Smart Innovation, Systems and Technologies

Volume 73

Series editors

Robert James Howlett, Bournemouth University and KES International,
Shoreham-by-sea, UK
e-mail: rjhowlett@kesinternational.org

Lakhmi C. Jain, University of Canberra, Canberra, Australia;
Bournemouth University, UK;
KES International, UK
e-mails: jainlc2002@yahoo.co.uk; Lakhmi.Jain@canberra.edu.au

About this Series

The Smart Innovation, Systems and Technologies book series encompasses the topics of knowledge, intelligence, innovation and sustainability. The aim of the series is to make available a platform for the publication of books on all aspects of single and multi-disciplinary research on these themes in order to make the latest results available in a readily-accessible form. Volumes on interdisciplinary research combining two or more of these areas is particularly sought.

The series covers systems and paradigms that employ knowledge and intelligence in a broad sense. Its scope is systems having embedded knowledge and intelligence, which may be applied to the solution of world problems in industry, the environment and the community. It also focusses on the knowledge-transfer methodologies and innovation strategies employed to make this happen effectively. The combination of intelligent systems tools and a broad range of applications introduces a need for a synergy of disciplines from science, technology, business and the humanities. The series will include conference proceedings, edited collections, monographs, handbooks, reference books, and other relevant types of book in areas of science and technology where smart systems and technologies can offer innovative solutions.

High quality content is an essential feature for all book proposals accepted for the series. It is expected that editors of all accepted volumes will ensure that contributions are subjected to an appropriate level of reviewing process and adhere to KES quality principles.

More information about this series at http://www.springer.com/series/8767

Ireneusz Czarnowski · Robert J. Howlett
Lakhmi C. Jain
Editors

Intelligent Decision Technologies 2017

Proceedings of the 9th KES International
Conference on Intelligent Decision
Technologies (KES-IDT 2017) – Part II

 Springer

Editors
Ireneusz Czarnowski
Maritime University
Gdynia
Poland

Robert J. Howlett
Bournemouth University
Poole
UK

and

KES International
Shoreham-by-Sea
UK

Lakhmi C. Jain
University of Canberra
Canberra, ACT
Australia

and

Bournemouth University
Poole
UK

and

KES International
Shoreham-by-Sea
UK

ISSN 2190-3018   ISSN 2190-3026   (electronic)
Smart Innovation, Systems and Technologies
ISBN 978-3-319-86622-2   ISBN 978-3-319-59424-8   (eBook)
DOI 10.1007/978-3-319-59424-8

Printed on acid-free paper

This Springer imprint is published by Springer Nature
The registered company is Springer International Publishing AG
The registered company address is: Gewerbestrasse 11, 6330 Cham, Switzerland

9th International KES Conference On Intelligent Decision Technologies (KES-IDT 2017), Proceedings, Part II

Preface

This volume contains the proceedings (Part II) of the 9th International KES Conference on Intelligent Decision Technologies (KES-IDT 2017), which will be held in Algarve, Portugal, on June 21–23, 2017.

The KES-IDT is a well-established international annual conference organized by KES International. The KES-IDT conference is a sub-series of the KES Conference series.

The KES-IDT is an interdisciplinary conference and provides excellent opportunities for the presentation of interesting new research results and discussion about them, leading to knowledge transfer and generation of new ideas.

This edition, KES-IDT 2017, attracted a number of researchers and practitioners from all over the world. The KES-IDT 2017 Program Committee received papers for the main track and 11 special sessions. Each paper has been reviewed by 2–3 members of the International Program Committee and International Reviewer Board. Following a review process, only the highest quality submissions were accepted for inclusion in the conference. The 63 best papers have been selected for oral presentation and publication in the two volumes of the KES-IDT 2017 proceedings.

We are very satisfied with the quality of the program and would like to thank the authors for choosing KES-IDT as the forum for presentation of their work. Also, we gratefully acknowledge the hard work of the KES-IDT international Program Committee members and of the additional reviewers for taking the time to review the submitted papers and selecting the best among them for presentation at the conference and inclusion in its proceedings.

We hope and intend that KES-IDT 2017 significantly contributes to the fulfillment of the academic excellence and leads to even greater successes of KES-IDT events in the future.

June 2017

Ireneusz Czarnowski
Robert J. Howlett
Lakhmi C. Jain

KES-IDT 2017 Conference Organization

Honorary Chairs

Lakhmi C. Jain   University of Canberra, Australia
and Bournemouth University, UK

Gloria Wren-Phillips   Loyola University, USA

Junzo Watada   Waseda University, Japan

General Chair

Ireneusz Czarnowski   Gdynia Maritime University, Poland

Executive Chair

Robert J. Howlett   KES International and Bournemouth University,
UK

Program Chair

Alfonso Mateos Caballero   Universidad Politécnica de Madrid, Spain

Publicity Chair

Izabela Wierzbowska   Gdynia Maritime University, Poland

Special Sessions

Decision Making Theory for Economics

Eizo Kinoshita   Meijo University, Japan
Takao Ohya   Kokushikan University, Japan

*Advances of Soft Computing in Industrial and Management Engineering:
New Trends and Applications*

Shing Chiang Tan   Multimedia University, Malaysia
Junzo Watada   Universiti Teknologi Petronas, Malaysia
Chee Peng Lim   Deakin University, Australia

Digital Architecture and Decision Management

Alfred Zimmermann   Reutlingen University, Germany
Rainer Schmidt   Munich University, Germany

*Specialized Decision Techniques for Data Mining, Transportation
and Project Management*

Piotr Jędrzejowicz   Gdynia Maritime University, Poland
Ireneusz Czarnowski   Gdynia Maritime University, Poland

Interdisciplinary Approaches in Business Intelligence Research and Practice

Ivan Luković   University of Novi Sad, Serbia
Ralf-Christian Härting   Hochschule Aalen, Germany

Eye Movement Data Processing and Analysis

Katarzyna Harezlak   Silesian University of Technology, Poland
Paweł Kasprowski   Silesian University of Technology, Poland

Decision Support Systems

Wojciech Froelich   University of Silesia, Poland

Pattern Recognition for Decision Making Systems

Paolo Crippa   Università Politecnica delle Marche, Italy
Claudio Turchetti   Università Politecnica delle Marche, Italy

Reasoning-Based Intelligent Systems

Kazumi Nakamatsu   University of Hyogo, Japan
Jair M. Abe   Paulista University, Brazil

Intelligent Data Analysis and Applications

Urszula Stańczyk   Silesian University of Technology, Gliwice,
                                        Poland
Beata Zielosko   University of Silesia, Katowice, Poland

Social Media Analysis and Mining

Clara Pizzuti   National Research Council of Italy (CNR), Italy

International Program Committee

Jair Minoro Abe   Paulista University and University of São Paulo,
                                        Brazil
Witold Abramowicz   Poznan University Economics, Poland
Stefan Aier   University of St Gallen, Germany
Rainer Alt   Universitat Leipzig, Germany
Colin Atkinson   University of Mannheim, Germany
Ahmad Taher Azar   Benha University, Egypt
Dariusz Barbucha   Gdynia Maritime University, Poland
Alina Barbulescu   Higher Colleges of Technology, Sharjah, UAE
Monica Bianchini   Dipartimento di Ingegneria dell'Informazione e
                                        Scienze Matematiche, Italy
Andreas Behrend   University of Bonn, Germany
Mokhtar Beldjehem   University of Ottawa, Canada
Gloria Bordogna   CNR IDPA, Italy
Oliver Bossert   McKinsey & Co. Inc., Germany

János Botzheim	Tokyo Metropolitan University, Japan
Lars Brehm	University of Munich, Germany
Wei Cao	School of Economics, HeFei University of Technology, China
Frantisek Capkovic	Slovak Academy of Sciences, Slovakia
Mario Giovanni C.A. Cimino	University of Pisa, Italy
Marco Cococcioni	University of Pisa, Italy
Angela Consoli	Defence Science and Technology Group, Australia
Paulo Cortez	University of Minho, Portugal
Paolo Crippa	Università Politecnica delle Marche, Ancona, Italy
Alfredo Cuzzocrea	University of Trieste, Italy
Ireneusz Czarnowski	Gdynia Maritime University, Poland
Eman El-Sheikh	University of West Florida, USA
Margarita N. Favorskaya	Siberian State Aerospace University, Russia
Michael Fellmann	University of Rostock, Germany
Raquel Florez-Lopez	University Pablo Olavide of Seville, Spain
Bogdan Franczyk	Universitat Leipzig, Germany
Ulrik Franke	KTH Royal Institute of Technology, Sweden
Wojciech Froelich	University of Silesia, Poland
Mauro Gaggero	National Research Council of Italy, Italy
Mauro Gaspari	Universita' di Bologna, Italy
Marina Gavrilova	University of Calgary, Canada
Michael Gebhart	Iteratec GmbH, Germany
Raffaele Gravina	University of Calabria, Italy
Christos Grecos	Central Washington University, USA
Foteini Grivokostopoulou	University of Patras, Greece
Jerzy W. Grzymala-Busse	University of Kansas, USA
Ralf-Christian Härting	Hochschule Aalen, Germany
Katarzyna Harezlak	Silesian University of Technology, Poland
Ioannis Hatzilygeroudis	University of Patras, Greece
Robert Hirschfeld	University of Potsdam, Germany
Dawn E.Holmes	University of California, USA
Katsuhiro Honda	Osaka Prefecture University, Japan
Daocheng Hong	Fudan University and Victoria University, China
Tzung-Pei Hong	National University of Kaohsiung, Taiwan
Yuh-Jong Hu	National Chengchi University, Taipei, Taiwan
Yuji Iwahori	Chubu University, Japan
Joanna Jedrzejowicz	University of Gdansk, Poland
Piotr Jędrzejowicz	Gdynia Maritime University, Poland
Björn Johansson	Lund University, Sweden
Nikos Karacapilidis	University of Patras, Greece
Dimitris Karagiannis	Universitat Wien, Germany
Pawel Kasprowski	Silesian University of Technology, Poland

Radosław Katarzyniak	Wrocław University of Technology and Science, Poland
Eizo Kinoshita	Meijo University, Japan
Frank Klawonn	Ostfalia University, Germany
Petia Koprinkova-Hristova	Bulgarian Academy of Sciences
Marek Kretowski	Bialystok University of Technology, Poland
Vladimir Kurbalija	University of Novi Sad, Serbia
Kazuhiro Kuwabara	Ritsumeikan University, Japan
Birger Lantow	University of Rostock, Germany
Frank Leymann	Universitat Stuttgart, Germany
Chee Peng Lim	Deakin University, Australia
Ivan Luković	University of Novi Sad, Serbia
Neel Mani	Dublin City University, Dublin, Ireland
Alfonso Mateos-Caballero	Universidad Politécnica de Madrid, Spain
Shimpei Matsumoto	Hiroshima Institute of Technology, Japan
Raimundas Matulevicius	University of Tartu, Estonia
Mohamed Arezki MELLAL	M'Hamed Bougara University, Algeria
Lyudmila Mihaylova	University of Sheffield, UK
Yasser Mohammad	ASSIUT University and EJUST, Egypt
Masoud Mohammadian	University of Canberra, Australia
Michael Möhring	University of Munich, Germany
Stefania Montani	University of Piemonte Orientale, Italy
Daniel Moldt	University of Hamburg, Germany
Mikhail Moshkov	KAUST, Saudi Arabia
Kazumi Nakamatsu	University of Hyogo, Japan
Selmin Nurcan	University of Paris, France
Marek Ogiela	AGH University of Science and Technology, Krakow, Poland
Takao Ohya	Kokushikan University, Japan
Isidoros Perikos	University of Patras, Greece
Petra Perner	Institute of Computer Vision and applied Computer Sciences IBaI, Germany
James F. Peters	University of Manitoba, Winnipeg, Canada
Gunther Piller	University of Mainz, Germany
Camelia-M. Pintea	UT Cluj-Napoca, Romania
Clara Pizzuti	National Research Council of Italy (CNR), Italy
Bhanu Prasad	Florida A&M University, USA
Jim Prentzas	Democritus University of Thrace, Greece
Radu-Emil Precup	Politehnica University of Timisoara, Romania
Georg Peters	Munich University of Applied Sciences
Marcos Quiles	Federal University of São Paulo - UNIFESP
Miloš Radovanović	University of Novi Sad, Serbia
Sheela Ramanna	The University of Winnipeg, Canada
Ewa Ratajczak-Ropel	Gdynia Maritime University, Poland
Manfred Reichert	Universitat Uulm, Germany

John Ronczka	SCOTTYNCC Independent research Scientist, Australia
Kurt Sandkuhl	University of Rostock, Germany
Mika Sato-Ilic	University of Tsukuba, Japan
Miloš Savić	University of Novi Sad, Serbia
Rafał Scherer	Częstochowa University of Technology, Poland
Rainer Schmidt	Munich University, Germany
Christian Schweda	Technical University Munich, Germany
Hirosato Seki	Osaka Institute of Technology, Japan
Bharat Singh	Big Data Labs, Hamburg, Germany
Urszula Stańczyk	Silesian University of Technology, Gliwice, Poland
Ulrike Steffens	Hamburg University of Applied Sciences, Germany
Janis Stirna	Stockholm University, Sweden
Catalin Stoean	University of Craiova, Romania
Ruxandra Stoean	University of Craiova, Romania
Mika Sulkava	Natural Resources Institute Finland
Keith Swenson	Fujitsu America Inc., USA
Shing Chiang Tan	Multimedia University
Mieko Tanaka-Yamawaki	Tottori University, Japan
Dilhan J. Thilakarathne	VU University Amsterdam, Netherlands
Claudio Turchetti	Università Politecnica delle Marche, Italy
Eiji Uchino	Yamaguchi University, Japan
Pandian Vasant	Universiti Teknologi PETRONAS, Tronoh, Malaysia
Ljubo Vlacic	Griffith University, Gold Coast Australia
Zeev Volkovich	Ort Braude College, Karmiel, Israel
Gottfried Vossen	WWU Munster, Germany
Junzo Watada	Waseda University, Japan
Fen Wang	Central Washington University, USA
Matthias Wissotzki	University of Rostock, Germany
Dmitry Zaitsev	International Humanitarian University, Odessa, Ukraine
Gian Piero Zarri	University Paris sorbonne, France
Beata Zielosko	University of Silesia, Katowice, Poland
Alfred Zimmermann	Reutlingen University, Germany

International Referee Board

Contents

Pattern Recognition for Decision Making Systems

A Development of Classification Model for Smartphone Addiction Recognition System Based on Smartphone Usage Data

Worawat Lawanont[✉] and Masahiro Inoue

Graduate School of Enginiering and Science,
Shibaura Intitute of Technology, Saitama, Japan
nb16501@shibaura-it.ac.jp, inouem@sic.shibaura-it.ac.jp

Abstract. The rapid growth of smartphone in recent years has resulted in many syndromes. Most of these syndromes are caused by excessive use of smartphone. In addition, people who tends to use smartphone excessively are also likely to have smartphone addiction. In this paper, we presented the system architecture for e-Health system. Not only we used the architecture for our smartphone addiction recognition system, but we also pointed out important benefits of the system architecture, which also can be adopted by other system. Later on, we presented a development of the classification model for recognizing likelihood of having smartphone addiction. We trained the classification model based on data retrieved from subjects' smartphone. The result showed that the best model can correctly classify the instance up to 78%.

Keywords: Smartphone addiction · Activity recognition · Data mining · e-Health system · Smartphone application

1 Introduction

Smartphone device has been increasing rapidly. A report [11] suggested that from a total population in South Korea, 88% of them own a smartphone and most developed countries remain above 50%. This information has proven that smartphone has become a part of people's modern lifestyle. Thus, consequences of excessive use of smartphone should be concerned.

In 2014, a medical research [7] presented the weight feel by cervical spine in each angle range of reading position, where reading position is also refer to the position when using smartphone. The work has shown that the more the head tilts forward, the more the weight feel by cervical spine will increase. This results in severe next pain, blurred vision, and headache.

Other well-known symptoms are, Computer Vision Syndrome (CVS) [2], pain in the wrist [14], and smartphone thumb (cellphone thumb) [8]. For example, CVS occurs from a low blink rate. The symptoms of CVS are dry eyes and headache. However, all this syndromes, including Text Neck, are results of excessive smartphone usage. Thus, people with smartphone addiction are likely to have one or more of these syndromes.

© Springer International Publishing AG 2018
I. Czarnowski et al. (eds.), *Intelligent Decision Technologies 2017*,
Smart Innovation, Systems and Technologies 73, DOI 10.1007/978-3-319-59424-8_1

In this paper, we present a development of classification model for smartphone addiction recognition system. The development includes the design of the system architecture and training of the classification model. For the experiment, we demonstrate the training processes of the classification model and how it is possible to use data from users' smartphone to predict whether they are likely to have smartphone addiction or not.

2   Related Works

In the past decades, there are many works that proposed solutions to overcome issues in healthcare system by using technologies. In this section, we outlined some of the important works that related to our proposal.

In 2014, Yang et al. [15] developed an intelligent medicine box. The purposes of the system were to monitor the patient behavior and to provide them with various of services, such as, reminder for taking medicines and a remote communication with a physician. The system showed a good example on using multiple Internet of Things (IoT) devices in one system. However, it did not contain or outline the uses of data mining in the work.

In term of smartphone addiction, several works have proposed statistical models what link the smartphone addiction to mental problems. In 2014, a work [1] shows a statistical model that presented the evidence that the use of smartphone for certain purposes and certain kind of smartphone addiction symptoms have significant impact on social capital building. Another work in 2015 [4] showed another perspective, and pointed out that social stress also has a positive influences on addictive smartphone behavior. However, in 2016, a research paper [12] presented a test on relation ship between smartphone addiction, stress level, and academic performance. The work showed a positive relationship between the smartphone addiction and stress level, but a negative relation ship between smartphone addiction and academic performance.

In term of a recognition system, Sano and Picard [13] presented a work to recognize stress by using wearable sensors and mobile phones. The work showed a possibility in recognizing mental issues by evaluating the model with perceived stress scale (PSS) [3]. The results showed that the system is capable of recognizing high or low perceived stress level with the highest accuracy of 75%.

For smartphone addiction, in 2014, Lee et al. presented a Smartphone Addiction Management System and Verification (SAMS) [10] to show the statistical analysis on the relationship between the application used on smartphone and possible smartphone addiction. The result of the SAMS showed a strong correlation between smartphone addiction and daily use count. Thus, we also considered this as one of the main attributes in our experiment.

3   Methodology

In this section, we discuss the techniques used in development of the classification model for smartphone addiction recognition system, which includes system architecture, and the smartphone application.

3.1   System Architecture

In recent years, there has been many development of e-health system to overcome technical limitations. Most of these developments started using IoT devices and data mining technique to provide a better system.

As in 2013, European Telecommunications Standards Institute (ETSI) has proposed a architecture that can be used by developers in building a service application [6]. As a result, a group of researchers has proposed a Next Generation e-Health Framework [5]. The framework adopted the concept from ETSI framework and extended the it further to match e-Health application requirements. From those proposed works, we have taken the idea and improved the design to better match the requirements of our system.

Figure 1 shows the design of the proposed system architecture. We took the idea of dividing data mining tasks into layers from a prior work [16] and separated the system into four main parts.

Body area network consists of sensor devices, which sense and transmit all raw sensor data to a body gateway. The Body gateway device must be capable of preprocessing raw data and send them over the Internet to the cloud services. By doing preprocessing at the body gateway, it also increases the abstraction level of the data. Thus, easier for personnel operating the cloud service to handle the data.

Mechanisms are the definition of how each part operates. In security mechanism, the architecture needs to specify how it handles the security issues of the system. For example, how the system will encrypt the data, which security protocol will it use to communicate between body area network and cloud services, and how will the system handle the privacy of users. On the other hand, sensor network mechanism specifies the protocol used between sensor devices and body gateway. This is different from the communication between the body gateway and cloud services, which is done over Internet, as there are more options to choose from. The chosen network protocol should consider the requirements of the system as well as sensor devices' capabilities.

Cloud services consists of several possible services. The cloud service should provide resource and services for data processing tasks, as most of e-health applications and systems have implemented with intelligence system, such as activity recognition. Thus, it is not suitable to perform those task in the body area network. Moreover, this will separate the tasks of data scientists and medical experts from handling the technical issues in body area network. Other services that the cloud service could provide are e-Health services, which are various services that need interaction between patient and medical personnel, and management service, which ease up the task of managing the whole system for administrators.

On the other hand, data processing does not concern the hardware nor the component of the system. However, data processing outlined four main tasks of data mining in e-health system and where it should be done. Raw data sensing (also known as data collection) and context management should be done at body area network level. While knowledge extraction (e.g., classification or clustering) and visualization and interaction should be done on the cloud services.

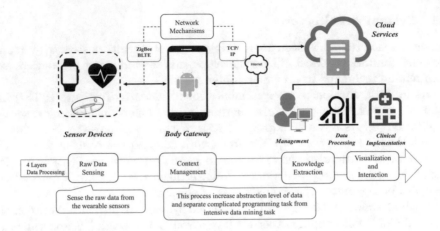

Fig. 1. Overview of the proposed system architecture for healthcare system

3.2 Smartphone Addiction Scale

Smartphone Addiction Scale (SAS) [9] is a self-diagnostic scale, which consists of 33 questions and each question is weighted equally on a 6-point scale. The SAS provides a score range between 0 to 188 where higher score indicates more serious smartphone addiction. As in this work, we used SAS as an evaluation tool. We recorded the SAS scores of all subjects and used the mean value of a total score as a separation point.

3.3 Application Overview

For the development of smartphone addiction recognition system, we developed an application to collect the data from the subjects. Figure 2 shows the 3 main application interfaces. The application consists of three main parts as follows.

The registration part is an interface for user to input important information, including, name, e-mail, date of birth, and gender. All information are kept in the cloud database, and it was not used publicly.

The purpose of the survey part is to collect the SAS score from subjects. After the application calculates the total SAS score, it sends them to cloud service to store them in database.

In monitoring part, the application handles all monitoring through Android service. The service allows the application to collect necessary data from the smartphone periodically without interfering users. The service runs for a total of 7 days and will stop itself after it finishes monitoring. The data collected are number of phone unlock, average phone usage time per phone unlock, maximum phone usage time per phone unlock, minimum phone usage time per phone unlock, total phone usage time, and total walking step count.

Fig. 2. Three main page of the developed application

4   Experiment

We conducted an experiment to prove that it is possible to predict the likelihood of having smartphone addiction by using data from smartphone sensor and logs. There were a total of 8 subjects participated in the experiment.

4.1   Experimental Design

We separated the experiment into three main stages. Figure 3 shows the overview of the experiment design. The first stage is to explain the detail and purpose of this work to the subject as well as allowing them to decide whether they want to participate in the experiment or not. If the subject chooses to participate in the experiment the application will be installed on the subject's smartphone. Then, the subject register their account to the system. In stage 2, all subject complete the SAS pre-survey, this survey score will be used later for evaluation. The detail of SAS is discussed later in this section. All pre-survey scores are stored in the cloud database. After the subject completes the survey, they can start the monitoring. The monitoring operates as a background service. Thus, it is possible for the subject to close the application and use their smartphone normally. In stage 3, after all subjects have completed the monitoring, which was lasted for 7 days, we retrieved all data and analyzed them. The attribute combinations and evaluation of each classification model is discussed later in this section.

4.2   Data Collection and Data Preprocessing

We developed the application on an Android 5.0 (API level 21) platforms. The application is responsible for three main features as mentioned earlier in Sect. 3.2.

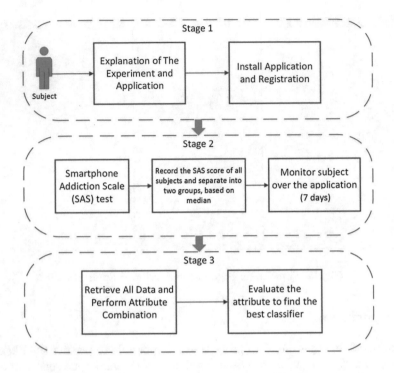

Fig. 3. Overview of the experiment design

The application collects the data periodically and preprocesses them before sending them to the cloud service. The application started the monitoring process and stopped itself after it reaches 7 days mark.

The application collected all the attributes mentioned earlier in the Sect. 3.3. Then, the application performs preprocessing by calculating the following values periodically:

1. Average Smartphone Usage per Unlock
2. Maximum Smartphone Usage per Unlock
3. Minimum Smartphone Usage per Unlock
4. Total Smartphone Usage Time
5. Amount of Walking Steps
6. Time Period (6.00–12.00, 12.00–18.00, 18.00–24.00PM, 24.00–6.00)
7. Phone Unlock Count

We set the application to update one instance of data to the cloud server every 30 min. Please note that, we set the period to 30 min in order to make sure that this data set can be used with any application or classification training, which require the period to be 30 min or longer, as well.

4.3   Modeling

The data retrieved from the cloud server is used in the training. We combine 2 instances into 1 instance. Therefore, an instance used in the modeling process is an instance with monitoring period of 60 min. All data were randomly sorted to avoid any biased in the training. The training set and test set were separated from all data with 70:30 ratio.

For performing supervised learning tasks, we labeled each instance as either 'High' or 'Low' for classification. However, as each subject has instance which was updated to the cloud during their inactive time. Thus, we labeled the instance where 'Total Smartphone Usage Time' equal to 0 as 'Inactive'. The lowest SAS for the experiment was 86 while the highest was 124 and the average score of all subjects was 110. Data instance from subject with score equal to or lower than 110 were labeled as 'Low' while the data instance from subject with score higher than 110 were labeled as 'High'.

In order to train the best performing classification model, we took all 7 attributes and calculated all possible combinations with at least two attributes. As a result we have a total of 120 attribute combinations. We used each combination as a training attributes with 4 classification algorithms, which are Naive Bayes, K-Nearest Neighbor (K-NN) (K = 5), Decision Tree (J48), and Support Vector Machine (SVM). We performed the training process using the same training set with 10-fold cross validation technique. Then, we tested the trained model on the same set, which we prepared earlier.

4.4   Experiment Result

The 20 most accurate results of all attribute combinations are discussed in this section. Table 1 shows the accuracy of each attribute combination. Please note that the number in attribute column represents the attribute according to the attribute list mentioned in Sect. 4.3.

The result shows that the combination of attributes 1, 3, 4, 5, 6, 7 and 2, 3, 4, 5, 6, 7 have the most accurate results when trained with Decision Tree (J48) algorithm, which the accuracy were equal at 78.74%. The two combinations also equal the accuracy at 68.11% with Naive Bayes algorithm. The accuracy with K-NN (K = 5) were 70.08% and 70.07% respectively and 61.42% and 61.45% respectively with SVM algorithm.

The least accurate attribute combination of this top 20 list was the combination of attributes 1, 2, 3, 4, 6. The accuracy was 65.76%, 70.47%, 73.62% and 58.27% with Naive Bayes, K-NN (K = 5), Decision Tree (J48), and SVM respectively.

From the algorithm perspective, Decision Tree (J48) has out performed all other algorithms in every attribute combinations. The most accurate combination with Decision Tree (J48) was 78.74% while the least accurate was 73.62%. For Naive Bayes, the most accurate was 68.50% and the least accurate was 65.76%. The most accurate for K-NN (K = 5) was 73.62% and the least accurate was 68.50%. For SVM, the performance was fairly poor as the most accurate combination was only 61.81% and the least accurate combination was 55.59%. The average accuracy of Naive Bayes, K-NN (K = 5), Decision Tree (J48), and SVM are 67.62%, 70.91%, 75.15%, and 58.80% respectively.

Table 1. Results of classifier evaluation on the test set

Attributes	Algorithm accuracy (%)			
	Naive Bayes	K-NN (K = 5)	Decision Tree (J48)	SVM
1, 3, 4, 5, 6, 7	68.11	70.08	78.74	61.42
2, 3, 4, 5, 6, 7	68.11	70.07	78.74	61.45
1, 2, 3, 5, 6, 7	68.11	69.69	78.34	61.42
2, 3, 5, 6, 7	68.11	69.29	78.30	60.23
3, 4, 5, 6, 7	68.11	68.50	78.29	61.02
1, 2, 3, 4, 5, 6, 7	68.11	70.47	76.77	61.41
1, 3, 5, 6, 7	66.93	68.50	76.38	61.42
1, 2, 4, 5, 7	68.11	72.44	75.98	59.94
1, 4, 5, 6, 7	67.72	69.69	75.98	59.06
1, 2, 3, 6, 7	67.72	70.87	75.59	61.42
1, 2, 4, 6, 7	68.50	69.29	75.59	59.84
1, 2, 5, 6, 7	67.71	70.87	75.59	61.4
1, 2, 3, 4, 5, 7	67.71	73.62	75.20	61.02
1, 2, 3, 4, 6, 7	68.5	71.26	75.20	61.81
1, 3, 4, 5, 7	68.11	73.62	75.20	61.02
1, 3, 4, 6, 7	67.72	70.87	75.20	55.60
2, 3, 4, 6, 7	68.5	72.05	75.2	55.59
2, 3, 4, 5, 7	68.5	73.23	75.19	61.02
1, 2, 4, 5, 6, 7	68.11	71.26	74.41	59.45
1, 2, 3, 4, 6	65.76	70.47	73.62	58.27

Table 2 showed the confusion matrix of the best performing attribute combination. From the result, the model correctly classified 37% of all instances in 'High' class or 70% of all 'High' instances. For instance in 'Low' class, the model correctly classified 16.54% of all instance or 76.36% of all 'Low' instances. The model correctly classified all instances in 'Inactive' class.

4.5   Discussion

The experiment has shown a satisfying results, and the best performing model has the accuracy of 78.74%. However, the results have also pointed out the room for improvements. In this experiment, we considered a total of 7 attributes. More attributes from other sensors could improve the accuracy of the model. Moreover,

Table 2. Confusion matrix of the best attribute combination

	High	Low	Inactive
High	37.00%	16.14%	0%
Low	5.12%	16.54%	0%
Inactive	0%	0%	25.20%

as another possible way to improve the result is to implement the application with artificial intelligence algorithm. By doing so, it is possible for the application to learn and study the smartphone usage behavior of each individual user and compare them with others. Nonetheless, the mentioned possibilities is out of the scope of this paper.

The limitation concerned in this experiment was that the size of the subject was small, and further experiment with larger subject group will provide a classification model that can handle more diverse usage characteristic. Nonetheless, the result of this experiment showed a positive side of using data from smartphone to recognize likelihood of having smartphone addiction.

5   Conclusion

In this paper, we showed the importance of recognizing smartphone addiction and how it could prevent users from suffering other syndromes related to smartphone usage. We presented the system architecture which is possible to implement in developing other e-health system. The architecture consists of four main parts which are clearly separate from each other. The abstraction level of the architecture made it easier for personnel from different fields to coordinate with each other.

Moreover, we presented a development of a classification model based on the data collected from the subjects smartphone. The result of the development was accurate up to 78.74% in recognizing whether their smartphone are Inactive or they have High or Low smartphone addiction likelihood. The results showed good opportunities for improvement and implementation in the smartphone.

Finally, despite the results are preliminary with the limited number of participants in the experiment. We have shown that the data from the smartphone can be used to recognize likelihood of smartphone addiction. In the future, by increasing the subject size and integrating multiple device into the system could provide us with a more accurate classification model. We will also attempt to work on a more robust and dynamic system in which the interaction between the system and users should be personalized to prevent smartphone addiction.

Acknowledgments. This work was supported by JSPS KAKENHI Grant number 15K00929.

References

1. Bian, M., Leung, L.: Linking loneliness, shyness, smartphone addiction symptoms, and patterns of smartphone use to social capital. Soc. Sci. Comput. Rev. **33**(1), 61–79 (2015)
2. Blehm, C., Vishnu, S., Khattak, A., Mitra, S., Yee, R.W.: Computer vision syndrome: a review. Surv. Ophthalmol. **50**(3), 253–262 (2005)
3. Cohen, S., Kamarck, T., Mermelstein, R.: A global measure of perceived stress. J. Health Soc. Behav. **24**, 385–396 (1983)
4. van Deursen, A.J., Bolle, C.L., Hegner, S.M., Kommers, P.A.: Modeling habitual and addictive smartphone behavior: the role of smartphone usage types, emotional intelligence, social stress, self-regulation, age, and gender. Comput. Hum. Behav. **45**, 411–420 (2015)
5. Fengou, M.A., Mantas, G., Lymberopoulos, D., Komninos, N., Fengos, S., Lazarou, N.: A new framework architecture for next generation e-health services. IEEE J. Biomed. Health Inform. **17**(1), 9–18 (2013)
6. ETSI ES 203 915-3 v1.2.1. ETSI/The Parlay Group: Open Service Access (OSA); Application Programming Interface (API); Part 3: Framework (Parlay5) (2007)
7. Hansraj, K.K.: Assessment of stresses in the cervical spine caused by posture and position of the head. Surg. Technol. Int. **25**, 277–279 (2014)
8. Karim, S.A.: From 'playstation thumb' to 'cellphone thumb': the new epidemic in teenagers. South African Med. J. (SAMJ) **99**(3), 161–162 (2009)
9. Kwon, M., Lee, J.Y., Won, W.Y., Park, J.W., Min, J.A., Hahn, C., Gu, X., Choi, J.H., Kim, D.J.: Development and validation of a Smartphone Addiction Scale (SAS). PloS One **8**(2), e56936 (2013)
10. Lee, H., Ahn, H., Choi, S., Choi, W.: The SAMS: smartphone addiction management system and verification. J. Med. Syst. **38**(1), 1–10 (2014)
11. Poushter, J.: Smartphone ownership and internet usage continues to climb in emerging economies. Global Attitudes & Trends, Pew Research Center (2016)
12. Samaha, M., Hawi, N.S.: Relationships among smartphone addiction, stress, academic performance, and satisfaction with life. Comput. Hum. Behav. **57**, 321–325 (2016)
13. Sano, A., Picard, R.W.: Stress recognition using wearable sensors and mobile phones. In: 2013 Humaine Association Conference on Affective Computing and Intelligent Interaction (ACII), pp. 671–676. IEEE (2013)
14. Jm, S.: The effect of carpal tunnel changes on smartphone users. J. Phys. Ther. Sci. **24**(12), 1251–1253 (2012)
15. Yang, G., Xie, L., Mäntysalo, M., Zhou, X., Pang, Z., Da Xu, L., Kao-Walter, S., Chen, Q., Zheng, L.R.: A health-IoT platform based on the integration of intelligent packaging, unobtrusive bio-sensor, and intelligent medicine box. IEEE Trans. Ind. Inform. **10**(4), 2180–2191 (2014)
16. Zhang, S., McCullagh, P., Nugent, C., Zheng, H., Black, N.: An ontological framework for activity monitoring and reminder reasoning in an assisted environment. J. Ambient Intell. Human. Comput. 4(2), 157–168 (2013)

Complex Object Recognition Based on Multi-shape Invariant Radon Transform

Ghassen Hammouda[1]($\boxtimes$), Atef Hammouda[2], and Dorra Sellami[1]

[1] National Engineering School of Sfax, University of Sfax, Sfax, Tunisia
gassenhammouda7@hotmail.fr, sellamimasmoudidorra@yahoo.com
[2] The Sciences Institute of Tunis, University of Tunis, Tunis, Tunisia
atef_hammouda@yahoo.fr

Abstract. Based on the properties of Template Matching and Radon Transform, a new Multi-Shape Invariant Radon Transform (MSIRT) is proposed in this paper. Unlike Radon Transform, integrating projections across lines, the MSIRT uses arbitrary given curves, which are derived from primitives contours. MSIRT leads to peaks once similar shapes are met in the projected image. For seek of genericity and invariance with respect to geometric transformations, we consider different primitives derived from MPEG7 dataset. Each object undergoes a series of pre-processing steps, segmentation and contour extraction, for generating the corresponding primitive. For each query object, the MSIRT is applied with respect to the different primitives and a vote approach will be used for object recognition. Validation of the proposed approach is done on the MPEG7 dataset, giving an accuracy of 94%. Comparison with some known approaches demonstrates the effectiveness of the proposed approach in detecting complex objects, even under geometric transformations.

Keywords: MSI Radon Transform · Primitives · Complex objects · Template matching · Vote

1 Introduction

Complex-shaped object detection is still an open problem in computer vision. Several works from the literature focus on the detection of objects with common geometric forms (line, square, circle) or parametric forms such as parabolas and hyperbolas. Only few approaches deals with detecting complex objects. However, most of them fail under geometric transformations.

In this paper, we introduce a new formalism for the generalisation of the Radon Transform to detect objects with complex shapes. By building a set of variable primitives, we made our approach invariant to geometric transformations. The remainder of this paper is organized as follows: Related works are described in Sect. 2. The proposed MSI Radon Transform is presented in Sect. 3. Experimental validation on MPEG 7 database and comparison results are given is Sect. 4. Finally, conclusions and perspectives are drawn in Sect. 5.

© Springer International Publishing AG 2018
I. Czarnowski et al. (eds.), *Intelligent Decision Technologies 2017*,
Smart Innovation, Systems and Technologies 73, DOI 10.1007/978-3-319-59424-8_2

2 Related Works

2.1 Template Matching Approaches

Template matching finds out appearance similarities between some template primitives and objects in the image. It ends at potentially locating template shapes in the image. Based on a template illustrating the most relevant traits of appearance of a focused pattern, a matching rate is computed to estimate the occurrence of the considered pattern in a set of images. It is a computational approach that has to deal with possible change in position, scale and rotation or any transformation in the image. The choice of the templates depends on the context and the constraints. To detect the similarity between a template image and a query image with equal dimensions, the cross-correlation approach can be adequate. This approach consists in summing the pairwise multiplications of corresponding pixel values of the images. However, one drawback of the cross correlation is that it cannot handle the change of brightness. Normalized cross-correlation (NCC) [1] is then introduced to improve the original approach. It subtracts the mean image brightness from each pixel value. NCC was adopted to recognize similar forms with a high precision but it is still sensitive to any change of scale or rotation.

2.2 Radon Transform

The Radon Transform (RT) is one of the oldest approach. In the literature, several variants of Radon Transforms have been developed [2,3]. Let f be a function defined on the Euclidean space. Each pixel has a (x, y) coordinate in a two dimensional cartesian system. So, the Radon Transform can be defined by:

$$R(x', \theta) = \int_{-\infty}^{\infty} \int_{-\infty}^{\infty} f(x, y)\, \delta\left(x' - x\cos(\theta) - y\sin(\theta)\right) \mathrm{d}x\, \mathrm{d}y \qquad (1)$$

Where δ is the Kronecker delta function that converts the two-dimensional integral to a line integral along the axis $x\cos(\theta) + y\sin(\theta) = x'$ and θ is the angle of orientation. Radon Transform offers a multitude of properties useful in resolving pattern recognition problems. The most relevant ones in the object recognition are:

– Symmetry:

$$R(x', \theta) = R(-x', \theta \pm \Pi) \qquad (2)$$

– Periodicity:

$$R(x', \theta) = R(x', \theta + 2k\Pi) \qquad (3)$$

where k is integer.
– Translation: a translation of f of $\bar{w} = (x, y)$ implies a translation of $\varpi = x_0 \cos(\theta) + y_0 \sin(\theta)$

$$R(x', \theta) = R(x' - x_0 \cos(\theta) - y_0 \sin(\theta), \theta) \qquad (4)$$

– Rotation: A rotation of the image by an angle θ_0 implies a shift of the Radon Transform in θ.

$$R(x', \theta) = R(x', \theta + \theta_0) \tag{5}$$

– Scaling: a zoom of $\alpha \neq 0$ in f involves a change of scale in Radon Transform:

$$R(x', \theta) = \frac{1}{\alpha} \times R(\alpha * x', \theta) \tag{6}$$

Radon Transform can be useful in pattern recognition. The projection of a pattern with RT is done without loss of information because only the non-null pixels are projected in the Radon matrix in order to retains the relevant information. The RT is also robust against noise. In fact, it can detect some scattered pixels without lack of accuracy. The relevant information detected as straight lines appears as a peak in the Radon space. Indeed, RT performs well in the detection of lines. Rojbani [4] propose an approach for object recognition called the GR-signature (GR). It is essentially based on the Radon Transform and the Gradient to measure the rectangularity of the form. This transform is robust to noise and it is discriminant even under deformation. It allows to estimate the shape of the object based on its characteristics.

S. Tabbone *et al.* [5,6] proposed an hybrid approach called the Histogram of the Transformed Radon (HTR). By statistically analysing the Radon Transform, this approach can detect lines. In other way, it offers a 2D histogram representing the length of the shape given at each direction. The HTR is invariant to translation and rotation but it still very sensitive to any noise or occlusion and detect exclusively lines.

The previous transforms are essentially concerned with straight lines in images. Recently, some works have focused on more complex shapes such as the Polynomial Discrete Radon Transform (PDRT) [7]. The PDRT offers the advantage of projecting a polynomial shape equation in all directions of an image to find it. The sum of pixels of the detected shape will be stored as a peak in the Radon space. In fact, this approach is limited to polynomial curves. The Generalized Radon Transform (GRT) [8] was also defined to project a 2D function over parametrized curves and provides a general solution for some complex forms and it is useful to detect parameterized shapes in an image. However, the GRT suffers from the absence of the multi directional criteria depriving the shape to be detected in different orientations.

To deal with this limitation, Elouedi *et al.* proposed the Generalized Multi-Directional Radon Transform (GMDRT) [9]. It allows to recognize multiple complex geometric curves presented as parametric equation such as circles, rectangles and parabolas in all directions. The GMDRT detects curves with any orientation of the initial shape. Even if the GMDRT offers a significant amelioration in the detection of geometric curves, it remains unable to detect any complex forms since there is no available parametric explicit description for these curves.

The application of various Radon Transform approaches has shown its efficiency in detecting straight lines and geometric forms with rectilinear shape.

An extension of the Radon Transform based on a parametric equation is used to identify the curve of different forms belonging to the same family. It improves the characteristics of the Radon Transform as a shape descriptor. However, application of Radon Transform was often considering specific forms as parabolas, polynomials, etc. The goal of the proposed approach called the Multi-Shape Invariant Radon Transform (MSI Radon Transform) here is to detect complex objects without need of a predefined parametric modeling. This is made possible by applying the MSI Radon Transform of the searched object on a number of primitives to detect its presence.

3   The MSI Radon Transform

3.1   General Brief Description

The MSI Radon Transform is a novel approach joining both features: Radon Transform and Template Matching. On the one hand, Radon Transform genericity inherited from considering variable primitives, and on the other hand accuracy of the Template Matching. The result of the application of MSI Radon Transform is some peaks in Radon space, in case of presence of specific shapes in the image. For seek of invariance, we consider in building the primitives different positions and sizes of the objects, we apply geometric transformations (scaling and rotation) to the different images in the dataset. In this way, MSI Radon Transform is made efficient under scaling and rotation. In a next step, similarity between images is computed from Radon space, the obtained peaks are analyzed for affecting each object to its correct class. All these steps are drawn in Fig. 1.

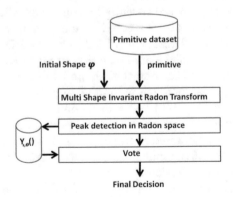

Fig. 1. The MSI Radon Transform based object detection steps

3.2   MSI Radon Transform Formalism

An MPEG7 dataset is used in validation. Let φ be an input initial primitive without any hypothesis made on its shape or size. Each image from the initial dataset is noted I_i and represents a two dimensional matrix which have undergone geometric transformations and deformations.

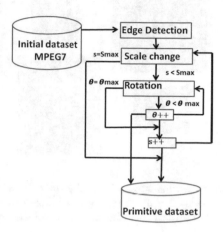

Fig. 2. Primitive generation steps

Primitive Generation. As shown in Fig. 2, for each Image I_i in the dataset, we apply a series of preprocessing steps including edge detection, scale change s and orientation θ, for sweeping them.

- Edge Detection: In order to reduce the computation complexity and to focus on the object shape, a contour extraction process is applied. The image is converted into a perceptual space HSV and the Split and Merge technique is applied. This process eliminates the shadow and keeps only the object relevant information. The Canny edge detector operator is then applied. Once the edge is extracted, a binary image is generated.
- Scale Change and Rotation: We apply a scaling of the image by a factor s ranging from $s_0 = 0.5$ to $s_{max} = 2$. For each scaled image of the k images of the dataset, we apply rotations by θ ranging from $\theta_0 = 0°$ to $\theta_{max} = 180°$. Let ns be the number of the scaling factors ($ns = 16$) and $n\theta$ the number of the rotations applied for each scale ($n\theta = 181$). The resultant images $I_{s,\theta}$ from these iterations constitute a bigger dataset of $k \times ns \times n\theta$ primitives.

Figure 3 illustrates some primitives generated from a bird image from the dataset MPEG7. $I_{s,\theta}$ is the result of the preprocessing steps and is given by Eq. (7).

$$
I_{s,\theta} = \begin{bmatrix}
I_{s,\theta}(-L,0) & I_{s,\theta}(-L,j) & \dots & I_{s,\theta}(-L,n-1) \\
\cdot & \cdot & \dots & \cdot \\
\cdot & \cdot & \dots & \cdot \\
\cdot & \cdot & \dots & \cdot \\
I_{s,\theta}(0,0) & I_{s,\theta}(0,j) & \dots & I_{s,\theta}(0,n-1) \\
\cdot & \cdot & \dots & \cdot \\
\cdot & \cdot & \dots & \cdot \\
\cdot & \cdot & \dots & \cdot \\
I_{s,\theta}(L,0) & I_{s,\theta}(L,j) & \dots & I_{s,\theta}(L,n-1)
\end{bmatrix} \tag{7}
$$

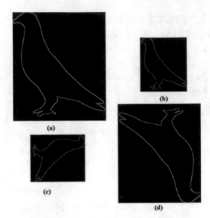

Fig. 3. Some primitives of a bird image from the dataset (a) original primitive (b) processed primitive rotated by $\theta = 0°$, scaled by s=0.5 (c) processed primitive rotated by $\theta = 90°$, scaled by s=0.5 (d) processed primitive rotated by $\theta = 180°$, scaled by s=0.9.

MSI Radon Transform. The MSI Radon Transform is given by:

$$y_\theta(n) = \sum_{m=-M}^{m=M} R_{m,\theta} \times I_{s,\theta}(n + m) \tag{8}$$

$y_\theta(n)$ is the resultant column of the matrix y_θ where $I_{s,\theta}$ the matrix of the primitive corresponding to an angle θ and the scale s starting on the column n is projected over φ. $I_{s,\theta}(n+m)$ is a fixed column of $I_{s,\theta}$. $R_{m,\theta}$ are $(2L+1) \times (2L+1)$ selection matrices introduced by Beylkin where are stored elements of $I_{s,\theta}(n+m)$ involved in the projection $y_\theta(n)$ [12]. Each row j, $-L < j < L$ in $R_{m,\theta}$ store the pixels from $I_{s,\theta}(n,m)$ belonging to φ starting at the position (j,n). The construction of the $R_{m,\theta}$ consists in presenting the shape of φ in a k position with $-L < k < L$. $y_\theta(n)$ is then the column resulting in the projection of φ starting in an initial coordinate (j,n) over the matrix of the primitive $I_{s,\theta}$. Each component $y_\theta(j,n)$ of this column is the sum of the pixels centered on the shape and started in the coordinate (j,n). M represents the number of columns $I_{s,\theta}$ involved in the computation of $y_\theta(n)$.

Peak Detection. Values of Radon peaks $y_\theta(n)$ are stored. They are arranged in a decreasing order for the further vote step.

Vote. Once the highest peaks are collected for each primitive, a vote is then used. The object class is then taken as the major class in the first primitives.

4   Experimental Results

In this section, an experimental set-up is provided in order to evaluate the performances of the MSI Radon Transform in complex form object detection. A comparison is done with the MPEG7 dataset.

4.1   MSI Radon Transform Performance Evaluation

To evaluate the MSI Radon Transform, a sequence of steps are undertaken and interact as a complex pattern recognition process. Below a brief description of the dataset is presented.

MPEG7 Datasets. The MPEG-7 standard Core Experiment CE-Shape-1 Part B [10,11]: Similarity based Retrieval dataset is available for the research community and is composed of 1400 images. In this dataset, 70 classes of different shapes are included with 20 images for each class. These images contain objects with complex forms. Figure 4 illustrates some images in this dataset.

Fig. 4. MPEG7 dataset.

Metric of Evaluation. The detection accuracy is used as a metric of evaluation. It is computed with the following equation:

$$R = \frac{TP}{TP + FN} \tag{9}$$

where TP is the total of the relevant images retrieved associated correctly to its original class and FN is the total of the objects affected to the wrong class. Each object of the dataset is compared to all the other objects of the other classes. The TPR also called sensitivity is the ratio of the true detected objects belonging to a specific class. The area under the curve is also used as a metric of evaluation. It is a common evaluation metric for binary classification problems used in order to evaluate a classifier. The area under the curve will be close to 1 in the case of a good classifier.

Comparison Result. To evaluate the performance of the MSI Radon Transform in the recognition of complex shape object, this approach is tested in the MPEG7 objects dataset. Moreover, comparison with other existing approaches is also achieved here in order to situate the proposed approach. For each approach, the recognition rate of a set of object forms in the dataset is estimated. All the previous approaches are implemented and represent each object by a shape descriptor specific to the approach used and is classified accordingly. The obtained accuracy rate is used as a metric of evaluation. Comparison Results as illustration, the sensitivity of some classes using several approaches is summarized in Table 1.

Table 1. Sensitivity and accuracy for some objects from the MPEG7 database.

	Apple	Circle	Fork	Symbol	Hammer	Fish	Glass	Bottle	Device	Results
GR [4]	0.15	0.1	0.1	0.4	0.2	0.95	0.2	0.45	0.35	0.33
RT [2]	0.9	0.95	0.55	0.95	0.65	0.7	0.7	1	0.8	0.78
GMDRT [9]	0.9	1	0.15	0.25	0.8	0.85	0.75	0.85	0.7	0.72
NCC [1]	0.95	1	0.8	0.9	0.95	1	0.75	1	0.75	0.91
MSIRT	0.95	0.95	0.9	0.9	0.95	0.9	0.95	0.9	0.85	0.94

The analysis of Table 1 reveals that the MSI Radon Transform approach and the NCC present the best detection rates (94% and 91% respectively) with a slight advantage for the MSI Radon Transform. These results concern all the forms evaluated and confirm the consistency of the proposed approach to distinguish any object with an acceptable recognition rate. The analysis by family of primitives for the nine classes described in Table 1 shows a stability of the results for each object class for this approach. For the other approaches (GR, RT, GMDRT), the results vary considerably from one primitive to another. This is the case of GMDRT which gives an acceptable rate for some primitives but is very limited in the recognition in other ones. Moreover, the MSI Radon Transform approach has a great ability to recognize irregular shapes. This is the case of the object fork where most approaches have provided a very low rate in its recognition while the MSI Radon Transform recognizes it at a rate of 90%. Although the results provided by the NCC are fairly close to MSI Radon Transform, it faces problems of scaling and orientation change. Indeed, the results are significantly affected by rotation and scale variations.

Comparative results given in Table 2 confirm that MSI Radon Transform approach remains stable against rotation and scale variation. The NCC performances are very low illustrating the sensitivity of this approach with respect to changes in these two parameters. However, MSI Radon Transform has some limitations that can be inherited from the contour detection approach and in case of bad contour detector, performance results can be tremendously affected. This latter is affected by the change of rotation and scale and can constitute a kind of limitation. This is the case of the object bottle for example that has

Complex Object Recognition Based on MSI Radon Transform   21

Table 2. Comparative study illustrating the performance of the proposed MSI Radon Transform and NCC in detecting primitives of the dataset with scale and rotation change.

	Apple	Circle	Fork	Symbol	Hammer	Fish	Glass	Bottle	Device	Results
NCC [1]	0.85	0	0.8	0	0.1	0	0	0.05	0	0.18
MSIRT	0.95	0.95	0.9	0.9	0.95	0.9	0.95	0.9	0.85	0.94

the lowest rate of 85% caused by the lost of some information in the contour detection. To illustrate the performance of the classifier in the MSI Radon Transform approach, the Receive Operating Characteristic (ROC) is used. We get the curve after sweeping the threshold separating between inter-class and intra-class distributions.

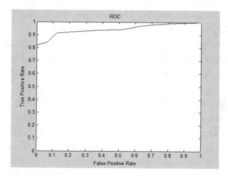

Fig. 5. Receive Operating Characteristic curve of the proposed approach.

Figure 5 illustrate an area under the curve (AUC) of 0.94. It denotes the ability of the approach to separate between objects.

5 Conclusion

A novel approach for the detection of objects has been proposed. It is a kind of Radon Transform. This transform focuses on the detection of complex shapes objects under geometric transformations changes. A dataset of primitives is obtained by applying preprocessing steps of edge detection, scale and orientation changes on the initial images (here the MPEG7 dataset). The MSI Radon Transform is applied for each query image in order to detect the presence of an initial input object in the dataset. A matrix of peaks in the Radon space revealing a possible presence of a primitive is set and a final vote allows to decide about the right object class. Experiments have been carried out. An area under the curve of 0.94 is obtained. Comparison results show also that the proposed approach outperformed existing ones, by presenting more accuracy and robustness to geometric transformations.

References

1. Raghavender Rao, Y., Prathapani, N., Nagabhooshanam, E.: Application of normalized cross correlation to image registration. Int. J. Res. Eng. Technol. **05**(3), 12–16 (2014)
2. Hasegawa, M., Tabbone, S.: Amplitude-only log radon transform for geometric invariant shape descriptor. Pattern Recogn. **47**(2), 643–658 (2014). doi:10.1016/j.patcog.2013.07.024. Elsevier
3. Tabbone, S., Wendling, L., Salmon, J.-P.: A new shape descriptor defined on the radon transform. Comput. Vis. Image Understand. **102**(1), 42–51 (2006). doi:10.1016/j.cviu.2005.06.005. Elsevier
4. Rojbani, H., Elouedi, I., Hamouda, A.: Rθ-signature: a new signature based on radon transform and its application in buildings extraction. In: IEEE International Symposium on Signal Processing and Information Technology (ISSPIT), pp. 490–495. IEEE (2011)
5. Hasegawa, M., Tabbone, S.: Histogram of radon transform with angle correlation matrix for distortion invariant shape descriptor. Neurocomputing **173**, 24–35 (2016). doi:10.1016/j.neucom.2015.04.100. Elsevier
6. Hasegawa, M., Tabbone, S.: A shape descriptor combining logarithmic scale histogram of radon transform and phase-only correlation function. In: International Conference on Document Analysis and Recognition (ICDAR), pp. 182–186. IEEE (2011)
7. Ines, E., Dhikra, H., Regis, F., Amine, N.-A., Atef, H.: Fingerprint recognition using polynomial discrete radon transform. In: 4th International Conference on Image Processing Theory, Tools and Applications (IPTA), pp. 1–6. IEEE (2014)
8. Hendriks, C.L.L., Van Ginkel, M., Verbeek, P.W., Van Vliet, L.J.: The generalized radon transform: sampling, accuracy and memory considerations. Pattern Recogn. **38**(12), 2494–2505 (2005). doi:10.1016/j.patcog.2005.04.018. Elsevier
9. Elouedi, I., Fournier, R., Nait-Ali, A., Hamouda, A.: Generalized multidirectional discrete radon transform. Sign. Process. **93**(1), 345–355 (2013). doi:10.1016/j.sigpro.2012.07.031. Elsevier
10. de Oliveira, A.B., da Silva, P.R., Barone, D.A.C.: A novel 2d shape signature method based on complex network spectrum. Pattern Recogn. Lett. **63**, 43–49 (2015). doi:10.1016/j.patrec.2015.05.018. Elsevier
11. Latecki, L.J., Lakamper, R., Eckhardt, T.: Shape descriptors for non-rigid shapes with a single closed contour. In: IEEE Conference on Computer Vision and Pattern Recognition, vol. 1, pp. 424–429. IEEE (2000)
12. Beylkin, G.: Discrete Radon transform. IEEE Trans. Acoustics Speech Sign. Process. **35**(2), 162–172 (1987)

Genetic Algorithms Based Resampling for the Classification of Unbalanced Datasets

Marco Vannucci[✉] and Valentina Colla

TeCIP Institute, Scuola Superiore Sant'Anna, Via G. Moruzzi, 1, 56124 Pisa, Italy
{mvannucci,colla}@sssup.it

Abstract. In the paper a resampling approach for unbalanced datasets classification is proposed. The method suitably combines undersampling and oversampling by means of genetic algorithms according to a set of criteria and determines the optimal unbalance rate. The method has been tested on industrial and literature datasets. The achieved results put into evidence a sensible increase of the rare patterns detection rate and an improvement of the classification performance.

1 Introduction

Many real world problems in different fields have to cope with the classification of unbalanced datasets. These tasks are characterized by the development of classifiers for the detection of particular rare patterns. In the context of binary classification problems, unbalanced datasets are characterized by the marked preponderance of samples belonging to the frequent class (C_F) with respect to the rare class (C_R). The correct identification of these patterns is fundamental in the context of the handled problems. For instance, in the industrial field machine faults detection [3] and product quality assessment [1] belong to this category, since both malfunctions and defects are far less frequent than normal situations but their correct and reliable identification would allow money saving and higher quality products delivered to customers. Further examples of this situation can be found in the medical field, as far as the rare disease concerns [2], in finance, for the identification of fraudulent credit card transactions, and in bioinformatics [4]. Another common point of these problems lies in the fact that the misclassification of frequent patterns (the so-called *false alarms*) is strongly preferable than the missed detection of rare ones.

The classification of unbalanced dataset is extremely complex due to a multitude of interacting factors. The main cause of the lower performance on this task of standard classifiers, such as those based on Artificial Neural Networks (ANN), Decision Trees (DT) and Support Vector Machines (SVM), is the basic assumption of even distribution of the training samples among classes. The goal of most machine learning algorithms is the achievement of an optimal *overall* performance, that is satisfactory in the case of balanced classes but, in the case of unbalanced datasets, the classifiers results to be biased toward the majority class and, as a consequence, the minority class is misclassified [5]. The complexity of

© Springer International Publishing AG 2018
I. Czarnowski et al. (eds.), *Intelligent Decision Technologies 2017*,
Smart Innovation, Systems and Technologies 73, DOI 10.1007/978-3-319-59424-8_3

the classification task reduces the predictive capabilities of classifiers as well. In effect, apart from cases where there is a clear spatial distinction among patterns belonging C_R and C_F, in the case of complex problems characterized by highly overlapping classes and complex boundaries, the mission of the learner becomes hard and standard classifier tend to solve conflicts in favour of majority class [6]. Other factors that contribute to complicate the task include the unbalance degree (the more unbalanced the dataset the more difficult is the detection of C_R samples) [6], and the poor quality (presence of noise and outliers) of data.

In this paper a method for the preprocessing of unbalanced datasets to be used for the training of classifiers (the so-called *resampling*) is proposed. This method is based on the optimal combination of the selection of a set of C_F samples to be removed from the original dataset and the generation of a set of synthetic C_R samples to be added by means of Genetic Algorithms (GA). The paper is organised as follows: Sect. 2 introduces the approaches designed for dealing with class imbalance. Section 3 describes the proposed method, whose performance are assessed in Sect. 4 through a set of tests involving unbalanced dataset coming from literature and industrial applications. Finally, in Sect. 5 conclusions are drawn and future perspectives of the proposed approach are discussed.

2   Classification of Unbalanced Datasets

Classifiers designed to cope with unbalanced datasets usually aim at the correct detection of rare patterns that are of interest for the specific applications. Two main classes of approaches can be distinguished: internal and external methods. **Internal methods** are the techniques based on the development or the modification of algorithms for the purpose of coping with unbalanced datasets. Since unbalance biases the classifiers toward C_F, these approaches directly promote the correct detection of C_R patterns despite the generation of an acceptable rate of *false positives*. A widely used family of internal methods is the so-called *Cost-Sensitive Learning* (CSL), which attributes a higher weight to the misclassification of rare patterns than the one of frequent samples. CSL is suitable for those problems where both class distribution and misclassification costs are unbalanced and can be easily coupled to most standard classifiers such as ANNs and DT. In [10] a CSL approach is used for the tuning of the output of an ANN devoted to the identification of machine faults. Different ANN-based methods are combined in [9] where the LASCUS method exploits a CSL approach within a fuzzy inference system to assign the clusters determined through a Self-Organizing Map to the C_R and C_F classes. An asymmetric cost matrix is used in [14] for the training of a DT for the choice of the split attribute of the nodes of the tree. Many existing algorithms have been adapted in order to successfully face unbalanced datasets. In [8] for instance an One-Class-Learning approach involving a modified version of SVM, the v-SVM, was proposed. Boosting belongs to the family of ensemble methods that combine the output of multiple weak learners, progressively added to the ensemble, to determine the overall output. In this framework the AdaCost method [7] adds at each step a new weak learner that is trained by using only those samples of the training dataset that are not correctly classified paying particular attention to those belonging to C_R so as to improve their detection at each step. Internal methods are, by their nature, often designed for facing particular problems and they generally achieve good results on the specific task they are developed for. However, their specificity limits their portability.

The **external methods** modify the training dataset by means of the *resampling* operation that changes the number of C_R and C_F samples in order to reduce the unbalance rate and allow the use of a standard classifier. In most of cases datasets are not totally rebalanced but rather the unbalance rate is decreased to a pre-determined ratio whose value varies for the different problems and data distributions without a rationale for its determination. The main advantage of external methods lies in their portability. They operate only on the training dataset and do not need any modification from the algorithms side, thus they can be combined with any type of classifier. There are basically two ways to reduce a dataset unbalance ratio: *under-sampling* operates by removing C_F samples from the dataset whilst *over-sampling* increases the number of C_R samples in the dataset. None of these two opposite techniques is prevalent and no optimal general purpose re-sampling strategy can be outlined [15]. Both for over and under sampling, the straightforward approach of randomly selecting the samples to be removed (undersampling) or replicated (oversampling) may lead to success in some cases but it is prone to significant counter effects. Random undersampling can exclude samples with high informative content, with a negative effect on the performance for the whole classifier. Random oversampling that casually replicates C_R samples can lead to the formation of compact *clusters* of minority samples that reduces, instead of expanding, the area of the input space associated to C_R and gives rise to over-fitting problems. In order to overcome these risks, several sophisticated approaches for the selection of the samples to remove or add to the training dataset have been developed. These approaches pursue the identification of the samples whose removal or addition maximizes the benefits in terms of the classifier performance. Examples of *focused* under-sampling techniques can be found in [11], where the selection is performed on the basis of noise in the data and their distance to class boundaries in order to reduce the dimension of domain areas associated to C_F patterns. In [12] data are undersampled by removing majority samples from compact homogeneous clusters of C_F samples in order to reduce redundancy. Focused over-sampling is proposed in [13], where the positive samples located in proximity of the boundary regions between minority and majority classes are replicated in order to spread the regions that the classifier associates to the C_R and to limit eventual classification conflicts.

In the framework of over-sampling approaches, advanced techniques are not limited to the replication of existing C_R samples that do not add information: SMOTE (Synthetic Minority Oversampling TEchnique) [16] is a well-known method that synthetically creates samples belonging to C_R and places them where they probably could be (i.e. along the lines connecting two existing minority samples). The creation of synthetic samples avoids the overfitting problem but the risk SMOTE assumes is the generation of misleading information due to the positioning of synthetic samples that could be placed close to existing C_F samples. This criticality is overcome in [17], where the procedure of generation of synthetic samples, inspired by SMOTE, takes into consideration both the existing samples belonging to C_R and C_F. In [18] the *Smart Undersampling* (SU) algorithms was proposed, which combines focused undersampling to the optimal determination of the training dataset unbalance ratio by selecting the samples to be removed on the basis of a set of criteria by means of the use of GA. The resampling method proposed in the present work associates the ideas behind the SU approach to a focused oversampling approach based on the creation of synthetic C_R samples.

3 Optimal GA-Based Resampling

The novel *Optimal GA-based Resampling* (OGAR) method combines focused undersampling and oversampling to the aim of overcoming the limitations described in Sect. 2 and maximizing the performance of classifier.

In terms of undersampling, the basic idea of OGAR is to remove from the training dataset only those C_F samples that prevent the correct characterization of minority class. As far as oversampling concerns, OGAR purpose is the addition of the most suitable subset of samples among a wide set of candidates synthetically generated by means of SMOTE. For the choice of the samples to be removed or added to the dataset multiple criteria related to classifiers performance are taken into consideration and managed within an optimization framework performed by means of GAs. GAs are used to determine the optimal resampling rate associated to each criterion (both for oversampling and undersampling). The OGAR algorithm is described in the following paragraphs. Without compromising the validity of the method all the variables of the dataset are normalized into the range [0; 1] and the original dataset DS is partitioned keeping the original unbalance rate into a training D_{TR} (50%), validation D_{VD} (25%) and a test D_{TS} (25%) datasets that are exploited as usual.

The first step consists in the creation of a set C_{SM} of candidate C_R samples by means of the SMOTE algorithm. The cardinality of C_{SM} is determined so as to fully balance the classes. Once C_{SM} is created, all C_{SM} and C_F samples are evaluated according to different criteria that rate them with a score $s \in [0, 1]$ that quantifies the detrimental and beneficial effect on classifier performance for C_F and C_{SM} samples respectively. In the case of C_{SM} samples, for instance, the higher s the better, whilst for C_F samples the higher s the more detrimental the sample. The criteria calculated for each samples $x \in C_F$ are the following:

1. **Distance to Closest Rare sample (UDCR)** computed as:

$$UDCR(x) = 1 - \min_{p \in C_R} \|x - p\| \tag{1}$$

where C_R represents the set of frequent patterns within the training dataset and the $\| \|$ operator represents the Euclidean distance. This value is higher (i.e. closer to 1) for those frequent patterns that are closest to rare ones thus their removal is advantageous.

2. **Distance to Closest Frequent sample (UDFR)** calculated as

$$UDFR(x) = \min_{p \in C_F} \|x - p\| \qquad (2)$$

that is higher for those x close to other frequent samples. The removal of this latter samples reduces unbalance rate without loosing informative content.

3. **Average Distance to Rare samples (UADR)** calculated as:

$$UADR(x) = 1 - \operatorname*{average}_{p \in C_F}(\|x - p\|) \qquad (3)$$

that is closer to 1 for those frequent samples whose neighbourhood includes a high number of rare samples.

The criteria calculated for each samples $x \in C_{SM}$ are the following:

1. **Distance to Closest Frequent sample (ODFR)** calculated as

$$ODFR(x) = \min_{p \in C_F} \|x - p\| \qquad (4)$$

that is higher for SMOTE-created samples far from existing frequent samples and thus do not generate conflicts with them (the major limit of SMOTE).

1. **Average Distance to Frequent samples (OADF)** calculated as:

$$OADF(x) = \operatorname*{average}_{p \in C_F}(\|x - p\|) \qquad (5)$$

that is higher for SMOTE-created samples whose neighbourhood includes a low number of frequent samples. Also in this case these latter samples are less prone to conflicts generations.

Once these ratings are calculated they are sorted in a decreasing order by creating 3 rankings for C_F samples and 2 for C_{SM} samples. The samples on top of the C_F rankings should be removed, while those ones on top of the C_{SM} ranking should be added. OGAR determines through GAs the optimal undersampling percentages R_{UDCR}, R_{UDFR} and R_{UADR} associated to the rankings related to UDCR, UDFR and UADR, respectively, and the optimal oversampling percentages R_{ODFR}, R_{OADF} associated to ODFR and OAFR rankings respectively. The undersampling percentages are used to pick from the top of the rankings the suitable amount of samples to be removed from the training dataset whilst the oversampling percentages are used to select in an analogous manner the synthetically generated samples to be actually added to the dataset.

The optimal resampling rates associated to each ranking are determined through a GAs-based optimization in order to maximize the benefits of the classifier in terms of performance. The GA candidate solutions are coded as 5 elements vectors ($[R_{UDCR}, R_{UDFR}, R_{UADR}, R_{ODFR}, R_{OADF}]$) where each element is associated to the resampling percentage related to the 5 involved rankings and varies in the range $[0\%, 100\%]$. GA candidate solutions are randomly initialised and evolved through the generations by means of a *single point*-type *crossover* and a *mutation* operator that varies in a range $[-5\%, +5\%]$ one random element within selected chromosomes. Selection operator is based on the *roulette wheel* technique, while the terminal condition is satisfied when an arbitrary number of generations of GAs has been completed or a stall condition is reached. Given a candidate solution, GAs minimize a fitness function as follows:

1. The candidate solution is used to build the resampled training dataset $\widehat{D_{TR}}$;
2. A DT classifier is trained by means of the C4.5 algorithm by exploiting $\widehat{D_{TR}}$;
3. The performance of the trained classifier is evaluated on both the not-resampled training dataset D_{TR} and the validation dataset D_{VD} according to the following expression proposed in [10]:

$$E = 1 - \frac{\gamma TPR - FPR}{ACC + \mu FPR} \qquad (6)$$

where TPR represents the True Positive Rate, FPR the False Positive Rate, ACC the overall accuracy while γ and μ are two empirical parameters. The calculated value E varies in the range $[0, 1]$ and is close to 0 for well performing classifiers. Equation (6) is used for the calculation of the performance E_{TR} on D_{TR} and E_{VD} on D_{VD}

4. The final fitness for the candidate is calculated as

$$Fitness = E_{TR} + |E_{TR} - E_{VD}| \qquad (7)$$

where the $|E_{TR} - E_{VD}|$ term is added to take into account the effect of overfitting.

The result of the GA-based optimization are the optimal oversampling and undersampling rates associated to the employed rankings. These rates are used to resample the original training dataset D_{TR} and generate $\widehat{D_{TR}}$ that is used for the final classifier training.

4   Test of the Method

The proposed approach has been tested on unbalanced datasets both related to industrial applications and taken from the UCI repository [19]. All the related applications are focused on the detection of rare patterns since they represent situations of interests. The datasets have different characteristics - summarized in Table 1.

The *Occlusion* dataset concerns the detection of clogs of nozzles during the continuous casting process in the steel-making industry. The detection of occlusions (1.2% of observations) can avoid product quality problems and machinery damages and is thus fundamental. The generation of false positives within this

Table 1. Main characteristics of the original datasets used in the tests.

Source	Dataset	Unb. rate	Samples	Vars
UCI rep.	CARDATA	3.8%	1728	6
	NURSERY	2.5%	1296	8
	SATELLITE	9.7%	6435	36
Industrial	OCCLUSION	1.2%	3756	6
	MSQ-1	24%	1915	11
	MSQ-2	0.3%	21784	9

classification problem is tolerable. The two *Metal Sheet Quality* (MSQ1, MSQ2) datasets refer to the automatic grading of metal sheets quality on the basis of the information provided by a set of sensors that inspect sheets surface. The main classification problem is to assess whether a metal sheet complies with quality standards. Since it is fundamental not to put into market defective products, the correct identification of non-complying sheets (the rare patterns) is aimed.

The results achieved by OGAR are compared to those obtained when no resampling is performed and to those of other resampling approaches: random oversampling and undersampling, SMOTE and SU. The classifier used within all the approaches is a DT for whose the C4.5 algorithm training was used. In the case of the methods that do not automatically reach an optimal unbalance rate the algorithms have been run testing different rates varying from 5% up to 50% and the best results have been considered. As far as the OGAR and SU settings are concerned, the GAs population cardinality is set to 100 whilst a terminal condition based on a maximum of 50 generations is adopted. The results obtained by all the involved methods on the test datasets are summarized in Table 2 in terms of ACC, TPR, FPR. In addition the optimal unbalance rate reached by SU and OGAR is reported and, in the case of the methods that do not automatically set it, the one for which the method performed best among all the tested rates. In Table 3 more information on the datasets resampled by means of the OGAR method and that led to the results depicted in Table 2 is provided.

The results put into evidence the validity of the proposed approach which creates a dataset that sensibly increases the classification performance. When no resampling is used the performance of the classifiers are poor: for all the datasets, the classifier is able to achieve an overall satisfactory rate of correct classifications but TPR is far lower with respect to other methods. When facing the literature datasets OGAR outperforms the other resampling techniques: it improves the TPR without sensibly increasing (or even decreasing) FPR. In the case of SATELLITE dataset OGAR gets a FPR 3% higher than SU (the best performer in terms of TPR) but is able to improve of 8% the TPR with respect to the best performing method. The OGAR behaviour on the industrial dataset is similar: it improves classification performance of all the tested datasets achieving higher TPR while keeping the FPR comparable or better than the other methods. It is worth to mention the case of MSQ-2 dataset which is strongly

Table 2. Results achieved by DT tested on datasets resampled by means of the proposed techniques. In the case of multiple tests characterized by different unbalance rates the best performing achievement is reported.

Database	Method	Unb. rate	ACC	TPR	FPR
CARDATA	No Resampling	-	98%	79%	1%
	Rand. Unders.	8%	98%	92%	2%
	SU	8%	98%	99%	2%
	Rand. Overs.	10%	99%	91%	1%
	SMOTE	25%	99%	81%	1%
	OGAR	24%	99%	100%	1%
NURSERY	No Resampling	-	99%	83%	1%
	Rand. Unders.	10%	97%	97%	3%
	SU	4%	97%	99%	3%
	Rand. Overs.	25%	99%	93%	1%
	SMOTE	10%	99%	84%	0
	OGAR	56%	98%	100%	2%
SATELLITE	No Resampling	-	91%	55%	5%
	Rand. Unders.	50%	82%	80%	18%
	SU	14%	87%	77%	12%
	Rand. Overs.	25%	91%	54%	5%
	SMOTE	45%	90%	62%	7%
	OGAR	40%	85%	88%	15%
OCCLUSION	No Resampling	-	98%	5%	2%
	Rand. Unders.	20%	90%	52%	10%
	SU	8%	88%	67%	11%
	Rand. Overs.	10%	99%	11%	1%
	SMOTE	25%	96%	0%	4%
	OGAR	36%	89%	73%	11%
MSQ-1	No Resampling	-	86%	65%	7%
	Rand. Unders.	30%	83%	68%	12%
	SU	26%	84%	73%	13%
	Rand. Overs.	25%	85%	67%	9%
	SMOTE	50%	84%	67%	11%
	OGAR	36%	86%	76%	10%
MSQ-2	No Resampling	-	99%	0%	0%
	Rand. Unders.	2%	99%	4%	1%
	SU	6%	90%	46%	10%
	Rand. Overs.	20%	99%	13%	1%
	SMOTE	5%	99%	10%	1%
	OGAR	53%	90%	65%	9%

Table 3. Details on the resampled dataset achieved by the OGAR method. The original total (Orig. #TR) and rare (Orig. $\#C_R$) number of samples are reported together with the number of removed samples (Rem.) and added (Add.) minority class samples.

Database	Orig. #TR	Orig. $\#C_R$	Rem.	Add.
CARDATA	865	33	298	140
NURSERY	6480	164	3567	3389
SATELLITE	3218	313	1096	907
OCCLUSION	1879	24	1126	385
MSQ-1	958	231	27	166
MSQ-2	5031	47	2611	2589

unbalanced, nevertheless OGAR is able to rise the TPR to 65%. Through all the tested datasets SU emerges as the best competitor of OGAR as it is able achieve high TPR and, more in general, its classification performance in the unbalanced dataset philosophy is satisfactory. The analysis of Table 3 shows that OGAR often strongly modifies the dataset unbalance rate either removing many C_F samples or adding a number of C_R samples: this behaviour is fruitful since in the tested problems drastically improves performance. The optimal unbalance rate reached by OGAR is generally higher than the one reached by SU due to the OGAR capability of adding minority samples to the dataset. The achieved performance compared to SMOTE puts into evidence that the intrinsic management of synthetic samples location pursued by OGAR is extremely fruitful.

5   Conclusions and Future Work

A method for the resampling of unbalanced datasets was proposed, which combines undersampling and oversampling finding in an automatical manner the optimal removal and addition rates of samples in order to maximize the classifier performance in accordance to the unbalanced dataset context. A GA optimization step determines simultaneously, according to a set of key-performance-indicators, the optimal undersampling and oversampling rates. The tests performed on literature and industrial unbalanced datasets put into evidence the superior performance of the proposed method with respect to other resampling approaches: it is in fact able to increase the rate of detected rare patterns without affecting the FPR and the overall accuracy. In the future more criteria will be involved in the GA-optimization process and a deeper connection among oversampling and undersampling will be investigated.

References

1. Borselli, A., Colla, V., Vannucci, M., Veroli, M.: A fuzzy inference system applied to defect detection in flat steel production. In: 2010 World Congress on Computational Intelligence, Barcelona, Spain, 18–23 July 2010, pp. 148–153 (2010)

2. Stepenosky, N., Polikar, R., Kounios, J., Clark, C.: Ensemble techniques with weighted combination rules for early diagnosis of alzheimer's disease. In: International Joint Conference on Neural Networks, IJCNN 2006 (2006)
3. Vannucci, M., Colla, V., Nastasi, G., Matarese, N.: Detection of rare events within industrial datasets by means of data resampling and specific algorithms. Int. J. Simul. Syst. Sci. Technol. **11**(3), 1–11 (2010)
4. García-Pedrajas, N., Ortiz-Boyer, D., García-Pedrajas, M.D., Fyfe, C.: Class imbalance methods for translation initiation site recognition. In: Proceedings of Trends in Applied Intelligent Systems, IEA/AIE 2010, Part I, pp. 327–336 (2010)
5. He, H., Garcia, E.A.: Learning from imbalanced data. IEEE Trans. Knowl. Data Eng. **21**(9), 1263–1284 (2009)
6. Estabrooks, A., Japkowicz, N.: A multiple resampling method for learning from imbalanced datasets. Comput. Intell. **20**(1), 18–36 (2004)
7. Fan, W., Stolfo, S.J., Zhang, J., Chan, P.K.: AdaCost: misclassification cost-sensitive boosting. In: Proceedings of the 16th International Conference on Machine Learning, ICML 1999, pp. 97–105 (1999)
8. Scholkopf, B., et al.: New support vector algorithms. Neural Comput. **12**, 1207–1245 (2000)
9. Vannucci, M., Colla, V.: Novel classification method for sensitive problems and uneven datasets based on neural networks and fuzzy logic. Appl. Soft Comput. **11**(2), 2383–2390 (2011)
10. Vannucci, M., Colla, V., Sgarbi, M., Toscanelli, O.: Thresholded neural networks for sensitive industrial classification tasks. In: Cabestany, J., Sandoval, F., Prieto, A., Corchado, J.M. (eds.) IWANN 2009. LNCS, vol. 5517, pp. 1320–1327. Springer, Heidelberg (2009)
11. Kubat, M., Matwin, S.: Addressing the curse of imbalanced training sets: one-sided selection. In: Proceedings of the Fourteenth International Conference on Machine Learning, pp. 179–186 (1997)
12. Laurikkala, J.: Improving identification of difficult small classes by balancing class distribution. In: Proceedings of Artificial Intelligence in Medicine: 8th Conference on Artificial Intelligence in Medicine in Europe, pp. 63–66 (2001)
13. Japkowicz, N.: The class imbalance problem: significance and strategies. In: International Conference on Artificial Intelligence, Las Vegas, Nevada, pp. 111–117 (2000)
14. Ling, C.X., Yang, Q., Wang, J., Zhang, S.: Decision trees with minimal costs. In: Proceedings of the Twenty-First International Conference on Machine Learning, pp. 69–76 (2004)
15. Batista, G.E.A.P.A., Prati, R.C., Monard, M.C.: A study of the behavior of several methods for balancing machine learning training data. SIGKDD Explor. Newsl. **6**(1), 20–29 (2004)
16. Chawla, N.V., Bowyer, K.W., Hall, L.O., Kegelmeyer, W.P.: SMOTE: Synthetic Minority Over-Sampling Technique. J. Artif. Intell. Res. **16**, 321–357 (2002)
17. Cateni, S., Colla, V., Vannucci, M.: A method for resampling imbalanced datasets in binary classification tasks for real-world problems. Neurocomputing **135**, 32–41 (2014)
18. Vannucci, M., Colla, V.: Smart under-sampling for the detection of rare patterns in unbalanced datasets. Smart Innov. Syst. Technol. **56**, 395–404 (2016)
19. Lichman, M.: UCI ML Repository. School of Information and Computer Science, University of California, Irvine (2013). http://archive.ics.uci.edu/ml

Analysis of Multiple Classifiers Performance for Discretized Data in Authorship Attribution

Grzegorz Baron[(✉)]

Silesian University of Technology, Akademicka 16, 44-100 Gliwice, Poland
grzegorz.baron@polsl.pl

Abstract. In authorship attribution domain single classifiers are often employed in research as elements of decision system. On the other hand, there is intuitive prediction that the use of multiple classifier with fusion of their outcomes may improve the quality of the investigated system. Additionally, discretization can be applied for input data which can be beneficial for the classification accuracy. The paper presents performance analysis of some multiple classifiers basing on the majority voting rule. Ensembles were composed from eight single well known classifiers. Influence of different discretization methods on the quality of the analyzed systems was also investigated.

Keywords: Multiple classifier · Ensemble classifier · Majority voting · Discretization · Authorship attribution

1 Introduction

One of the typical approaches to classification tasks involves application of selected inducer and executing supervised learning process. In many cases it allows to obtain a good decision system. Some solutions in data mining domain proposed use of a set of classifiers with fusion of their outputs. It is obvious that some classifiers perform better than others in certain subspaces of the problem domain. In other words, for complex problems different subsystems of the decision system can find best solution. Analysis of this in the context of authorship attribution constituted the main objective for the presented work.

On the other hand, analysis of the influence of discretization on overall efficiency of the decision systems, was executed in authorship attribution belonging to the stylometric domain [1]. Generally speaking discretization is the process which allows to convert data in continuous form into discrete space. It can be employed when subsequent elements of a decision system cannot operate on continuous numbers. Applying discretization for data can bring other advantages like improvement of performance of the decision system. It happens because discretization changes the nature of data, the way of expressing knowledge contained in analyzed data, and decreases its volume.

Another problem related to discretization of input data sets must be considered, namely the way of discretization of test datasets. It has been shown [2] that

© Springer International Publishing AG 2018
I. Czarnowski et al. (eds.), *Intelligent Decision Technologies 2017*,
Smart Innovation, Systems and Technologies 73, DOI 10.1007/978-3-319-59424-8_4

employment of specially prepared test datasets is more reliable for evaluation of decision systems, such as presented in the paper. In [4] three approaches were proposed and investigated: "independent", "test on learn", and "glued". In the first method learning and test datasets are discretized independently, given the same parameters. For "test on learn" approach test sets are discretized basing on the information calculated for training data. In the "glued" approach learning and test data is concatenated for the discretization process and then split in order to obtain again training and test sets.

The paper presents research results of combined influence of application of multiple classifiers and discretization on the efficiency of the decision system. Different approaches to test sets discretization were taken into consideration. To the best of author's knowledge no other published research addressed this subject for authorship attribution domain.

The paper is organized as follows. Section 2 gives the theoretical background and an overview of methods employed in the research. Section 3 presents the experimental setup, datasets used, and techniques employed. The results and their discussion are given in Sect. 4. Section 5 provides conclusions.

2   Theoretical Background

The following paragraphs present background of the presented research including multiple classifier systems, discretization, and employed inducers.

2.1   Multiple Classifier

Multiple classifier is considered as a set of classifiers with their results fused in some way to obtain better decisions. There is a lot of names concerning that methodology, such as ensemble methods, classifier fusion, aggregation, decision combination, hybrid methods, cooperative agents [11].

The reason to use more than one classifier to improve the classification accuracy seems to be intuitively correct. Yet the question arises, how to properly select classifiers for a specific task. In [10] authors mentioned that accuracy and diversity of the individual members of an ensemble of classifiers constitute necessary and sufficient conditions to obtain better results. The classifier is considered as accurate when its error rate is better than random guessing for new input. The diversity of two classifiers can be defined as ability to make different errors for new data. The short example explains this issue [6]. Lets assume a multiple classifier composed of three classifiers $\{c_1, c_2, c_3\}$. If they are similar, i.e. not diverse, then for a new hypothesis h if $c_1(h)$ is wrong, also $c_2(h)$ and $c_3(h)$ are wrong. When the diversity condition is satisfied, if $c_1(h)$ is wrong, $c_2(h)$ and $c_3(h)$ may be correct, which means that employment of majority vote rule can deliver better results.

According to [6] there are three reasons why ensemble of classifiers might perform better: statistical, computational, and representational. The statistical reason is based on the idea of averaging outputs of individual classifiers in an ensemble. For example it is possible to train a set of classifiers with re-substitution error equal to 0. So their performance is similar taking training data into consideration. On the other hand, they may behave differently for new data. Therefore fusion of outputs of classifiers belonging to the ensemble seems to be a better solution than selecting only one classifier.

The computational reason is related to the problem of local optima, which is vital for many classifiers. Running the optimum search from different starting points in multiple classifier may provide better results, than the individual ones.

The representational reason concerns the properties of classifiers in respect to the problem to be solved. For example single linear classifiers may not be suitable for solving some stated problem, while the ensemble of linear classifiers can approximate the decision area. Of course it is possible to employ a single, but more complex classifier, such as neural network, to solve the same problem. However, it is likely that computational cost will be higher [14].

Depending on the types of classifiers output different methods of fusing their results can be applied. According to [20] three general output levels can be listed:

- the abstract level – classifier produces only the class label; there is no information about certainty of the predicted classes,
- the rank level – the outputs are ranked in the queue; order is determined by the plausibility of each output,
- the measurement level – classifiers produce outputs, where for each class there is provided information about support for the hypothesis, that input vector belongs to that class.

There are two general architectures of multiple classifiers [18]:

- homogeneous classifier - the same algorithm is applied for diversified data: bagging, boosting, multiple partitioned data, multi-class specialized systems;
- heterogeneous classifiers - different learning algorithms utilize the same data: voting, rule-fixed aggregation, stacked generalization or meta-learning

In the presented research the voting algorithm was investigated. Let us assume that $\Omega = \{\omega_1, \ldots \omega_c\}$ is a set of c class labels, and classifiers outputs are described by vectors $[d_{i,1}(x), \ldots, d_{i,c}(x)]^T \in \{0,1\}^c$, $i = 1, \ldots, L$, where L is the number of classifiers in the ensemble and x is the input. $d_{i,j}(x) = 1$ if individual classifier D_i labels input x with ω_j, otherwise it is set to 0. If the following formula is satisfied for $k \in \Omega$,

$$\sum_{i=1}^{L} d_{i,k}(x) = \max_{j=1}^{c} \sum_{i=1}^{L} d_{i,j}(x) \tag{1}$$

it means that multiple classifier declares ensemble result for class ω_k. Formally Eq. (1) describes the plurality vote often called also majority vote.

To examine properties of such classifier let us assume that for each classifier p describes the probability of correct answer for a given input $x \in \Re^n$, and the number of classifiers L is odd. Their outputs are assumed to be independent, which means that the joint probability can be described as:

$$P(D_{i_1} = S_{i_1}, \ldots, D_{i_K} = S_{i_K}) = P(D_{i_1} = s_{i_1}) \times \cdots \times P(D_{i_K} = s_{i_K}) \quad (2)$$

for any subset of classifiers $A \subseteq \mathbf{D}, A = \{D_{i_1}, \ldots, D_{i_K}\}$ where s_{i_j} is the label output of classifier $D_{i,j}$. The accuracy of the multiple classifier is as follows [14]:

$$P_{\mathrm{maj}} = \sum_{m=\lfloor L/2 \rfloor + 1}^{L} \binom{L}{m} p^m (1-p)^{L-m} \quad (3)$$

The analysis of behavior of P_{maj} in relation to p leads to the following conclusions:

- if $p > 0.5$ then $P_{\mathrm{maj}} \longrightarrow 1$ when $L \longrightarrow \infty$
- if $p < 0.5$ then $P_{\mathrm{maj}} \longrightarrow 0$ when $L \longrightarrow \infty$
- if $p = 0.5$ then $P_{\mathrm{maj}} = 0.5$ for any L

So if classifiers in the ensemble have $p > 0.5$, improvement of quality of the multiple classifier can be expected, comparing to the results of single inducers.

2.2   Discretization

In many research areas input data is given as numerical values, which is obvious because the nature of real world is also continuous. Yet many algorithms and methods used in data mining can not operate at all for real valued features, or perform better for discrete data. In such cases discretization can be used as a way of converting continuous data into nominal representation. Such processing can improve the overall quality of a decision system.

There are many discretization algorithms. One way of categorization splits these methods into supervised and unsupervised. Supervised algorithms such as Fayyad and Irani's MDL [7] or Kononenko MDL [13], take into consideration information about class of each instance in the data set. Because the nature of data is analyzed in some way, they can deliver better, more accurate results. Unsupervised methods perform discretization process basing only on given parameters and values of attributes in all instances. Class information is omitted.

The most popular unsupervised methods are equal width and equal frequency binning. These two are simple algorithms which determine minimum and maximum values of any given attribute, and then search for cut points in order to obtains bins of the equal width or bins filled with the same number of instances in case of equal frequency binning. There is also a modification of equal width method, available in WEKA [9], which performs a kind of bin number optimization using the leave-one-out estimation of estimated entropy.

Evaluation of analyzed decision systems was performed using test datasets. The previous research in that area proved that it is the most reliable method of assessing the quality of the system [2]. When discretization of input data is taken into account a new problem arises, how to discretize test datasets. Some earlier research in that area allowed to formulate three approaches to that problem: "independent" – *Id* – where learning and test data are discretized separately, given the same parameters; "test on learn" – *ToLd* – where test data is discretized using cut points calculated for training data, and "glued" – *Gd* – where test and learn datasets are concatenated, discretized and finally separated to obtain training and test datasets again. Deeper discussion of these methods, their advantages and disadvantages can be found in [4].

2.3   Classifiers Used in Research

Construction of voting multiple classifier requires selection of different classifiers. As aforementioned, it is a good idea to choose a diversified set of inducers, taking into consideration their way of data processing and performed algorithm. On the other hand, candidates should have satisfactory performance for individual tasks. Basing on such criteria the following list of classifiers was created: Naive Bayes (NB), Bayesian networks (BNet), C4.5, PART, Random Forest (RF), k-Nearest Neighbors (k-NN), Multilayer Perceptron (MLP), Radial Basis Function (RBF) network, all implemented in WEKA.

3   Experimental Setup

The following steps were performed in the experiments:

- input data preparation including splitting of data into learning and test sets,
- discretization of input sets applying all chosen methods,
- training of each of selected classifiers,
- evaluation of individual and multiple classifiers

Selection of attributes is the big research area and many different methods can be applied [15–17]. Also the principal components analysis (PCA) must be mentioned as useful technique in authorship attribution [5,8,12,19]. However, it was not investigated because the feature extraction was not in the main focus of the research. Basing on the preliminary experiments the set of function words containing two-letter words taken from the list of the most frequently used words in English has been created. The personal pronouns have been excluded. The input datasets were prepared basing on the works of male and female English authors. Texts were grouped into pairs in respect to gender, frequencies of occurrence of selected descriptors were calculated for each part, and datasets containing instances belonging to one of two classes representing authors were built. Similar volume of texts of each author and their splitting into the same number of parts guarantee that obtained datasets contain balanced classes. Finally sets were divided into learning and test datasets basing on the disjunctive works of authors.

During research all discretization methods described in Sect. 2 were utilized using *Id*, *Gd*, and *ToLd* approaches to discretization of test datasets. Finally, discrete descriptors of bins were converted to their ordinal numbers, and attributes type of data was declared as numeric. It allowed to avoid problems related to inequality of numbers of bins in learning and test datasets, which can occur for the *Id* approach. Analysis and investigation of this problem was conducted in [3].

Multiple classifiers were built basing on the individual classifiers presented in Sect. 2 using the majority vote rule. The general idea was to create three-member ensembles from classifiers of different nature. In this way the following multiple classifiers were constructed and tested (abbreviations as presented in Sect. 2): NB_MLP_C4.5, BNet_RBF_PART, NB_MLP_k-NN, MLP_RBF_k-NN, NB_MLP_RF. As can be noticed each ensemble contains at least one neural classifier and some combinations of Bayesian, decision tree and k-Nearest Neighbors classifiers. Observations of preliminary results of experiments, performed for aforementioned setups, indicated that some other combinations also could be interesting, so the following ensembles were created: BNet_C4.5_PART and NB_k-NN_RF.

4   Results and Discussion

Experiments were executed separately for male and female authors datasets, but all results are presented as average of outcomes. In further discussion whenever experiment or result is mentioned it is considered as a pair of experiments or averaged result for male and female authors. For MLP and RBF classifiers the multistarting criterion was applied and the best result from 20 rounds of network training and evaluation was taken for further analysis. For k-NN there were experiments for $k = 1, \ldots, 9$ neighbors and also the best result was taken.

Preliminary set of experiments was performed for nondiscretized data to obtain reference points for discussion. The evaluation results of individual and multiple classifiers are presented in Fig. 1 as *raw data* bars. As evaluation quality measure there was taken the percentage of correctly classified instances.

The next experiments were performed for discretized data, with three different approaches to discretization of test datasets. Supervised methods do not require any parameters so for each experiment the single result was obtained. Figure 1 presents outcomes for supervised discretization algorithms.

Unsupervised methods need additional parameters like the required numbers of bins. In the research the values of these parameters varied from 2 to 10. Such range was defined basing on the earlier experiments [1]. In such case for each experiment the set of 9 values was delivered as result. In order to show these results clearly the boxplot diagrams were chosen. Figure 2 presents all results with respect to discretization methods and the way of test sets discretization.

Diagrams presenting the results for "independent" way of test sets discretization were omitted because of extremely poor results for supervised methods, comparing to the reference ones. The possible reason lies in the nature of data. Independent discretization of learning and test sets leads to calculation of so different cut points in both sets that coherency between them is lost. It is very interesting observation which warns against uncritical use of *Id* method. Also results presented in Fig. 1b show poorer performance for discretized data comparing to the *raw data*. Even though some relations between individual and multiple classifiers can be observed there, it is better to perform such observations for Gd method presented in Fig. 1a.

The main research task was an analysis of performance of multiple classifiers. The analysis can base on observations of single classifiers quality (Fig. 1a). It is easy to notice that MLP classifier is the best one. Others perform slightly poorer, but almost all work better for discretized data than for nondiscretized. Taking the multiple classifiers into consideration, one general observation is obvious: none of them is better than MLP. It is interesting that all ensembles containing MLP (NB_MLP_C4.5, NB_MLP_k-NN, MLP_RBF_k-NN, NB_MLP_RF) cannot perform better than a single MLP classifier. On the contrary, fusion with other classifiers decreases the overall quality. The probable explanation of this fact is that MLP, as a very good classifier, utilized exhaustively all information included in the learning data, and other worse classifiers in the ensemble decrease the overall quality. The similar conclusion can be stated for BNet_RBF_PART ensemble where for discretized data PART classifier is the best one. On the other hand, ensemble B_k-NN_RF performs better than each individual classifier for data discretized using Kononenko MDL.

a)

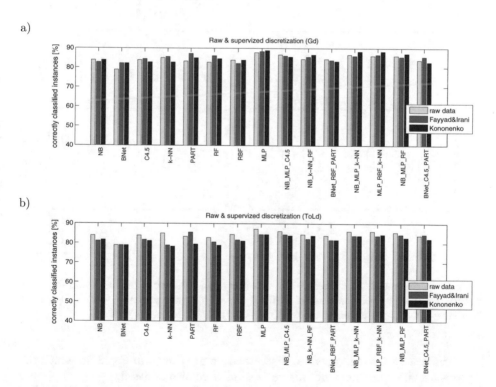

b)

Fig. 1. Performance of individual and multiple classifiers for nondiscretized data and for supervised Fayyad and Irani and Kononenko discretization. Test sets discretized using methods: (a) "glued" (Gd), (b) "test on learn" (TLd).

a)

b)

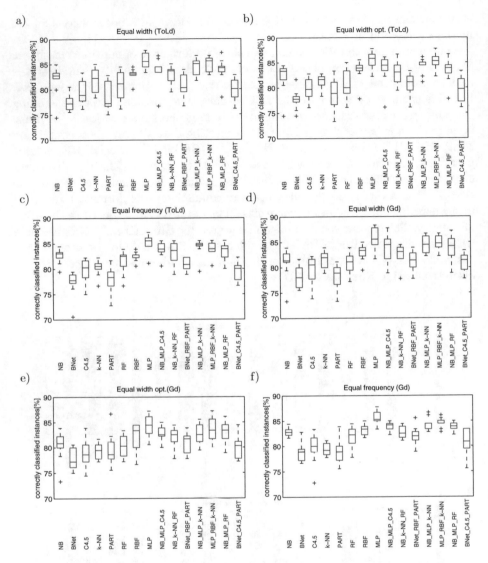

c)

d)

e)

f)

Fig. 2. Performance of classifiers for unsupervised discretization methods: (a) equal width – *ToLd*, (b) optimized equal width – *ToLd*, (c) equal frequency – *ToLd*, (d) equal width – *Gd*, (e) optimized equal width – *Gd*, f) equal frequency – *Gd*

The results for unsupervised discretization are shown in Fig. 2, in which diagrams for the "independent" discretization were also omitted.

The general observations are similar to those formulated for previous results – the performance of MLP classifier exceeds all other individual and multiple solutions. But some positive comments can be made also for multiple classifier containing simpler components. For example BNet_C4.5_PART or NB_k-NN_RF perform better than their members individually. They are relatively simple and do not perform well comparing to the winners of the ranking. In such situation multiple classifier can improve the overall quality of the decision system.

It is worth to mention, that the discussion is based on analysis of boxplots' medians. Observation of results allows to expect that deeper analysis of outcomes in relation to parameters of discretization process may deliver better results.

5   Conclusions

The paper presents results of research on application of multiple classifiers in authorship attribution domain for discretized data. Eight single classifiers, Naive Bayes, Bayesian network, C4.5, Random Forest, PART, k-NN, multilayer perceptron, and radial basis function network, were selected and composed into some ensemble classifiers. Experiments were performed for different discretization methods, and three approaches to discretization of test datasets were additionally taken into consideration.

The general conclusion is that, if ensemble contains a well performing classifier which explores deeply the input data space, it is difficult to obtain better results using multiple classifier performing the majority voting fusion. When classifiers in the ensemble are of average quality, the improvement is possible.

An important observation concerns the discretization of test datasets using "independent" approach. As it was noticed during the research, such way of discretization can deliver unexpectable results. It is strongly visible for supervised discretization where algorithm decides the number and positions of the cut points. Test datasets processed in such way may lose their relation to the original data. A system which performed well for nondiscretized data, completely lost its quality for the discretized input.

To summarize, the experiments show that application of multiple classifier under conditions presented in the paper can be beneficial, especially when the ensemble is composed of simple classifiers. In such situations short learning time or extensive use of resources can be the advantage. On the other hand, employment of a powerful classifier like multilayer perceptron can solve the problem using a single decision unit.

Acknowledgments. The research described was performed at the Silesian University of Technology, Gliwice, Poland, in the framework of the project BK/RAu2/2017. All experiments were performed using WEKA workbench [9].

References

1. Baron, G.: Influence of data discretization on efficiency of Bayesian classifier for authorship attribution. Procedia Comput. Sci. **35**, 1112–1121 (2014)
2. Baron, G.: Comparison of cross-validation and test sets approaches to evaluation of classifiers in authorship attribution domain. In: Czachórski, T., Gelenbe, E., Grochla, K., Lent, R. (eds.) Computer and Information Sciences: 31st International Symposium, ISCIS 2016, Kraków, Poland, October 27–28, 2016, Proceedings, pp. 81–89. Springer International Publishing, Cham (2016)

3. Baron, G.: On influence of representations of discretized data on performance of a decision system. Procedia Comput. Sci. **96**(c), 1418–1427 (2016)
4. Baron, G., Haężlak, K.: On approaches to discretization of datasets used for evaluation of decision systems. In: Czarnowski, I., Caballero, M.A., Howlett, J.R., Jain, C.L. (eds.) Intelligent Decision Technologies 2016: Proceedings of the 8th KES International Conference on Intelligent Decision Technologies (KES-IDT 2016) - Part II, pp. 149–159. Springer International Publishing, Cham (2016)
5. Crippa, P., Curzi, A., Falaschetti, L., Turchetti, C.: Multi-class ECG beat classification based on a gaussian mixture model of Karhunen-Loéve transform. Int. J. Simul. Syst. Sci. Technol. **16**(1), 2.1–2.10 (2015)
6. Dietterich, T.G.: Ensemble methods in machine learning. In: Proceedings of the 1st International Workshop on Multiple Classifier Systems, MCS 2000, pp. 1–15. Springer-Verlag, London (2000)
7. Fayyad, U.M., Irani, K.B.: Multi-interval discretization of continuousvalued attributes for classification learning. In: 13th International Joint Conference on Articial Intelligence, vol. 2, pp. 1022–1027. Morgan Kaufmann Publishers (1993)
8. Gianfelici, F., Turchetti, C., Crippa, P.: A non-probabilistic recognizer of stochastic signals based on KLT. Sig. Process. **89**(4), 422–437 (2009)
9. Hall, M., Frank, E., Holmes, G., Pfahringer, B., Reutemann, P., Witten, I.H.: The weka data mining software: an update. SIGKDD Explor. **11**(1), 10–18 (2009)
10. Hansen, L.K., Salamon, P.: Neural network ensembles. IEEE Trans. Pattern Anal. Mach. Intell. **12**(10), 993–1001 (1990)
11. Ho, T.K.: Multiple classifier combination: lessons and next steps. In: Kandel, A., Bunke, H. (eds.) Hybrid Methods in Pattern Recognition, pp. 171–198. World Scientific, Singapore (2011)
12. Jamak, A., Savatić, A., Can, M.: Principal component analysis for authorship attribution. Bus. Syst. Res. **3**, 49–56 (2012). Proceedings of 11th International Conference Symposium on Operational Research in Slovenia
13. Kononenko, I.: On biases in estimating multi-valued attributes. In: 14th International Joint Conference on Articial Intelligence, pp. 1034–1040 (1995)
14. Kuncheva, L.I.: Combining Pattern Classifiers: Methods and Algorithms. Wiley-Interscience, Hoboken (2004)
15. Stamatatos, E.: A survey of modern authorship attribution methods. J. Am. Soc. Inf. Sci. Technol. **60**(3), 538–556 (2009)
16. Stańczyk, U.: Feature evaluation by filter, wrapper, and embedded approaches. In: Stańczyk, U., Jain, L.C. (eds.) Feature Selection for Data and Pattern Recognition, pp. 29–44. Springer, Heidelberg (2015)
17. Stańczyk, U.: Ranking of characteristic features in combined wrapper approaches to selection. Neural Comput. Appl. **26**(2), 329–344 (2015)
18. Stefanowski, J.: Multiple classifiers (2009). http://www.cs.put.poznan.pl/jstefanowski/aed/DMmultipleclassifiers.pdf. Accessed 27 Jan 2017
19. Turchetti, C., Biagetti, G., Gianfelici, F., Crippa, P.: Nonlinear system identification: an effective framework based on the Karhunen-Loéve transform. IEEE Trans. Signal Process. **57**(2), 536–550 (2009)
20. Xu, L., Krzyzak, A., Suen, C.Y.: Methods of combining multiple classifiers and their applications to handwriting recognition. IEEE Trans. Syst. Man Cyber. **22**(3), 418–435 (1992)

Speaker Identification in Noisy Conditions Using Short Sequences of Speech Frames

Giorgio Biagetti, Paolo Crippa$^{(\boxtimes)}$, Laura Falaschetti, Simone Orcioni, and Claudio Turchetti

DII – Dipartimento di Ingegneria dell'Informazione,
Università Politecnica delle Marche, Via Brecce Bianche 12, 60131 Ancona, Italy
{g.biagetti,p.crippa,l.falaschetti,s.orcioni,c.turchetti}@univpm.it

Abstract. The application of speaker recognition technologies on domotic systems, cars, or mobile devices such as tablets, smartphones and smartwatches faces with the problem of ambient noise. This paper studies the robustness of a speaker identification system when the speech signal is corrupted by the environmental noise. In the everyday scenarios the noise sources are highly time-varying and potentially unknown. Therefore the noise robustness must be investigated in the absence of information about the noise. To this end the performance of speaker identification using short sequences of speech frames was evaluated using a database with simulated noisy speech data. This database is derived from the TIMIT database by rerecording the data in the presence of various noise types, and is used to test the model for speaker identification with a focus on the varieties of noise. Additionally, in order to optimize the recognition performance, in the training stage the white noise has been added as a first step towards the generation of multicondition training data to model speech corrupted by noise with unknown temporal-spectral characteristics. The experimental results demonstrated the validity of the proposed algorithm for speaker identification using short portions of speech also in realistic conditions when the ambient noise is not negligible.

Keywords: Cepstral analysis · Classification · Digitized voice samples · Discrete Karhunen-Loéve transform (DKLT) · Expectation Maximization (EM) algorithm · Feature extraction · Gaussian Mixture Model (GMM) · Short sequences · Robust speaker identification · Noisy conditions · Speech · Speech frames · TIMIT · Noise · Environmental noise

1 Introduction

The voice is the most natural signal to produce and the simplest to acquire using technologies on handheld devices or the Internet [1,18] for recognizing, identifying or verifying individuals [19,25,27]. Basically there are two fundamental modes of operations: identification and verification.

In identification systems, the issue is to detect which speaker from a given set the unknown speech is derived from, while in verification systems the speech of

© Springer International Publishing AG 2018
I. Czarnowski et al. (eds.), *Intelligent Decision Technologies 2017*,
Smart Innovation, Systems and Technologies 73, DOI 10.1007/978-3-319-59424-8_5

the unknown person is compared against both the claimed identity and against all other speakers (the imposter or background model) [9,10,14,16,17]. As a result, in speaker identification the classification is based on a set of S models (one for each speaker), while, in speaker verification, two models (one for the hypothesized speaker and one for the background model), have to be derived during training.

Some recent interesting applications are vehicle access control, telephone services for transaction authorization, personalized domotic control, speaker diarization, personalized smart training [5,6,8].

Due to the mobile nature of Internet related systems, this work investigates the problem of speaker identification in noisy conditions, assuming that speech signals are corrupted by environmental noise. Because the noise sources are inherently highly time-varying and potentially unknown, the noise robustness in the absence of information about the noise is a key requirement [21,23]. In particular the performance of a speaker identification framework presented in [3,4] for the clean speech, has been further investigated here by adding several environmental noises to the clean speakers' utterances in the testing stage. More in detail, the algorithm was tested using the TIMIT database with simulated noisy speech data. The TIMIT database has been redeveloped by synthetically adding the data in the presence of various noise types, used to test the model for speaker identification with a focus on the varieties of noise.

This paper is also aimed towards the implementation of a new model for real-world applications. This include the generation of multicondition training data to model noisy speech, the combination of different training data to optimize the recognition performance, and the reduction of the model's complexity.

This paper addresses the problem of speaker identification with short sequences of speech frames, that is with utterance duration of less than 1 s, for real-world applications. In particular, as this is a very severe test for speaker identification, we want to investigate for feature representations of voice sample that guarantees the best performance in terms of classification accuracy for noisy speech. It is well known that, among linear transforms which can be used for feature extraction and dimensionality reduction, the best linear feature extractor is the Karhunen-Loève transform (KLT) expansion [15,28,29]. In addition, as robust speaker recognition remains an important problem in speaker identification [20,22,26,30,31], in a recent paper [24] it has been shown that Principal Component Analysis (PCA) transformation minimizes the effect of noise and improves the speaker identification rate as compared to the conventional Mel-frequency cepstral coefficient (MFCC) features. In particular here we exploit some convergence properties of the truncated version of the discrete Karhunen-Loève transform (DKLT) [2,11], that guarantee good performance in terms of speaker classification accuracy [3,4] extending its applicability also in noisy conditions.

This paper is organized as follows. Section 2 describes the speaker identification framework. Section 3 reports the experimental results of the classifier on the database signals and comments the performance. Finally, conclusions are drawn in Sect. 4.

2   Identification Method

In this section we briefly summarize the mathematical formulation of the overall short-sequence speaker identification framework.

2.1   Single Frame Identification

Let us refer to a frame $y = [y_1 \cdots y_N]$, representing the power spectrum of speech signal, extracted from the time domain waveform of the utterance under consideration, through a pre-processing algorithm including pre-emphasis, framing and log-spectrum. The duration of frames is 25 ms and each frame is generated every 10 ms: as a result consecutive frames overlap by 15 ms.

For Bayesian speaker identification, a group of S speakers is represented by the probability density functions (pdf's) $p_s(y) = p(y \mid \theta_s)$, $s = 1, 2, \cdots, S$ where θ_s are the parameters to be estimated during training. The objective of classification is to find the speaker model θ_s that has the maximum a posteriori probability for a given frame y:

$$\hat{s}(y) = \operatorname*{argmax}_{1 \leq s \leq S} \{p_r(\theta_s \mid y)\} = \operatorname*{argmax}_{1 \leq s \leq S} \left\{ \frac{p(y \mid \theta_s) p_r(\theta_s)}{p(y)} \right\}. \tag{1}$$

Considering $p_r(\theta_s) = 1/S$ (equally likely speakers) and $p(y)$ being the same for all speakers models, the Bayesian classification is equivalent to

$$\hat{s}(y) = \operatorname*{argmax}_{1 \leq s \leq S} \{p(y \mid \theta_s)\}. \tag{2}$$

Adopting the Gaussian Mixture Model (GMM) for the single speaker, the pdf is a weighted sum of F components densities and given by the equation

$$p(y \mid \theta_s) = \sum_{i=1}^{F} \alpha_{(s),i} \frac{(2\pi)^{-N/2}}{\sqrt{|C_{(s),i}|}} \exp\left\{ -\frac{1}{2}(y - \mu_{(s),i})^T C_{(s),i}^{-1}(y - \mu_{(s),i}) \right\} \tag{3}$$

where $\alpha_{(s),i}$, $i = 1, \ldots, F$ are the mixing weights, and $\mu_{(s),i}$ and $C_{(s),i}$ are the mean and covariance matrix values, respectively, of the s-th mixture. As a result

$$\theta_s = \{\alpha_{(s),1}, \mu_{(s),1}, C_{(s),1}, \ldots, \alpha_{(s),F}, \mu_{(s),F}, C_{(s),F}\} \tag{4}$$

is the set of parameters needed to specify the s-th Gaussian mixture.

Finally the maximum likelihood (ML) estimate of θ_s,

$$\hat{\theta}_{s,\mathrm{ML}} = \operatorname*{argmax}_{\theta_s}\{\log p(\mathcal{W} \mid \theta_s\}, \tag{5}$$

where $\mathcal{W}$ represents the training data, has been obtained using an optimized version of the Expectation Maximization (EM) algorithm [12].

2.2 Multiframe Frame Identification

In order to face the problem of dimensionality [7], to reduce y to a vector k_T of lower dimension, the DKLT of y, defined as $k = \Phi^T y$, where Φ is the solution of the eigenvector matrix equation $R_{yy} = E\{yy^T\} = \Phi \Lambda \Phi^T$ has been used in its truncated version to $M < N$ orthonormal basis functions, because among the linear transforms is the one that ensures the minimum mean square error.

The accuracy of speaker identification can be considerably improved using a sequence of frames instead of a single frame alone. To this end let us refer to a sequence of frames defined as $Y = \{y^{(1)}, \ldots, y^{(V)}\}$ where $y^{(v)}$ represents the v-th frame. Applying (2) to the truncated DKLT of y, we can determine the class each frame $y^{(v)}$ belongs to. Thus the S sets $\mathcal{Z}_s = \{y^{(v)} \mid y^{(v)}$ belongs to class $s\}$, $s = 1, \ldots S$, are univocally determined.

Given Y, we define the score for each class s as $r_s(Y) = \text{card}\{\mathcal{Z}_s\}$, where the cardinality operator $\text{card}\{\cdot\}$ extracts the number of elements belonging to $\mathcal{Z}_s$.

Finally the multi-frame speaker identification is based on:

$$\hat{s}(Y) = \underset{1 \leq s \leq S}{\text{argmax}} \{r_s(Y)\} \ . \tag{6}$$

3 Experimental Results

To get a better understanding of the performance achieved with the proposed approach, several experiments were conducted on increasing population sizes, and considering the English language.

To this end the evaluations were carried out using the widely available TIMIT speech corpus [13]. The TIMIT corpus is a collection of phonetically balanced sentences, 10 sentence utterances from 630 speakers across 8 dialect regions in the USA. From this corpus the first 100 speakers of the test set (64 male and 36 female) spanning the dialect of New England (Dialect 1), Northern (Dialect 2), North Midland (Dialect 3), South Midland (Dialect 4), and Southern (Dialect 5) regions, have been chosen.

Table 1 reports the consistency of the database used for this TIMIT-based experimental evaluation in terms of dialect and sex of the speakers.

Table 1. Consistency of the database used for the TIMIT-based experimental evaluation.

Total speakers		Dialect 1		Dialect 2		Dialect 3		Dialect 4		Dialect 5	
M	F	M	F	M	F	M	F	M	F	M	F
64	36	7	4	18	8	23	3	16	16	0	5

For each speaker, the 10 sentences have an average length of 3.10 s (minimum length: 1.09 s, maximum length: 7.57 s) at a sample frequency of 16 kHz

(16-bit samples). These sentences have been split into overlapping frames of 25 ms (400 samples), with a frame shift of 10 ms (160 samples) thus obtaining from each sentence an average number of about 480 voiced frames (ranging from a minimum value of 169 to a maximum of 1170). Finally, from each speaker, an average number of 4807 voiced speech frames have been extracted.

Then the full database so obtained was divided in two datasets containing for each speaker 80% of speech frames for training (model evaluation) and 20% for testing (performance evaluation).

Table 2. Identifier overall sensitivity with $M = 20$ DKLT components and using TIMIT corpus in different noisy conditions and SNRs.

Noise condition	SNR [dB]	Speech sequence length		
		1 frame	100 frames	350 frames
Clean speech		21.01%	80.63%	97.24%
White	15	6.84%	35.12%	58.99%
Cell phone ringing	20	19.32%	79.43%	96.77%
	15	17.90%	75.93%	95.39%
	10	15.91%	70.02%	94.47%
Jet engine	20	19.17%	77.57%	97.24%
	15	17.68%	72.98%	96.31%
	10	15.19%	67.61%	93.09%
Office	20	15.52%	69.37%	93.09%
	15	12.35%	59.19%	89.55%
	10	9.07%	43.33%	76.12%
Pop song female	20	17.99%	77.79%	95.85%
	15	15.24%	72.32%	94.47%
	10	11.60%	59.96%	87.56%
Rain	20	18.38%	75.49%	94.93%
	15	16.30%	68.60%	92.17%
	10	13.28%	51.75%	73.73%
Restaurant	20	12.43%	56.35%	87.10%
	15	9.10%	42.23%	68.66%
	10	6.54%	28.56%	51.15%
Street	20	19.46%	77.90%	97.24%
	15	18.38%	74.95%	95.85%
	10	17.07%	70.68%	91.71%
Train ride	20	19.59%	78.67%	97.70%
	15	18.67%	78.23%	96.77%
	10	17.17%	72.32%	93.55%

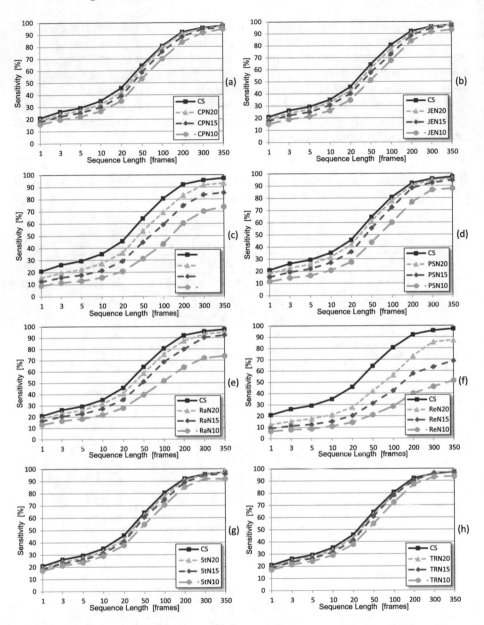

Fig. 1. Classifier performance as a function of sequence length, with $M = 20$ DKLT components and using TIMIT corpus in eight different noisy conditions: *(a)* cellular phone ringing noise, *(b)* jet engine noise, *(c)* office noise, *(d)* pop song female noise, *(e)* rain noise, *(f)* restaurant noise, *(g)* street noise, *(h)* train ride noise, and various SNRs (SNR = 20, 15, 10 dB).

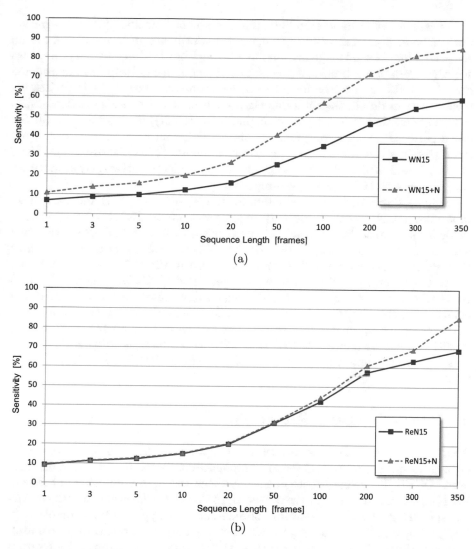

Fig. 2. Classifier performance as a function of sequence length, with $M = 20$ DKLT components and using TIMIT corpus in presence of *(a)* white noise (WN) and *(b)* restaurant noise (ReN) for the noiseless (solid line) and noised (dashed line) model (SNR = 15 dB).

Table 2 shows the micro-averaged overall sensitivity obtained on the described dataset, synthetically adding to the testing speech frames different type of artificial (white noise) and natural environmental noises, for three selected segment durations (single frame, 100 frames, and 350 frames). As can be seen, results are very consistent with all the tested natural noises.

To gain a better understanding on how segment duration affects recognition performance, Fig. 1 reports the framework performance in terms of the overall sensitivity for different sequence lengths ranging from 1 (25 ms) to 350 (3.5 s) consecutive overlapping frames. $M = 20$ DKLT components have been considered and the TIMIT corpus in different noisy conditions has been used in the testing stage. More in detail, the following eight environmental noises have been taken into consideration: cellular phone ringing noise (CPN), jet engine noise (JEN), office noise (OfN), pop song female noise (PSN), rain noise (RaN), restaurant noise (ReN), street noise (StN), and train ride noise (TRN), with varying SNRs of 10, 15, and 20 dB. For reference, the speaker identification performance for the clean speech (CS) case is also reported.

Figure 2 shows the same evaluation for the white noise (WN) and the restaurant noise (ReN) cases whose performance was worst. Here we tried another modeling system adding a variety of white noises with varying SNRs to the train frames (SNR $= 10, 12, 14, 16, 18, 20$ dB) in the modeling phase, which helped increase resilience to noise. The results are seen in the WN15+N trace, where the test was done by adding white noise to get a 15 dB SNR, and in the ReN15+N trace, where the test was done by adding restaurant noise to get a 15 dB SNR. For comparison, the WN15 and the ReN15 traces show the same results obtained with the non-noise-augmented model. For sequences of 1 s (3.5 s) the overall sensitivity increased from 35.12% to 57.33% (from 58.99% to 85.25%) for the white noise case and from 42.23% to 44.31% (from 68.66% to 85.25%) for the restaurant noise case.

4   Conclusions and Future Work

This paper investigates the robustness against the environmental noise of a speaker identification system for short sequences of speech frames based on truncated DKLT. The performance of the algorithm was evaluated using in the testing phase, a database synthesized from the TIMIT speech corpus by rerecording the speech signal with the addition of different noise types (cellular phone ringing, jet engine, office, pop song female, rain, restaurant, street, train ride noise). The experimental results demonstrated the validity of the proposed algorithm for speaker identification using short portions of speech also in realistic conditions when the ambient noise is not negligible. More in detail for sequences of 1 s (3.5 s) overall sensitivities ranging from 28.56% to 79.43% (from 51.15% to 97.70%) have been obtained in different environmental noisy conditions with SNR of 10, 15 and 20 dB. Additionally in order to optimize the recognition performance a white noise at different levels of energy has been added to the speech frames used in the training stage as a first step towards the generation of multicondition training data to model speech corrupted by noise with unknown temporal-spectral characteristics. Finally, future efforts will be invested in building a realistic noise model that definitively could improve the performance of the identifier as the preliminary results reported for the white noise model have been demonstrated.

References

1. Bhardwaj, S., Srivastava, S., Hanmandlu, M., Gupta, J.: GFM-based methods for speaker identification. IEEE Trans. Cybern. **43**(3), 1047–1058 (2013)
2. Biagetti, G., Crippa, P., Curzi, A., Orcioni, S., Turchetti, C.: A multi-class ECG beat classifier based on the truncated KLT representation. In: UKSim-AMSS 8th European Modelling Symposium on Computer Modelling and Simulation (EMS 2014), pp. 93–98, October 2014
3. Biagetti, G., Crippa, P., Falaschetti, L., Orcioni, S., Turchetti, C.: An investigation on the accuracy of truncated DKLT representation for speaker identification with short sequences of speech frames. IEEE Trans. Cybern. (in press). doi:10.1109/TCYB.2016.2603146
4. Biagetti, G., Crippa, P., Curzi, A., Orcioni, S., Turchetti, C.: Speaker identification with short sequences of speech frames. In: 4th International Conference on Pattern Recognition Applications and Methods (ICPRAM 2015), Lisbon, Portugal, pp. 178–185. January 2015
5. Biagetti, G., Crippa, P., Falaschetti, L., Orcioni, S., Turchetti, C.: A rule based framework for smart training using sEMG signal. In: Neves-Silva, R., Jain, L.C., Howlett, R.J. (eds.) Intelligent Decision Technologies, Smart Innovation, Systems and Technologies, vol. 39, pp. 89–99. Springer, Cham (2015)
6. Biagetti, G., Crippa, P., Falaschetti, L., Orcioni, S., Turchetti, C.: Distributed speech and speaker identification system for personalized domotic control. In: Conti, M., Martínez Madrid, N., Seepold, R., Orcioni, S. (eds.) Mobile Networks for Biometric Data Analysis, pp. 159–170. Springer, Cham (2016)
7. Biagetti, G., Crippa, P., Falaschetti, L., Orcioni, S., Turchetti, C.: Multivariate direction scoring for dimensionality reduction in classification problems. In: Czarnowski, I., Caballero, A.M., Howlett, R.J., Jain, L.C. (eds.) Intelligent Decision Technologies 2016: Proceedings of the 8th KES International Conference on Intelligent Decision Technologies (KES-IDT 2016) - Part I, pp. 413–423. Springer, Cham (2016)
8. Biagetti, G., Crippa, P., Falaschetti, L., Orcioni, S., Turchetti, C.: Robust speaker identification in a meeting with short audio segments. In: Czarnowski, I., Caballero, A.M., Howlett, R.J., Jain, L.C. (eds.) Intelligent Decision Technologies 2016: Proceedings of the 8th KES International Conference on Intelligent Decision Technologies (KES-IDT 2016) - Part II, pp. 465–477. Springer, Cham (2016)
9. Bimbot, F., Bonastre, J.F., Fredouille, C., Gravier, G., Magrin-Chagnolleau, I., Meignier, S., Merlin, T., Ortega-García, J., Petrovska-Delacrétaz, D., Reynolds, D.A.: A tutorial on text-independent speaker verification. EURASIP J. Appl. Sig. Process. **2004**, 430–451 (2004)
10. Campbell, J.P.: J.: Speaker recognition: a tutorial. Proc. IEEE **85**(9), 1437–1462 (1997)
11. Crippa, P., Curzi, A., Falaschetti, L., Turchetti, C.: Multi-class ECG beat classification based on a Gaussian mixture model of Karhunen-Loève transform. Int. J. Simul. Syst. Sci. Technol. **16**(1), 2.1–2.10 (2015)
12. Figueiredo, M.A.F., Jain, A.K.: Unsupervised learning of finite mixture models. IEEE Trans. Pattern Anal. Mach. Intell. **24**(3), 381–396 (2002)

13. Garofolo, J.S., Lamel, L.F., Fisher, W.M., Fiscus, J.G., Pallett, D.S.: DARPA TIMIT acoustic-phonetic continous speech corpus CD-ROM. NIST speech disc 1-1.1. NASA STI/Recon Technical report N 93, 27403 (1993)
14. Gianfelici, F., Biagetti, G., Crippa, P., Turchetti, C.: AM-FM decomposition of speech signals: an asymptotically exact approach based on the iterated Hilbert transform. In: IEEE/SP 13th Workshop on Statistical Signal Processing 2005, pp. 333–338, July 2005
15. Gianfelici, F., Turchetti, C., Crippa, P.: A non-probabilistic recognizer of stochastic signals based on KLT. Sig. Process. **89**(4), 422–437 (2009)
16. Gish, H., Schmidt, M.: Text-independent speaker identification. IEEE Sig. Process. Mag. **11**(4), 18–32 (1994)
17. Jain, A., Duin, R.P.W., Mao, J.: Statistical pattern recognition: a review. IEEE Trans. Pattern Anal. Mach. Intell. **22**(1), 4–37 (2000)
18. Jain, A., Ross, A., Prabhakar, S.: An introduction to biometric recognition. IEEE Trans. Circ. Syst. Video Technol. **14**(1), 4–20 (2004)
19. Kinnunen, T., Li, H.: An overview of text-independent speaker recognition: From features to supervectors. Speech Commun. **52**(1), 12–40 (2010)
20. Maina, C., Walsh, J.: Joint speech enhancement and speaker identification using approximate Bayesian inference. IEEE Trans. Audio Speech Lang. Process. **19**(6), 1517–1529 (2011)
21. McLaughlin, N., Ming, J., Crookes, D.: Speaker recognition in noisy conditions with limited training data. In: 2011 19th European Signal Processing Conference, pp. 1294–1298, August 2011
22. McLaughlin, N., Ming, J., Crookes, D.: Robust multimodal person identification with limited training data. IEEE Trans. Hum. Mach. Syst. **43**(2), 214–224 (2013)
23. Ming, J., Hazen, T.J., Glass, J.R., Reynolds, D.A.: Robust speaker recognition in noisy conditions. IEEE Trans. Audio Speech Lang. Process. **15**(5), 1711–1723 (2007)
24. Patra, S., Acharya, S.: Dimension reduction of feature vectors using WPCA for robust speaker identification system. In: 2011 International Conference on Recent Trends in Information Technology (ICRTIT), pp. 28–32, June 2011
25. Reynolds, D.: An overview of automatic speaker recognition technology. In: 2002 IEEE International Conference on Acoustics, Speech, and Signal Processing (ICASSP), vol. 4, pp. IV–4072–IV–4075, May 2002
26. Sadjadi, S., Hansen, J.: Blind spectral weighting for robust speaker identification under reverberation mismatch. IEEE/ACM Trans. Audio Speech Lang. Process. **22**(5), 937–945 (2014)
27. Togneri, R., Pullella, D.: An overview of speaker identification: Accuracy and robustness issues. IEEE Circ. Syst. Mag. **11**(2), 23–61 (2011)
28. Turchetti, C., Biagetti, G., Gianfelici, F., Crippa, P.: Nonlinear system identification: an effective framework based on the Karhunen Loève transform. IEEE Trans. Signal Process. **57**(2), 536–550 (2009)
29. Turchetti, C., Crippa, P., Pirani, M., Biagetti, G.: Representation of nonlinear random transformations by non-Gaussian stochastic neural networks. IEEE Trans. Neural Netw. **19**(6), 1033–1060 (2008)
30. Zhao, X., Shao, Y., Wang, D.: CASA-based robust speaker identification. IEEE Trans. Audio Speech Lang. Process. **20**(5), 1608–1616 (2012)
31. Zhao, X., Wang, Y., Wang, D.: Robust speaker identification in noisy and reverberant conditions. IEEE/ACM Trans. Audio Speech Lang. Process. **22**(4), 836–845 (2014)

Human Activity Recognition Using Accelerometer and Photoplethysmographic Signals

Giorgio Biagetti, Paolo Crippa$^{(\boxtimes)}$, Laura Falaschetti, Simone Orcioni, and Claudio Turchetti

DII – Dipartimento di Ingegneria dell'Informazione,
Università Politecnica delle Marche, Via Brecce Bianche 12, 60131 Ancona, Italy
{g.biagetti,p.crippa,l.falaschetti,s.orcioni,c.turchetti}@univpm.it

Abstract. This paper presents an efficient technique for real-time recognition of human activities by using accelerometer and photoplethysmography (PPG) data. It is based on singular value decomposition (SVD) and truncated Karhunen-Loève transform (KLT) for feature extraction and reduction, and Bayesian classification for class recognition. Due to the nature of signals, and being the algorithm independent from the orientation of the inertial sensor, this technique is particularly suitable for implementation in smartwatches in order to both recognize the exercise being performed and improve the motion artifact (MA) removal from PPG signal for accurate heart rate (HR) estimation. In order to demonstrate the validity of this methodology, it has been successfully applied to a database of accelerometer and PPG data derived from four dynamic activities.

Keywords: Activity recognition · Accelerometer · Photoplethysmography (PPG) · Motion artifact reduction · Real-time · Health · Fitness · Bayesian classification · Singular Value Decomposition (SVD) · Expectation Maximization (EM) · Karhunen-Loève Transform (KLT) · Feature extraction

1 Introduction

Human activity recognition using wearable sensors, i.e. sensors that are positioned directly or indirectly on the human body, is one of the most interesting topics in the healthcare, ambient assisted living, sport and fitness research areas.

These sensors, which can be embedded into clothes, shoes, belts, sunglasses, smartwatches and smartphones, or positioned directly on the body generate signals (accelerometric, photoplethysmography (PPG) [2,5,23], electrocardiography (ECG) [3,14], surface electromyography (sEMG) [9], ...) that can be used to collect information such as body position and movement, heart rate (HR), muscle fatigue of the user performing activities [2,4,11]. In particular, exercise routines and repetitions can be counted in order to track a workout routine as well as determine the

© Springer International Publishing AG 2018
I. Czarnowski et al. (eds.), *Intelligent Decision Technologies 2017*,
Smart Innovation, Systems and Technologies 73, DOI 10.1007/978-3-319-59424-8_6

energy expenditure of individual movements. Indeed, mobile fitness coaching has involved topics ranging from quality of performing such sports actions to detection of the specific sports activity [7].

On the one hand, among wearable sensors, accelerometers are probably the most frequently used for activity monitoring. In particular, they are effective in monitoring actions that involve repetitive body motions, such as walking, running, cycling, climbing stairs [8,13,18,20,21]. Because smartphones and smartwatches have become very popular, their accelerometer sensors can be used for providing accurate and reliable information on peoples activities and behaviors, thus ensuring a safe and sound living environment [1,15,17]. There are several techniques based on signal processing and neural networks for representing nonlinear transformations derived from stochastic systems such as the human body [22].

On the other hand, PPG is a well-known noninvasive method for monitoring the HR that shines light into the body and measures the amount of light that is reflected back to measure the blood flow. Unlike the ECG and the sEMG monitoring that need sticky metal electrodes across the body skin in order to monitor electrical activity from heart and muscles [6,10], PPG monitoring can be performed at peripheral sites on the body and needs a simpler body contact. As a result, PPG sensors are more and more used in wearable devices (smartwatches), as the preferred modality for HR monitoring in everyday activities by non-specialist users. However accurate estimation of PPG signal recorded from subject wrist, when the subject is performing various physical exercises, is often a challenging problem as the raw PPG signal is severely corrupted by motion artifacts (MAs). These are principally due to the relative movement between the PPG light source/detector and the wrist skin of the subject during motion. In order to reduce the MAs, a number of signal processing techniques based on data derived from the smartphone built-in triaxial accelerometer have been proven to be very useful [5,23].

This paper proposes an efficient technique for real-time recognition of human activities, by using data gathered from accelerometer and PPG sensors. The proposed technique is based on singular value decomposition (SVD) and truncated Karhunen-Loève transform (KLT) for feature extraction and reduction, and Bayesian classification for class recognition. The algorithm is independent of the orientation of the accelerometer sensor making it particularly suitable for implementation in wearable devices such as smartphones where the orientation of the sensors can be unknown or their placement could be not always correct. The algorithm could be used to both recognize the exercise being performed and improve the tracking of the PPG signal for accurate HR estimation. In order to demonstrate the validity of this technique, it has been successfully applied to a database of accelerometer and PPG data derived from four dynamic activities.

The paper is organized as follows. Section 2 provides a brief overview of the human activity recognition algorithm. Section 3 presents the experimental results carried out on a public domain data set in order to show the effectiveness of the proposed technique. Finally, the conclusions of this work are drawn in Sect. 4.

2   Recognition Algorithm

In this section both the model generation and the recognition algorithm will be presented. They exploit a Bayesian classifier based on a Gaussian mixture model (GMM) [16] of the probability density functions of the dimensionality-reduced feature vectors. The feature vectors themselves are built from the normalized singular value spectrums of both the accelerometer and the PPG Hankel data matrices. A schematic diagram of the activity detection algorithm is shown in Fig. 1.

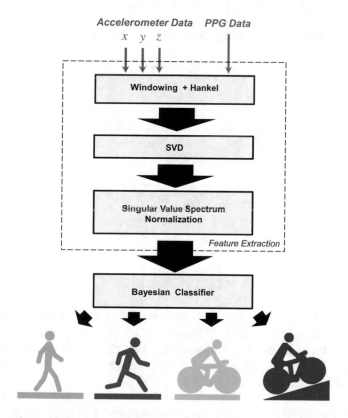

Fig. 1. Flow chart of the proposed framework for human activity classification (x, y, z are the 3-axial accelerometer signals).

In order to recognize the activity being performed at a given time instant t, the incoming signals are first windowed into relatively short windows (8 s in our sample application) wherein the activity can reasonably be considered invariant. From these slices of signals, a compact set of features ξ_t is extracted as detailed next.

Let x, y, z be the accelerometer signals and p the PPG signal, sliced into windows $N + L - 1$ samples long, indicated as $p_t = [p(t) \ldots p(t + N - 1)]^T$.

Then let $P_t = [p_t^{(1)} \dots p_t^{(L)}]$, with $p_t^{(i)} = p_{t+i-1}$, be the Hankel matrix derived from the PPG signal. Analogously, let $X_t = [x_t^{(1)} \dots x_t^{(L)}]$, $Y_t = [y_t^{(1)} \dots y_t^{(L)}]$, and $Z_t = [z_t^{(1)} \dots z_t^{(L)}]$ be the Hankel data matrices for the three accelerometer signals, where $x_t^{(i)}$, $y_t^{(i)}$, $z_t^{(i)}$, $i = 1, \dots, L$, represent the observations achieved from a three-axes accelerometer, each shifted in time by i samples, just as $p_t^{(i)}$.

As in [8], the accelerometer signals are grouped together to avoid dependence on sensor orientation, but the PPG data is left on its own as it is not commensurable with acceleration. So, the matrices

$$H_t^{\mathrm{A}} = [X_t\, Y_t\, Z_t] \in \mathbb{R}^{N \times 3L} \qquad H_t^{\mathrm{P}} = [P_t] \in \mathbb{R}^{N \times L} \tag{1}$$

can be represented by their singular value decomposition (SVD) as

$$H_t^{\mathrm{A}} = S_t^{\mathrm{A}} \Lambda_t^{\mathrm{A}} R_t^{\mathrm{AT}} = \sum_{i=1}^{N} \lambda_i^{\mathrm{A}} s_i^{\mathrm{A}} r_i^{\mathrm{AT}}, \tag{2}$$

where, if $N < 3L$, $S_t^{\mathrm{A}} = [s_1^{\mathrm{A}} \dots s_N^{\mathrm{A}}]$, $R_t^{\mathrm{A}} = [r_1^{\mathrm{A}} \dots r_N^{\mathrm{A}}]$, with s_i^{A}, r_i^{A} being the corresponding left and right singular vectors, and λ_i^{A} are the singular values in decreasing order $\lambda_1^{\mathrm{A}} \geq \lambda_2^{\mathrm{A}} \geq \dots \geq \lambda_N^{\mathrm{A}}$. Similarly, from the PPG signal p, we compute the SVD of its Hankel matrix as

$$H_t^{\mathrm{P}} = S_t^{\mathrm{P}} \Lambda_t^{\mathrm{P}} R_t^{\mathrm{PT}} = \sum_{i=1}^{N} \lambda_i^{\mathrm{P}} s_i^{\mathrm{P}} r_i^{\mathrm{PT}}, \tag{3}$$

where λ_i^{P} are the singular values in decreasing order $\lambda_1^{\mathrm{P}} \geq \lambda_2^{\mathrm{P}} \geq \dots \geq \lambda_N^{\mathrm{P}}$.

The chosen feature vector is defined as

$$\xi_t = \begin{bmatrix} \Lambda_t^{\mathrm{A}}/\|\Lambda_t^{\mathrm{A}}\| \\ w\, \Lambda_t^{\mathrm{P}}/\|\Lambda_t^{\mathrm{P}}\| \end{bmatrix} \in \mathbb{R}^{2N} \tag{4}$$

where $\| \cdot \|$ represents the norm of a vector and w is a weighting factor to be determined. Since the dimension $2N$ is usually too high to directly model a GMM on it, it is reduced by means of a truncated Karhunen-Loève transformation to a vector k_{tM} of lower dimension by a linear application Ψ such that $\mathrm{k}_{tM} = \Psi\, \xi_t$ where $\xi_t \in \mathbb{R}^{2N}$, $\mathrm{k}_{tM} \in \mathbb{R}^{M}$, $\Psi \in \mathbb{R}^{M \times 2N}$, and $M \ll 2N$.

Let us refer to a frame $\mathrm{k}_{tM}[n]$, $n = 0, \dots, M-1$, containing compacted features extracted from both the accelerometer and PPG signals. For Bayesian classification, a group of Γ activities is represented by the probability density functions (pdfs) $p_\gamma(\mathrm{k}_{tM}) = p(\mathrm{k}_{tM} \mid \theta_\gamma)$, $\gamma = 1, 2, \cdots, \Gamma$, where θ_γ are the parameters to be estimated during training.

The objective of classification is to find the activity $\widehat{\gamma}$ which has the maximum *a posteriori* probability for a given frame k_{tM}, i.e.

$$\widehat{\gamma}(\mathrm{k}_{tM}) = \underset{1 \leq \gamma \leq \Gamma}{\mathrm{argmax}} \left\{ \frac{p(\mathrm{k}_{tM} \mid \theta_\gamma) p(\theta_\gamma)}{p(\mathrm{k}_{tM})} \right\} = \underset{1 \leq \gamma \leq \Gamma}{\mathrm{argmax}} \{p(\mathrm{k}_{tM} \mid \theta_\gamma)\}, \tag{5}$$

assuming equally likely activities (i.e. $p(\theta_\gamma) = 1/\Gamma$) and noting that $p(\mathrm{k}_{tM})$ is the same for all activity models.

The statistical model we adopted for $p(\mathbf{k}_{tM} \mid \theta_\gamma)$ of the γ-th exercise is the GMM [19] given by the equation

$$p(\mathbf{k}_{tM} \mid \theta_\gamma) = \sum_{i=1}^{F} \alpha_i \, \mathcal{N}(\mathbf{k}_{tM} \mid \mu_i, \mathbf{C}_i) \qquad (6)$$

where α_i, $i = 1, \ldots, F$ are the mixing weights, and $\mathcal{N}(\mathbf{k}_{tM} \mid \mu_i, \mathbf{C}_i)$ represents the density of a Gaussian distribution with mean μ_i and covariance matrix $\mathbf{C}_i$, and θ_γ is the set of parameters defined as $\theta_\gamma = \{\alpha_1, \mu_1, \mathbf{C}_1, \ldots, \alpha_F, \mu_F, \mathbf{C}_F\}$.

To obtain an estimate of the mixture parameters we used a variant of the expectation maximization (EM) algorithm [16], which integrates both model estimation and component selection, i.e. the ability of choosing the best number of mixture components F according to a predefined minimization criterion, in a single framework.

3   Experimental Results

PPG recordings were collected from 8 subjects, for approximately 5 min each, as they undertook a range of different physical activities on a treadmill and on an exercise bike. The activities performed by each person are detailed in Table 1. All of the signals (including a simultaneous ECG for gold standard heart rate) belong to the Physionet database in https://physionet.org/works/ WristPPGduringexercise/ [12]. We follow the original Authors' naming of the subjects, i.e., s1, ..., s6, s8, s9, and we splitted the available signals between a training set and a testing set, so as to include exactly two signals for each activity type in the training, as detailed in the same table.

Table 1. Distribution of the activities performed by the various subjects. The database was also split in a training set (*), comprising subjects s1, s2, s4, and s8 (8 signals in total), and a testing set (+), comprising subjects s3, s5, s6, and s9 (10 signals in total, all of the remaining).

Activities	s1	s2	s3	s4	s5	s6	s8	s9
High resistance bike (H)	*	*	+					
Low resistance bike (L)	*	*	+		+	+		
Run (R)			+	*	+	+	*	
Walk (W)	*		+			+	*	+

The available data was sampled at 256 Hz, and it was sliced in overlapping windows of 2048 samples (8 s long), each window being shifted by 512 samples (2 s). These windows were used to build four $N \times L$ Hankel matrices, one for each acceleration direction and one for the PPG signal, with $N = 400$ and $L = 1649$. The acceleration-derived matrices are then fed together to the SVD, to remove

sensor orientation effects, and the PPG-derived matrix is fed to an independent
SVD block. The resulting normalized singular values are concatenated, with a
weight w applied to the PPG-derived values, and are used as the feature vectors
for the classifier, after KLT-based dimensionality reduction to $M = 10$ principal
components.

As an example, the average of these feature vectors, within each activity, is
shown in Fig. 2. It is apparent that different types of motion (bike, run, walk)
produce differing distribution of the singular values, making recognition of the
class of activity quite easy. Unfortunately, there is almost no way to differentiate
the resistance level on the bike.

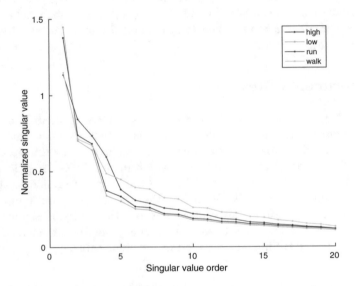

Fig. 2. Per-class average of a portion (extracted from the accelerometer data) of the
feature vectors.

This is confirmed by a recognition experiment, whose results are reported
in Table 2. Here the test was performed disregarding the PPG signal (i.e., by
setting $w = 0$), and the first two classes are often confused. The PPG signal can
help differentiate between these two cases, that differs only in the effort and not
in the type of motion.

Unfortunately, as can be seen in Table 3, which reports the results of an
experiment made disregarding the accelerometer signal, PPG alone is not a good
candidate to recognize the physical activity. To see if it can help in discriminating
the effort it must be combined with the accelerometer signal.

In order to appropriately combine the information coming from the
accelerometer and from the PPG for identification purposes, a 4-fold cross-
validation was performed on the training set, leaving out one of the subjects
at a time and averaging the overall recognition accuracy on the three remaining

Table 2. Confusion matrix obtained using only the accelerometer data as the feature vector. The resulting overall accuracy is 65.7%.

Input	Recognized			
	H	L	R	W
H	127	9	0	1
L	220	170	16	9
R	8	30	408	1
W	95	63	32	220

Table 3. Confusion matrix obtained using only the PPG data as the feature vector. The resulting overall accuracy is 44.7%.

Input	Recognized			
	H	L	R	W
H	20	117	0	0
L	5	140	172	98
R	0	0	157	290
W	0	0	97	313

Table 4. Confusion matrix obtained using both the PPG and the accelerometer data as the feature vector, with the PPG features being optimally scaled with $w = 0.155$. The resulting overall accuracy is 78.0%.

Input	Recognized			
	H	L	R	W
H	120	17	0	0
L	65	232	49	69
R	0	0	435	12
W	0	9	89	312

Table 5. Summary of the performance obtained using both the PPG and the accelerometer data as the feature vector, with the PPG features being optimally scaled with $w = 0.155$.

Activities	Sensitivity [%]	Precision [%]	F1-score [%]
H	87.59	64.86	74.53
L	55.90	89.92	68.95
R	97.32	75.92	85.29
W	76.10	79.39	77.71

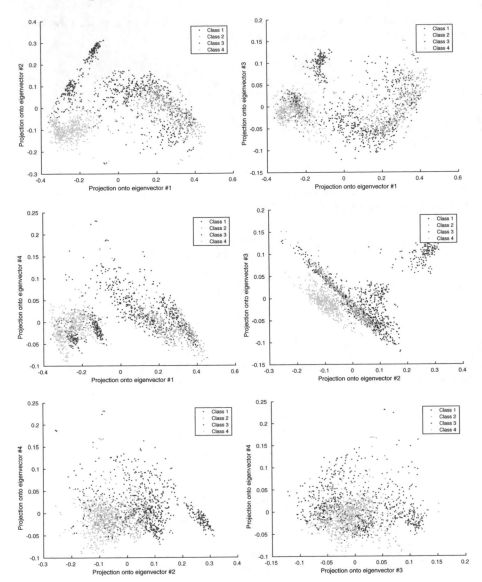

Fig. 3. Projections of the training set over the first four eigenvectors.

subjects, while varying the weight w used to combine the two signals. The optimum value was found to be $w = 0.155$. The results of this experiment are shown in Table 4, and it clearly improved the overall accuracy from 65.7% to 78.0%. For a better view of the performance, a few indices are also reported in Table 5.

Finally, Fig. 3 shows projections of the feature vectors over the first few eigenvectors. As can be seen, on the strongest projections the first two classes (bike) have a significant overlap, but they can be separated by higher-order projection thanks to the introduction of the PPG signal.

4 Conclusion

In this paper we extended the results presented in [8] to also take into account PPG data, besides acceleration data, to help in human activity identification. In fact, there are sometimes different activities that can usefully be differentiated but that involves essentially the same movements, differing only in the amount of effort exerted, like pedaling on a bike at different resistance levels. These activities are difficult to identify with acceleration alone, as the results here presented show. By adding information from a PPG sensor, recognition accuracy can be considerably improved provided the two types of signals are appropriately combined. To this end, a cross-validation approach was employed, resulting in an improvement from an initial 65.7% to 78.0% overall accuracy on experiments conducted on a publicly-available data set.

References

1. Anguita, D., Ghio, A., Oneto, L., Parra, X., Reyes-Ortiz, J.L.: Energy efficient smartphone-based activity recognition using fixed-point arithmetic. J. Univ. Comput. Sci. **19**(9), 1295–1314 (2013)
2. Bacà, A., Biagetti, G., Camilletti, M., Crippa, P., Falaschetti, L., Orcioni, S., Rossini, L., Tonelli, D., Turchetti, C.: CARMA: a robust motion artifact reduction algorithm for heart rate monitoring from PPG signals. In: 23rd European Signal Processing Conference, pp. 2696–2700, September 2015
3. Biagetti, G., Crippa, P., Curzi, A., Orcioni, S., Turchetti, C.: A multi-class ECG beat classifier based on the truncated KLT representation. In: UKSim-AMSS 8th European Modelling Symposium on Computer Modelling and Simulation (EMS 2014), pp. 93–98, October 2014
4. Biagetti, G., Crippa, P., Curzi, A., Orcioni, S., Turchetti, C.: Analysis of the EMG signal during cyclic movements using multicomponent AM-FM decomposition. IEEE J. Biomed. Health Inf. **19**(5), 1672–1681 (2015)
5. Biagetti, G., Crippa, P., Falaschetti, L., Orcioni, S., Turchetti, C.: Artifact reduction in photoplethysmography using Bayesian classification for physical exercise identification. In: Proceedings of the 5th International Conference on Pattern Recognition Applications and Methods, Rome, Italy, pp. 467–474, February 2016
6. Biagetti, G., Crippa, P., Falaschetti, L., Orcioni, S., Turchetti, C.: Wireless surface electromyograph and electrocardiograph system on 802.15.4. IEEE Trans. Consum. Electron. **62**(3), 258–266 (2016)
7. Biagetti, G., Crippa, P., Falaschetti, L., Orcioni, S., Turchetti, C.: A rule based framework for smart training using sEMG signal. In: Neves-Silva, R., Jain, L.C., Howlett, R.J. (eds.) Intelligent Decision Technologies, Smart Innovation, Systems and Technologies, vol. 39, pp. 89–99. Springer, Cham (2015)
8. Biagetti, G., Crippa, P., Falaschetti, L., Orcioni, S., Turchetti, C.: An efficient technique for real-time human activity classification using accelerometer data. In: Czarnowski, I., Caballero, A.M., Howlett, R.J., Jain, L.C. (eds.) Intelligent Decision Technologies 2016: Proceedings of the 8th KES International Conference on Intelligent Decision Technologies - Part I, pp. 425–434. Springer, Cham (2016)

9. Biagetti, G., Crippa, P., Falaschetti, L., Orcioni, S., Turchetti, C.: Homomorphic deconvolution for MUAP estimation from surface EMG signals. IEEE J. Biomed. Health Inf. **21**(2), 328–338 (2017)

10. Biagetti, G., Crippa, P., Orcioni, S., Turchetti, C.: An analog front-end for combined EMG/ECG wireless sensors. In: Conti, M., Martínez Madrid, N., Seepold, R., Orcioni, S. (eds.) Mobile Networks for Biometric Data Analysis, pp. 215–224. Springer, Cham (2016)

11. Biagetti, G., Crippa, P., Orcioni, S., Turchetti, C.: Surface EMG fatigue analysis by means of homomorphic deconvolution. In: Conti, M., Martínez Madrid, N., Seepold, R., Orcioni, S. (eds.) Mobile Networks for Biometric Data Analysis, pp. 173–188. Springer, Cham (2016)

12. Casson, A.J., Galvez, A.V., Jarchi, D.: Gyroscope vs. accelerometer measurements of motion from wrist PPG during physical exercise. ICT Expr. **2**(4), 175–179 (2016)

13. Catal, C., Tufekci, S., Pirmit, E., Kocabag, G.: On the use of ensemble of classifiers for accelerometer-based activity recognition. Appl. Soft. Comput. **37**, 1018–1022 (2015)

14. Crippa, P., Curzi, A., Falaschetti, L., Turchetti, C.: Multi-class ECG beat classification based on a Gaussian mixture model of Karhunen-Loève transform. Int. J. Simul. Syst. Sci. Technol. **16**(1), 2.1–2.10 (2015)

15. Dernbach, S., Das, B., Krishnan, N.C., Thomas, B.L., Cook, D.J.: Simple and complex activity recognition through smart phones. In: 8th International Conference on Intelligent Environments, pp. 214–221, June 2012

16. Figueiredo, M.A.F., Jain, A.K.: Unsupervised learning of finite mixture models. IEEE Trans. Pattern Anal. Mach. Intell. **24**(3), 381–396 (2002)

17. Khan, A., Lee, Y.K., Lee, S., Kim, T.S.: Human activity recognition via an accelerometer-enabled-smartphone using kernel discriminant analysis. In: 2010 5th International Conference on Future Information Technology, pp. 1–6, May 2010

18. Mannini, A., Intille, S.S., Rosenberger, M., Sabatini, A.M., Haskell, W.: Activity recognition using a single accelerometer placed at the wrist or ankle. Med. Sci. Sports Exerc. **45**(11), 2193–2203 (2013)

19. Reynolds, D.A., Rose, R.C.: Robust text-independent speaker identification using Gaussian mixture speaker models. IEEE Trans. Speech Audio Process. **3**(1), 72–83 (1995)

20. Rodriguez-Martin, D., Samà, A., Perez-Lopez, C., Català, A., Cabestany, J., Rodriguez-Molinero, A.: SVM-based posture identification with a single waist-located triaxial accelerometer. Expert Syst. Appl. **40**(18), 7203–7211 (2013)

21. Torres-Huitzil, C., Nuno-Maganda, M.: Robust smartphone-based human activity recognition using a tri-axial accelerometer. In: 2015 IEEE 6th Latin American Symposium on Circuits Systems, pp. 1–4, February 2015

22. Turchetti, C., Crippa, P., Pirani, M., Biagetti, G.: Representation of nonlinear random transformations by non-Gaussian stochastic neural networks. IEEE Trans. Neural Networks **19**(6), 1033–1060 (2008)

23. Zhang, Z., Pi, Z., Liu, B.: TROIKA: A general framework for heart rate monitoring using wrist-type photoplethysmographic signals during intensive physical exercise. IEEE Trans. Biomed. Eng. **62**(2), 522–531 (2015)

Digital Architecture
and Decision Management

When Energy Revolution Meets Digital Transformation

Shéhérazade Benzerga[1,2(✉)], Dominik Hauf[1], Michael Pretz[1], and Ahmed Bounfour[2]

[1] Daimler AG, Stuttgart, Germany
sheherazade.benzerga@daimler.com
[2] Paris-Sud University, Paris, France

Abstract. Digital transformation can be observed in business as well as in broader society, but we are far from having identified and defined all the issues and implications of this ongoing transformation and how it impacts on a traditional company. In an attempt to better understand this phenomenon and the changing role of IT that it brings with it, this article presents the field of energy efficiency as an example of the strategic implementation in the context of digital transformation.

In the first section, the Digital Economy trend will be presented in order to better understand to what extent this trend could be a key to the future of traditional industries on the macro level. The paper goes into more detail in the second section with the definition of digital transformation. Its dimensions will be addressed in order to better understand the inherent strategic impacts. Coming to the micro level of the plant in a manufacturing company then, the specific case of energy efficiency will be analyzed, which is supported by a patented process innovation based on digital technologies such as self-learning algorithms and hence on an increasing role of IT in the production area.

Finally, in order to encourage researchers and practitioners to advance in the field, future areas of interest are identified and further application fields are proposed as a conclusion to this article.

Keywords: Digital transformation · Digital economy · IT function · Industrie 4.0 · Energy efficiency · Smart factory

1 Introduction: Digital Economy – Is It the Key for the Future in Automated Industry?

As of today, the use of technology as a means of dramatically improving the performance or scope of business of digital processing is of major interest for any type of organization. This is in line with a study conducted among 400 companies by Cap Gemini Consulting in collaboration with MIT showing that the digitally most advanced companies are 26% more profitable compared with others [1]. With information technologies being such a powerful tool for companies, adapting to this new economy is essential. Even though the Digital Economy[1] is rapidly growing, its economic potential is still widely underestimated [2].

---

[1] The term Digital Economy was introduced in [6].

© Springer International Publishing AG 2018
I. Czarnowski et al. (eds.), *Intelligent Decision Technologies 2017*,
Smart Innovation, Systems and Technologies 73, DOI 10.1007/978-3-319-59424-8_7

1.1   What Defines the Digital Economy?

Since the development of the internet in the twentieth century, an evolution in the economy is taking place or in other words: a new economy is born, the Digital Economy [3]. But what is this economy about? Simply said, it is an economy that is based on digital technologies. In fact these technologies, e.g. the internet, mobile devices, big data, analytics, 3D printing, social media or cloud computing, are changing the way of doing business and thereby also influencing the relationship between companies and their customers [4, 5].

The Digital Economy is part of a global change in which companies as well as the whole of society are becoming more collaborative, more open and thus having a greater reach (Fig. 1).

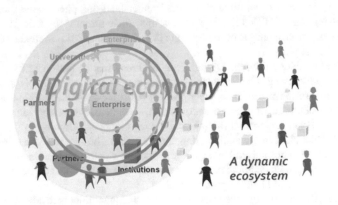

Fig. 1. The digital economy ecosystem

In consequence, the economic players have to deal with a highly complex ecosystem that is becoming more and more informational and immaterial and is also omnipresent in every aspect of daily life.

Given such omnipresence, for any company the Digital Economy inevitably forms an important aspect in business evolution.

1.2   The Digital Transformation as a Consequence of the Digital Economy

Today, society as such is facing a digital transformation. But this transformation is also an important topic for industry, and it represents a genuine challenge for traditional manufacturing companies in particular. As a consequence, a continuous emergence and growth of political actions is inherent to this new area. For example, on April 14th, 2015, the French President launched the project "Industrie du future," which is playing a very important role in the transformation of the economic model in the digital domain, equivalent to the German "Industrie 4.0" concept. Both initiatives are forming the basis for a

Franco-German cooperation, as was seen at the French-German conference on the Digital Economy that took place on October 27th, 2015, at the Elysée Palace in Paris[2].

1.3   Research Goal

However, as digital transformation gains importance, there is still a lack of research into the management of information systems, as Yoo et al. remark in their publication *"Digital technology's transformative impact on industrial-age products has remained surprisingly unnoticed in the Information System (IS) literature"* [2].

One of the challenges for research still is to understand how traditional industry is influenced by this digital transformation. This is what it will be dealt with in this paper, with a special focus in the production area.

The aim of this paper is to investigate the following research questions:

- How to understand digital transformation and how does it take place in a traditional OEM industry?
- Based on a use case study – the energy efficiency use case – how the use and implementation of new technology is revolutionizing our industry and how the IT role is impacted?
- Can production facilities be automated and optimized online for energy efficiency?
- How can a generally valid method be developed (i.e. automated optimization)?

The following section contextualizes the concept of digital transformation, the concept of Industrie 4.0, its challenges in the production area as well as the vision for the specific case of energy efficiency. Further, a case study in the area of energy efficiency focusing on an implementation in a plant environment will be provided, in order to evaluate the strategic impact of such an industry revolution. A conclusion and an outlook on future work complete the article.

2   Digital Transformation

2.1   Definition and Dimension

Since the commercial introduction of the internet through the World Wide Web in the middle of the 1990s, we can observe a fast and global transformation of our society. Our private and business lives have been modified, as has the way we communicate. The most recent OECD publication on the Digital Economy reported that 80% of adults in OECD countries are internet users and 77% of OECD companies are present in the internet through a website or social media activities [7].

This technical revolution brought by the internet is just at the beginning and every day new innovations strengthening this transformation can be observed. But what is the digital transformation?

---

[2] http://www.elysee.fr/communiques-de-presse/article/conference-numerique-franco-alle-mande/.

As of today, there is no standard definition of the digital transformation. In the literature and in practice, the phenomenon so far is approached very differently, leading to a multitude of definitions (see Table 1) and showing that digitalization is not considered on an abstract level, but is rather approached on a case-by-case basis. While finding an abstract definition remains an academic challenge, with this article we aim to add a new perspective to what has been thought so far about digital transformation by analyzing on the basis of a case study how digitalization can and will change classical approaches to plant energy efficiency.

Table 1. Diverse definitions of digital transformation

Definition	Reference
It as "an organizational change through the use of digital technologies to materially improve performance"	[8]
"The digital transformation can be understood as the changes that the digital technology causes or influences in all aspects of human life"	[9]
Digitization – i.e. the transformation of continuous variables into zeros and ones[a]	[10, 11]
Digital transformation is a "global megatrend that is fundamentally changing existing value chains across industries and public sectors"	[12]
"Digital transformation is a new development in the use of digital artifacts, systems and symbols within and around organisations"	[13]

[a]Own translation from german: "Digitalisierung – also die Transformation kontinuierlicher Größen in Nullen und Einsen"

2.2 Dimensions

In the literature, it can be observed that different dimensions of the digital transformation have been proposed and defined in order to provide a better understanding of digitalization. In this regard, in a special issue of MIS Quarterly from 2013 building on an analysis of five expert papers in the field of digitalization, [4, 14], four dimensions of a company's digital business strategy were highlighted as key attributes for digitalization:

- Its scope, which needs to extend beyond the traditional boundaries of the firm;
- Its scale, with the emergence of platforms that create important network effects in a context of data abundance;
- Its speed, whether to launch products or services, to take decisions, etc.;
- The source of value creation and capture (data, networks, digital architecture).

Bounfour [13] explains that when taking into account the scope dimension, there is a need to develop new approaches to ecosystems. In terms of scale, the power of digital markets lies in platforms that compete to leverage the key resource of the intangible economy: data. In this context, speed plays a critical role for success. Acceleration calls for the development of a new form of capital: the capital of digital systems is defined as the cumulative knowledge of the design of IT artifacts, which should be distinguished from, for example, discrete strategies used to enlarge the cumulative stock of knowledge [13]. It will be seen in the empirical study under the section "Energy revolution meets digital transformation – an example of use", how the scope, scale and speed in the specific case of energy efficiency could be impacted. But before turning to the possible impacts, an analysis will be made to identify the enabler of such a transformation in the organization.

2.3   ICT as Enabler

On the basis of a systemic view of the topic of digital transformation, the introduction of the internet in the twentieth century and the development of ICT (information and communication technology) can be considered as the key drivers of digitalization. [15] highlighted data, processes, structures, systems, software and IT infrastructures as key subjects in the digital world. How can we support the shift that technologies are bringing? In answering this question, [4] introduces the term "digital business strategy," being defined as follows: *"Digital business strategy is different from traditional IT strategy in the sense that it is much more than a cross-functional strategy, and it transcends traditional functional areas (such as marketing, procurement, logistics, operations, or others) and various IT-enabled business processes (such as order management, customer service, and others). Therefore, digital business strategy can be viewed as being inherently transfunctional. All of the functional and process strategies are encompassed under the umbrella of digital business strategy with digital resources serving as the connective tissue."*

In fact, it can be observed in practice that companies are undergoing a paradigm shift from "IT following the business" to "IT as an enabler for new digital business," which is why closer collaboration between the IT department and other departments is a cornerstone for the success of a digital transformation.

2.4   Challenges

Based on the result of the international Information Systems Dynamics (ISD) Research Program sponsored by CIGREF and large international companies [16], the three following propositions were made to describe potential challenges in the context of digital transformation:

1. Digital induces tensions between regulation and freedom, and between privacy and the freedom to do business. Copyright, in particular, is endangered by the development of open-source and open-access practices.
2. Digital is a generative machine that produces spontaneous, unexpected innovations, often contributed by external developers.
3. Network abundance multiplies the tensions involved in organizing the enterprise (fly-by-wire vs. decision support; security vs. privacy; ownership vs. profitability; public goods vs. private goods) [13].

We can thus understand that digitalization is a very complex phenomenon that combines both technological and organizational challenges.

To better understand the challenges that the digital area brings in the manufacturing industry and more specifically in our plants, let's understand first the fourth industrial revolution then the challenge of energy efficiency to finally come to our vision and challenge in the area where energy efficiency meets digital transformation.

3 Industrie 4.0

3.1 Definition

Given that historically the industrial sector has undergone several industrial revolutions in Germany, the governmental initiative on the digital transformation in production is called Industrie 4.0 (Industry 4.0). For the German government, Industrie 4.0 is a strategic initiative to support the fourth industrial revolution and was adopted in 2011 as part of the "High-Tech Strategy 2020 Action Plan" [17, 18].

The fourth industrial revolution, from a historical perspective, builds on the following evolutional background:

- The starting point was the first industrial revolution (which lasted for almost 100 years), characterized by the use of mechanical production facilities with the support of water and steam power;
- This was followed by the second industrial revolution (which lasted for 60 years), characterized by the use of division of labor and mass production with the support of electrification (use of electrical energy);
- The subsequent third industrial revolution (which lasted 50 years) can be characterized by the use of advanced electronics and information technologies, resulting in the automation of production processes [18, 19].

The ongoing fourth revolution is characterized by the introduction and deployment of internet technology and is based on the use of cyber physical systems (CPSs) [18, 19], which form the merging element between the physical domain and the digital or virtual domain. New possibilities such as the Internet of Things and Services match with the production and manufacturing skills and form the basis of this latest revolution. The already existing plants with their production machines as well as new planned plants will be transformed into smart factories. Such smart factories consist of several CPSs, like smart products, smart logistics or smart buildings [18] (Fig. 2).

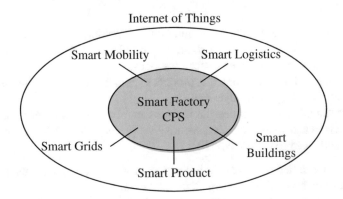

Fig. 2. Industry 4.0 and smart factories as part of the internet of things and services, based on [18]

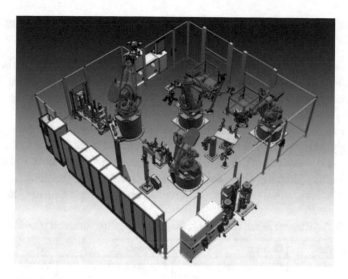

Fig. 3. AREUS demonstration cell. System setup with four KUKA KR210 2700 prime robots with four different subordinated technologies: handling, riveting, gluing and spot welding [20].

Smart grids are an important element of the described smart factory within the Industrie 4.0 initiative. Energy supply systems in the future will have to be more flexible and intelligent. In this regard, new energy supply technologies in the factory automation layer also have to be developed. The approach of the EU research project AREUS (Automation and Robotics for European Sustainable Manufacturing) describes such a new "smart automation grid" [20].

In the field of energy-flexible production systems, a lot of results were published as part of a German research project called "FOREnergy – Energy flexibility in production" [21].

Both energy flexibility and intelligent grids are main topics in research and the key for modern production systems in the context of Industrie 4.0. One benefit that such consideration brings is a much better and more efficient solution to achieve sustainable production.

Another key factor for the successful realization of Industrie 4.0 is the integrated digital process chain during the engineering phase. Before a production line exists in reality, a digital model has to be engineered. The challenge for the future is to use this data to build a so-called digital twin of such an engineered production line for a virtual runtime environment, where the runtime can be supplied by input data from real manufacturing in near real time. The digital or cyber physical twin in this context refers to a Primary IT Trend[3] that is central to our research into bringing together ICT, manufacturing production with its robots, and the challenge of energy efficiency.

---

[3] You can find a definition of the Cyber Physical Twins and other Primary IT Trends in [22].

3.2   Smart Automation and Digital Engineering - Enabling Technologies for Energy Efficiency

Energy efficiency represents a necessary and logical consequence of the basic idea of Industrie 4.0. In the future, there will be intelligent smart automation grids and integrated digital process chains. Both parts are essential for generating a benefit in terms of financial savings as well as in terms of sustainability.

Up to now, process energy efficiency has been considered in the context of digital planning only peripherally. If the targets of the European Union are to be reached by 2020 (20% energy reduction), energy (efficiency) considerations have to be part of the digital production planning [23]. Table 2 shows an excerpt of the parts and systems of the digital process chain during the planning and engineering of a production unit.

Table 2.   Excerpt of the digital process chain for production cell planning [26]

Application area	Robots	Technology	Cell
Subject	Off-line robot programming	Process models	Virtual commissioning
Energy simulation	Possible	Not possible	Not possible
Maturity	Prototype	None	Concept

Prototype software tools are already available for the area of off-line robot programming (OLP) to simulate the energy consumption of the robot based on an individual trajectory. Meike [24] and Paryanto et al. [25] describe how the energy consumption of industrial robots (IRs) can be simulated as a function of a given trajectory. Further activities are taking place to accommodate the topic of energy simulation in the standard for robot simulation [24]. [26] describes how the complete energy simulation of an entire production cell can be integrated into the virtual commissioning tool chain.

The described process top-down from digital transformation to smart grid, integrated digital process chain and finally energy efficiency shows the challenge of how the traditional OEM industry is being revolutionized by the use and implementation of new technology.

For a better understanding of how digitalization through programming and the cloud helps to increase performance in the area of energy efficiency, let us have a look at an example.

The digitalization example described in the next section of our paper is based on practical experiences and the implementation of energy consumption simulation and optimization in the digital process chain.

4   Energy Revolution Meets Digital Transformation – An Example of Use

Today, all robots in body-in-white manufacturing run with a motion speed of 100%. But in fact the robots work together in teams; they interact. That means that not all robots have to work at full motion speed, because they have to wait for each other. These waiting times provide the opportunity to slow down non-cycle-time-critical robots. The result of slower motion is a reduction in energy consumption.

This paper looks especially at the cooperation of robots and the opportunity to save energy during this process, through the use and implementation of digital technologies (self-learning algorithms, cloud, CPS, software, etc.). Experiments have shown that a reduction in robot velocity can decrease energy consumption considerably. The decrease in energy consumption takes place until 40% of the velocity of a robot movement is reached. Below 40%, the effect reverts and leads to an increase in energy consumption [26].

Simulations with the OLP system *Process Simulate* (Siemens) also demonstrate this effect (see Fig. 4). The system setup for the simulation was a demonstration cell with four *KUKA KR210 2700 prime robots* with four different subordinated technologies: handling, riveting, gluing and spot welding (see Fig. 3). The graph shows the average of different trajectories with different tools.

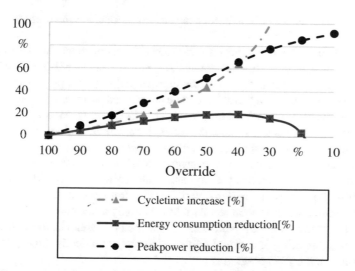

Fig. 4. Comparison of energy consumption, peakpower reduction and cycletime extension regarding to override adaption

The energy consumption reduction (ordinate) as a function of override decrease (abscissa) can be determined until a value of 40%. These are the same dimensions as described by [27]. For the cycle time increase and the peak power reduction, the simulated values are also feasible. An override reduction causes a cycle time increase and a peak power reduction.

Another investigation determines the differences between the velocity, acceleration and override adaption (see Fig. 5). The override and velocity adaption have nearly the same effect on energy reduction. A saving of up to 20% is possible (value of 40% override or velocity). A further decrease in the parameters causes a contrary effect: The energy consumption increases. The acceleration adaption also shows the impact on energy reduction, but the effect is not as large as for override and velocity. In practice, adjusting the override is a very useful self-learning method that can be implemented as an algorithm in the PLC (programmable logic controller) code of a production cell.

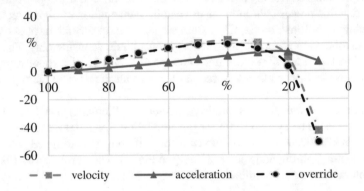

Fig. 5. Energy reduction as a function of velocity, acceleration and override adaption

As described above, the occurrence of waiting times of robots in multi-cooperative robot cells is ubiquitous; slowing down the speed of robots will result in shortened waiting times and reduced energy consumption. To realize this approach, the two criteria "unchanged cycle time" and "costs" are to be considered, with "costs" meaning that the implementation costs of a method must not exceed the savings that originate from lower energy consumption. As a first step, this paper therefore investigates an approach in a self-learning PLC program that drives a dynamic velocity adjustment.

The first question that has to be answered is where in the robot program creation process a method can be realized for dynamic velocity adjustment. After that, it can be clarified how the method is implemented. For the first step, an overview of the robot creation program process is obtained. As it has to be a dynamic automatic method, which has to adapt to changes on the production side, the method cannot be implemented during the offline programming, the virtual commissioning, or manually during the operation of a robot cell in the PLC program.

For this reason, only digital add-ons like a self-learning algorithm as a function in the PLC program or an entirely new system with a self-learning algorithm communicating with the PLC program and the robot controller remain. Hence, based on current

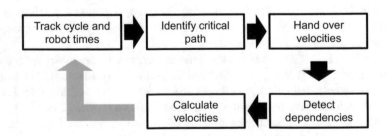

Fig. 6. General logic of an iterative method for dynamic velocity adjustment

experience with a dynamic velocity adjustment function in a PLC program, a general logic can be designed. It consists of five function blocks and is visualized in Fig. 6.

Function block one is responsible for tracking robot and cycle times. The next block identifies the critical path. After that a new block, which is not implemented in the PLC program yet, detects the dependencies among the robots. This block should be responsible for blocking the velocity value of a specific robot program if a change would lead to an extension of the cycle time. Following that, the modified velocities can be calculated. In a last step, the velocities are handed over to the PLC. It is important that this method is iterative, meaning that it always tracks the new cycle and robot times and starts the dynamic velocity adjustment from the beginning. This ensures that the velocity values are updated consistently. This has the advantage that if a modification needs to be made in the robot programs, the velocities are computed again and the cycle time stays almost the same.

The application and validation of the method was executed on a demonstration cell, which was set up at the Daimler plant in Sindelfingen as part of the EU research project AREUS. The demonstration cell consists of three sequences. Sequence 1 starts with picking up the workpieces and moving to the gluing station. In sequence 2, the workpiece is placed on the clamping fixture and a second robot executes the rivet operation. The third sequence comprises the stamping and the final placing of the workpiece.

In Fig. 7, the overall cycle time is marked in sequence 2. That is the situation after normal programming of the robots with 100% override. As a constraint, the overall cycle may not be modified. The algorithm is therefore only allowed to adapt the velocities of robots in sequences 1 and 3.

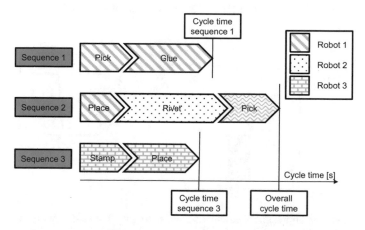

Fig. 7. Temporal production sequence before dynamic velocity scaling

After application of the dynamic velocity adjustment method, the velocities of the robots in sequences 1 and 3 decreases and the cycle time thus increases. The cycle graph has now changed (see Fig. 8). All sequences are harmonized and fully adapted, and the overall cycle time has not been modified.

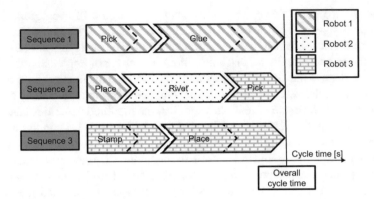

Fig. 8. Production sequence after dynamic velocity scaling

To achieve adaption and harmonization of more than one production cell, for example of an entire production line, in the future we will have to use new opportunities such as the cloud and big data in the factory layer. Such new systems can collect all the recorded data from all intelligent sensors and actors of a production system.

The automatic adaption of velocities should be executed in a big data system, where all motion parameters of the system are saved to a global cloud. The algorithms of the velocity calculation are executed in this additional system. To close this control loop, the adapted velocities have to be played back. The big challenge of this new system is the interface setup. Production-related systems (especially industrial bus systems) have to be interconnected with an IT system for cloud computing (see Fig. 9).

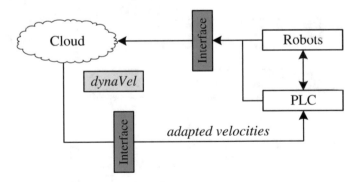

Fig. 9. Cloud system approach for calculating and setting up dynamic velocity in an automated production cell

The main advantage of such a superordinate system is that it is possible to optimize in terms of energy efficiency not just a single production cell but also a complete production line consisting of different cells (see Fig. 10). The harmonization of production speed is not only executed within a single cell. This can be very useful for adjusting line override as a function of the fluctuating number of cars to be produced (for example night shift with a smaller number of units).

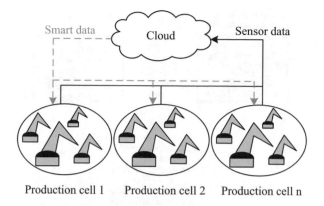

Production cell 1   Production cell 2   Production cell n

Fig. 10. Intelligent connection between production cells and cloud applications

A patent application on this innovative energy efficiency process is pending. The application describes not only the adjustment of override but also the direct setting of velocities and accelerations in the robot program. A further interesting approach for this configuration is to set energy modes of robots and other electric consumers by using an intelligent superordinate system. In practice, a robot can be operated in five energy modes (for example "normal", "stand-by," etc.). The control of these energy modes for all production cells can be executed by the described system. Another possible use of the system is the intelligent and dynamic scheduling of production processes to save energy [28].

5   Changing Role of IT Function

Through the example of energy efficiency in production plants of a classical industry[4] like the car manufacturing industry, some lessons are learned. In fact, this use case is one example among many that are happening in the digitalization context. We can see in Fig. 11 that the impacting trends for the digital transformation are linked to a major change in the IT role inside the company. In fact, the IT is the indispensable enabler for digital transformation. We see that we are undergoing a paradigm change from "IT following the business" to "IT being the strategic partner for business." This is impacting the work with the business departments and requires a more transversal integration, agility and in some cases a swarm organization.

Through the use of IT technologies as a disruptive driver for business (like in our use case cloud technology, big data, digital process chain and smart automation), new skills requiring considerable IT know-how are needed in the production area.

These changes are also dramatically impacting the role of the Chief Information Officer (CIO), who is more and more involved in strategic decisions being made in the

---

[4] Also called "latecomer industries" by Kohli and Johnson in R. Kohli, S. Johnson, Digital Transformation in Latecomer Industries: CIO and CEO Leadership Lessons from Encana Oil & Gas (USA) Inc., MISQ Executive Vol. 10 No. 4, December 2011.

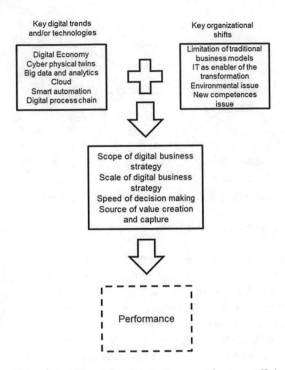

Fig. 11. Summary of the digital transformation in the case of energy efficiency, figure adapted and issue of [29] and inspired by [4].

different business units. The IT and business departments are thus becoming more and more intertwined, forming new business models this way.

Moreover, agility as a methodology originally stemming from software development is being used more and more in all business units to speed up the processes. Building on the success of the two-speed IT approach, which seems to have become a role model for other business units as well, IT and business collaboration and co-creation are strengthened by using a mix of classical and agile project management.

6   Conclusion and Perspectives

This paper attempted to provide a better understanding of the phenomenon of digital transformation taking place in traditional manufacturing companies.

The paper aims to contribute to both research and practice.

In terms of research, the paper is intended to fit in the growing knowledge base on key digital trends (like the Digital Economy) and key digital technologies (like CPS, big data and analytics, the cloud, smart automation or the digital process chain). It tries to give some answers to the key organizational shift appearing in the context of this digitization. On the basis of the observation that Bharadwaj et al. made in the MISQ Special Issue of 2013 [4], it can be ascertained that taking into account every defined dimension (scope, scale, speed and source) helps to improve performance factors.

On the other hand, practicing entities like traditional manufacturing companies may benefit from this research by gaining an overview of the potentials offered by an innovative combination of IT and business.

More specifically, this paper shows in the empirical study part that a dynamic automatic velocity adjustment in the PLC program does work and that it reduces the waiting time of robots, ultimately leading to energy savings in the robot cell. The present status of the PLC function further shows that the research in this particular field still is at an early stage. The missing dependencies among the robots further need to be taken into account in the calculation of robot speeds in order to avoid increases in cycle time. Future studies have to prove if an approach in a PLC program is feasible and if this PLC approach has chances of success. If not, a completely new system has to be developed to perform a dynamic velocity adjustment in a robot cell, such as that it has to interact with the PLC and KRC controller in order to evaluate the generated values and conduct a dynamic velocity adjustment with them. Such a method can, for example, help car manufacturers like Daimler AG to reach achieve the governmental energy-saving targets.

Further, this paper aims to show the tremendous impact that digital transformation is having on the producing industry. As [25] explains, the integration of software in manufactured products at a radical pace challenges a company's established processes. It could be interesting for researchers and practitioners to better understand this new challenge by further analyzing similar case studies.

Understanding the new key capabilities needed by companies will be one of the key elements for success in the digital era, beginning with new skills for business leaders and a new role for the CIO [30], thus promoting closer collaboration between both of them.

Such a transformation has to be seen in more detail and brings new fields of research to light, such as the question of how to handle and how to create new business assets from data in the companies' digitalization context.

The winners in the Digital Economy will be those companies that are able to operate under new assumptions and that are willing to explore new approaches. In other words: "the global competitive battle will not be won in the real world alone, but also in the digital one" [31].

References

1. Fitzgerald, M., Kruschwitz, N., Bonnet, D., Welch, M.: Embracing digital technology. MIT Sloan Manag. Rev. **55**, 1–12 (2013)
2. Yoo, Y., Henfridsson, O., Lyytinen, K.: The new organizing logic of digital innovation: an agenda for information systems research. Inf. Syst. Res. **21**(4), 724–735 (2010)
3. Porter, M.: Strategy and the internet. Harvard Bus. Rev. **79**(3), 62–78 (2001)
4. Bharadwaj, A., El Sawy, O., Pavlou, P.A., Venkatraman, N.: Digital business strategy: toward a next generation of insights. MIS Q. **37**, 471–482 (2013)
5. Lucas, H.C., Agarwal, R., Clemons, E.K., El Sawy, O., Weber, B.: Impactful research on transformational information technology: an opportunity to inform new audiences. MIS Q. **37**(2), 371–382 (2013)

6. Tapscott, D.: The Digital Economy: Promise and Peril in the Age of Networked Intelligence. McGraw-Hill, New York (1997)
7. OECD: Measuring the Digital Economy: A New Perspective. OECD Publishing, Paris (2014)
8. Marchand, D.A., Wade, M.: IMD Discovery Event, 14 May 2014
9. Stolterman, E., Croon Fors, A.: Information technology and the good life. In: Information Systems Research: Relevant Theory and Informed Practice, pp. 687–692 (2004)
10. Keuper, F., Hamidian, K., Verwaayen, E., Kalinowski, T., Kraijo, C.: Digitalisierung und Innovation: Planung – Entstehung – Entwicklungsperspektiven. Springer Gabler, Wiesbaden (2013)
11. von Leibniz, G.: Explication de l'Arithmétique Binaire. Des Sciences, Académie royale des sciences (1703). https://hal.archives-ouvertes.fr/ads-00104781/document
12. Collin, J., Hiekkanen, K., Korhonen, J.J., Halén, M., Itälä, T., Helenius, M.: IT leadership in transition - the impact of digitalization on finnish organizations, Research rapport. Aalto University Publication Series (2015). https://aaltodoc.aalto.fi/bitstream/handle/123456789/16540/isbn9789526062433.pdf?sequence=1&isAllowed=y
13. Bounfour, A.: Digital Futures, Digital Transformation, From Lean Production to Acceluction. Springer, New York (2015)
14. Bharadwaj, A., El Sawy, O., Pavlou, P.A., Venkatraman, N.: Visions and voices on emerging challenges in digital business strategy. MIS Q. 37(2), 633–634 (2013)
15. Brenner, W., Karagiannis, D., Kolbe, L., Krüger, J., Lamberti, H.-J., Leifer, L., Leimeister, J.M., Österle, H., Petrie, C., Plattner, H., Schwabe, G., Uebernickel, F., Winter, R., Zarnekow, R.: User, use and utility research. Res. Notes Bus. Inf. Syst. Eng. 6(1), 55–61 (2014)
16. ISD (Information Systems Dynamics): Project description. www.fondation-cigref.org/programme-isd/
17. Kagermann, H., Lukas, W.D., Wahlster, W.: Industrie 4.0: mit dem internet der dinge auf dem weg zur 4. industriellen revolution. VDI Nachrichten, issue 13, 1 April 2011
18. Kagermann, H., Wahlster, W., Helbig, J.: Recommendations for Implementing the Strategic Initiative Industrie 4.0: Final Report of the Industrie 4.0 Working Group (2013). http://www.acatech.de/fileadmin/user_upload/Baumstruktur_nach_Website/Acatech/root/de/Material_fuer_Sonderseiten/Industrie_4.0/Final_report__Industrie_4.0_accessible.pdf
19. Drath, R., Horch, A.: Industrie 4.0: hit or hype? [Industry forum]. IEEE Ind. Electron. Mag. 8, 56–58 (2014)
20. AREUS (Automation and Robotics for European Sustainable Manufacturing): Project description. www.areus-project.eu
21. Graßl, M., Vikdahl, E., Reinhart, G.: A petri-net based approach for evaluating energy flexibility of production machines. In: Zäh, M.F. (ed.) Enabling Manufacturing Competitiveness and Economic Sustainability, pp. 303–308. Springer, New York (2014). ISBN 978-3-319-02053-2
22. Benzerga, S., Pretz, M., Riegg, A., Reimann, W., Bounfour, A.: Appflation – a phenomenon to be considered for future digital services. In: Zimmermann, A., Rossmann, A. (eds.) Digital Enterprise Computing 2015. LNI, pp. 39–50. Gesellschafts für Informatik, Bonn (2015)
23. European Union: Directive 2012/27/EU of the European Parliament and of the Council, Brussels (2012)
24. Meike, D.: Increasing energy efficiency of robotized production systems in automobile manufacturing. Ph.D. dissertation, Riga Technical University, Riga (2013). ISBN 978-9934-10-500-5. https://ortus.rtu.lv/science/en/publications/17487/summary
25. Paryanto, P., Brossog, M., Bornschlegl, M., Franke, J.: Reducing the energy consumption of industrial robots in manufacturing systems. Int. J. Adv. Manuf. Technol. 1–14 (2015)

26. Hauf, D., Meike, D., Paryanto, P., Franke, J.: Energy consumption modeling within the virtual commissioning tool chain. In: IEEE International Conference on Automation Science and Engineering (CASE), pp. 1357–1362 (2015)
27. Meike, D., Pellicciari, M., Berselli, G.: Energy efficient use of multirobot production lines in the automotive industry: detailed system modeling and opimization. IEEE Trans. Autom. Sci. Eng. (TASE) **11**(3), 798–809 (2013)
28. Hauf, D., Meike, D., Lebrecht, M.: Verfahren zum betreiben einer mehrzahl von robotern einer produktionsanlage. Patent application, Stuttgart, German Patent and Trade Mark Office, 31 March 2016
29. Benzerga, S.: Dissertation concept working paper, Digital business patterns and leveraging data (digital assets and digital transformation) for innovation and value creation. RITM, Paris-Sud University (2014)
30. Harvard Business Review analytic services report. Driving digital transformation: new skills for leaders, new role for the CIO (2015)
31. Zetsche, D.: The digital transformation of industry. A European study commissioned by the Bundesverband der Deutschen Industrie e.V

Decision-Controlled Digitization Architecture for Internet of Things and Microservices

Alfred Zimmermann[1(✉)], Rainer Schmidt[2], Kurt Sandkuhl[3],
Dierk Jugel[1,3], Justus Bogner[1,4], and Michael Möhring[2]

[1] Herman Hollerith Center, Reutlingen University,
Danziger Str. 6, 71034 Böblingen, Germany
{alfred.zimmermann,
dierk.jugel}@reutlingen-university.de
[2] Munich University of Applied Sciences, Lothstrasse 64,
80335 Munich, Germany
{rainer.schmidt,michael.moehring}@hm.edu
[3] University of Rostock, Albert Einstein Str. 22, 18059 Rostock, Germany
kurt.sandkuhl@uni-rostock.de
[4] Hewlett Packard Enterprise, Herrenberger Str. 140,
71034 Böblingen, Germany
justus.bogner@hpe.com

Abstract. Digitization of societies changes the way we live, work, learn, communicate, and collaborate. In the age of digital transformation IT environments with a large number of rather small structures like Internet of Things (IoT), Microservices, or mobility systems are emerging to support flexible and agile digitized products and services. Adaptable ecosystems with service-oriented enterprise architectures are the foundation for self-optimizing, resilient run-time environments and distributed information systems. The resulting business disruptions affect almost all new information processes and systems in the context of digitization. Our aim are more flexible and agile transformations of both business and information technology domains with more flexible enterprise information systems through adaptation and evolution of digital enterprise architectures. The present research paper investigates mechanisms for decision-controlled digitization architectures for Internet of Things and Microservices by evolving enterprise architecture reference models and state of the art elements for architectural engineering for micro-granular systems.

Keywords: Digitization architecture · Architectural evolution · Internet of Things · Microservices · Decision analytics and management

1 Introduction

Smart connected products and services expand physical components from their traditional core by adding information and connectivity services using the Internet. Digitized products and services amplify the basic value and capabilities and offer exponentially expanding opportunities [1]. Digitization enables human beings and autonomous objects to collaborate beyond their local context using digital technologies [2].

© Springer International Publishing AG 2018
I. Czarnowski et al. (eds.), *Intelligent Decision Technologies 2017*,
Smart Innovation, Systems and Technologies 73, DOI 10.1007/978-3-319-59424-8_8

Information, data, and knowledge become more important as fundamental concepts of our everyday activities [2]. The exchange of information enables more far-reaching and better decisions of human beings, and intelligent objects. Social networks, smart devices, and intelligent cars are part of a wave of digital economy with digital products, services, and processes driving an information-driven vision [1, 2].

The Internet of Things (IoT) [3–5] connects a large number of physical devices to each other using wireless data communication and interaction based on the Internet as a global communication environment. Additionally, we have to consider some challenging aspects of the overall architecture [6, 7] from base technologies: cyber-physical systems, social networks, big data with analytics, services, and cloud computing. Typical examples for the next wave of digitization are smart enterprise networks, smart cars, smart industries, and smart portable devices.

The fast moving process of digitization [2] demands flexibility to adapt to rapidly changing business requirements and newly emerging business opportunities. To be able to handle the increased velocity and pressure, a lot of software developing companies have switched to integrate Microservice Architectures (MSA) [8]. Applications built this way consist of several fine-grained services that are independently scalable and deployable. Using Microservice Architectures, organizations can increase agility and flexibility for business and IT systems, which fits better with small-sized integrated systems and is vital in the age of digital transformation.

Digitization [2] requires the appropriate alignment of business models and digital technologies for new digital strategies and solutions, as same as for their digital transformation. Unfortunately, the current state of art and practice of enterprise architecture lacks an integral understanding and decision management when integrating a huge amount of micro-granular systems and services, like Microservices and Internet of Things, in the context of digital transformation and evolution of architectures. Our goal is to extend previous approaches of quite static enterprise architecture to fit for flexible and adaptive digitization of new products and services. This goal shall be achieved by introducing suitable mechanisms for collaborative architectural engineering and integration of micro-granular architectures.

Our current research in progress paper investigates the research questions, which are answered by following main sections applying a design science methodology [9]:

RQ1: *How should the digital architecture be holistically tailored to integrate a huge amount of Internet of Things and Microservices architectures, researching the hypotheses that these micro-granular structures can be integrated into a consistent view into a digital enterprise architecture?*

RQ2: *How can we architect a huge amount of the Internet of Things and Microservices to support the digitization of products and processes?*

RQ3: *What are architectural implications for a decision-controlled composition of micro-granular elements, like Internet of Things and Microservices?*

The following Sect. 2 explains the setting of a digital enterprise architecture and links it with specific architectural integration mechanisms for micro-granular systems and services. Section 3 focusses on architecting the Internet of Things for supporting the digital transformation. Section 4 presents an architectural approach to integrating micro-granular systems and services architectures using Microservices. In Sect. 5 we are investigating concepts and mechanisms for analyzing and decision management of multi-perspective digital architectures with a huge amount of micro-granular systems and services. Finally, we summarize in Sect. 6 our research findings and limitations, and our ongoing and next work in academia and practice.

2 Digitization Architecture

Today, Enterprise Architecture Management [10, 11] defines a quite large set of different views and perspectives with frameworks, standards, tools, and practical expertise. An architecture management approach for digital enterprises should support digitization of products and services and should be both holistic [2, 14] and easily adaptable [6]. It should also support digital transformation using new business models and technologies that are based on a large number of micro-structured digitization systems with their own micro-granular architectures like IoT, mobility devices, or with Microservices.

In this paper, we are extending our previous service-oriented enterprise architecture reference model for the context of digital transformations with Microservices and Internet of Things with decision making [15], which are supported by interactive functions of an EA cockpit [16]. Enterprise Services Architecture Reference Cube (ESARC) [14] is our improved architectural reference model for an extended view on evolved micro-granular enterprise architectures (Fig. 1).

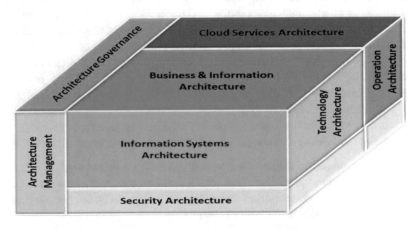

Fig. 1. Enterprise Services Architecture Reference Cube [6, 14]

The new ESARC for digital products and services is more specific than existing architectural standards of EAM [12, 13] and uses eight integral architectural domains to provide a holistic classification model. While it is applicable for concrete architectural instantiations to support digital transformations, it still abstracts from a concrete business scenario or technologies. The Open Group Architecture Framework [12] provides the basic blueprint and structure for our extended service-oriented enterprise architecture domains.

Our research extends an existing metamodel-based model extraction and integration approach from [14] for digital enterprise architecture viewpoints, models, standards, frameworks and tools. The approach supports the adaptable integration of micro-granular architecture. Currently, we are working on the idea of continuously integrating small architectural descriptions for relevant objects of a digital architecture. It is a huge challenge to continuously integrate numerous dynamically growing architectural descriptions from different microstructures with micro-granular architecture into a consistent digital architecture. To address this problem, we are currently formalizing small-decentralized mini-metamodels, models, and data of architectural microstructures, like Microservices and IoT into DEA-Mini-Models (Digital Enterprise Architecture Mini Model).

DEA-Mini-Models consists of partial DEA-Data, partial DEA-Models, and partial EA-Metamodel. They are associated with Microservices and/or objects from the Internet of Things. These structures are based on the Meta Object Facility (MOF) standard [17] of the Object Management Group (OMG). The highest layer M3 represents abstract language concepts used in the lower M2 layer and is, therefore, the meta-metamodel layer. The next layer M2 is the metamodel integration layer and defines the language entities for M1 (e.g. models from UML, ArchiMate [13], or OWL [18]). These models are a structured representation of the lowest layer M0 that is formed by collected concrete data from real-world use cases.

By integrating DEA-Mini-Models micro-granular architectural cells (Fig. 2) for each relevant IoT object or Microservice, the integrated overall architectural metamodel becomes adaptable and can mostly be automatically synthesized by considering the integration context from a growing number of previous similar integrations. In the case of new integration patterns, we have to consider additional manual support.

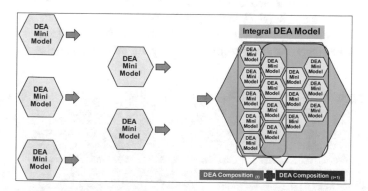

Fig. 2. Federation by composition of DEA-Mini-Models [8]

A DEA-Mini-Model covers partial EA-IoT-Data, partial EA-IoT-Models, and partial EA-IoT-Metamodels associated with main IoT objects like IoT-Resource, IoT-Device, and IoT-Software-Component [3, 19]. The challenge of our current research is to federate these DEA-Mini-Models to an integral and dynamically growing DEA model and information base by promoting a mixed automatic and collaborative decision process [15, 16]. We are currently extending model federation and transformation approaches [20, 21] by introducing semantic-supported architectural representations, from partial and federated ontologies [18, 22] and associate mapping rules with special inference mechanisms.

Fast changing technologies and markets usually drive the evolution of ecosystems. Therefore, we have extracted the idea of digital ecosystems from [23] and linked this with main strategic drivers for system development and their evolution. Adaptation drives the survival of digital architectures, platforms and application ecosystems.

3   Internet of Things Architecture

The Internet of Things [19] connects a large number of physical devices to each other using wireless data communication and interaction, based on the Internet as a global communication environment. Real world objects are mapped into the virtual world. The interaction with mobile systems, collaboration support systems, and systems and services for big data and cloud environments is extended. Furthermore, the Internet of Things is an important foundation of Industry 4.0 [24] and adaptable digital enterprise architectures [14]. The Internet of Things, supports smart products as well as their production enables enterprises to create customer-oriented products in a flexible manner. Devices, as well as human and software agents, interact and transmit data to perform specific tasks part of sophisticated business or technical processes [3, 4].

The Internet of Things embraces not only a things-oriented vision [5] but also an Internet-oriented and a Semantic-oriented one. A cloud-centric vision for architectural thinking of a ubiquitous sensing environment is provided by [25]. The typical setting includes a cloud-based server architecture, which enables interaction and supports remote data management and calculations. By these means, the Internet of Things integrates software and services into digitized value chains.

A layered Reference Architecture for the Internet of Things is described in [19] and (Fig. 3), where layers can be implemented using suitable technologies.

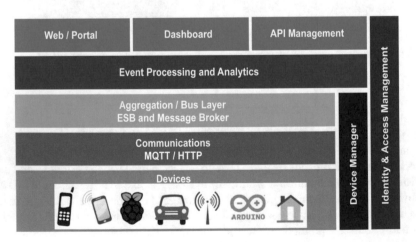

Fig. 3. Internet of Things reference architecture [19]

The main question is, how the Internet of Things architecture fits in a context of a service-based enterprise computing environment? A service-oriented integration approach for the Internet of Things is referenced in [26]. The core issue is, how millions of devices can be flexibly connected to establish useful advanced collaborations within business processes. The service-oriented architecture abstracts the heterogeneity of embedded systems, their hardware devices, software, data formats and communication protocols.

From the inherent connection of a magnitude of devices, which are crossing the Internet over firewalls and other obstacles, are resulting a set of generic requirements [26]. Because of so many and dynamically growing numbers of devices we need an architecture for scalability. Typically, we additionally need a high-availability approach in a 24 × 7 timeframe, with deployment and auto-switching across cooperating data-centers in the case of disasters and high scalable processing demands. The Internet of Thing architecture has to support automatically managed updates and remotely managed devices. Typically, often connected devices collect and analyze personal or security relevant data. Therefore, it should be mandatory to support identity management, access control and security management on different levels: from the connected devices through the holistic controlled environment.

The contribution from [3] considers a role-specific development methodology and a development framework for the Internet of Things. The development framework specifies a set of modeling languages for a vocabulary language to be able to describe domain-specific features of an IoT-application, besides an architecture language for describing application-specific functionality and a deployment language for deployment features. Associated with programming language aspects are suitable automation techniques for code generation, and linking, to reduce the effort for developing and operating device-specific code. The metamodel for Internet of Things applications from [3] specifies elements of an Internet of Things architectural reference model like IoT resources of type: sensor, actuator, storage, and user interface. Base functionalities of IoT resources are handled by components in a service-oriented way by using computational services. Further Internet of Thing resources and their associated physical devices are differentiated in the context of locations and regions.

4 Microservices Architecture

The Microservices approach is spreading quickly. Defined by James Lewis and Martin Fowler, as in [8], it is a fine-grained, service-oriented architecture style combined with several DevOps elements. A single application is created from a set of services. Each of them is running in its own process. Microservices communicate using lightweight mechanisms. Often, Microservices are combined with NoSQL databases from on-premise and optional Cloud environments.

Microservices are built implement business capabilities and are independently deployable, using an automated deployment pipeline. The centralized management elements of these services are reduced to a minimum. Microservices are implemented using different programming languages. Different data storage technologies may be used. As opposed to big monolithic applications, a single Microservice tries to represent a unit of functionality that is as small and coherent as possible. This unit of functionality or business capability is often referred to as a bounded context, a term that originates from Domain-Driven Design (DDD) [27].

Microservices need a strong DevOps culture [28] to handle the increased distribution level and deployment frequency. Moreover, while the single Microservice may be of reasonably low complexity, the overall complexity of the system has not been reduced at all. Gary Olliffe [28] distinguishes between the inner architecture and the outer architecture of Microservices (Fig. 4).

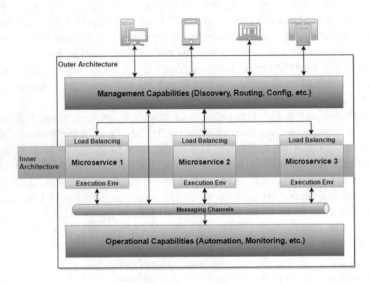

Fig. 4. Microservices inner and outer architecture, based on [28]

Using fine-grained independent services, the hindering complexity is shifted from the inner architecture to the outer architecture. There, inter-service communication, service discovery, or operational capabilities are handled. An important advantage of the Microservices architectures is the possibility to apply a best-of-breed approach for each bounded context [29]. Typical examples are: increased application resilience, independent and efficient scalability and faster and easier deployment. Especially the last advantage increases the agility of business and IT systems.

Microservices enable technological heterogeneity and thus reduce the possibility of lock-ins by outdated technology. Unfortunately, classical enterprise architecture approaches are not flexible enough for the kind of diversity and distribution present in a Microservice Architecture.

5   Decision Analytics

We are exploring in our current research, which extends the more fundamentally approach of a decision dashboard for Enterprise Architecture [30, 31], how an Architecture Management Cockpit [15, 16] can be leveraged and extended to a Decision Support System (DSS) [31] for digital architecture management. An architectural cockpit in Fig. 5 implements a facility, which enables analytics and optimizations using multi-perspective interrelated viewpoints on the system under consideration. Each stakeholder taking part in a cockpit meeting can utilize a viewpoint that displays the relevant information. Viewpoints, which are applied simultaneously, are linked to each other in a such manner that the impact of a change performed in one view can be visualized in other views as well.

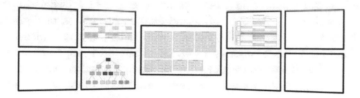

Fig. 5. Architecture Management Cockpit [15, 16]

Jugel et al. [15] present a collaborative approach for decision-making for architecture management. They identify decision making in such complex environment as a knowledge-intensive process reflecting the balance between decentral and central architectural decisions. Therefore, the collaborative approach presented is built based on methods and techniques of adaptive case management (ACM), as defined in [32].

A decision-making step is based on case data consisting of an architectural model and additional insights elicited in previous steps. Consequently, the insights gained during each step contribute to the case file (CaseFile) of the decision-making case. Derived values, like the values of KPIs are thereby not considered additional information, but only a different way of representing and aggregating existing information. If stakeholders based on the values of a KPI decide on affected architecture elements, these decisions and considerations represent new information, which is added to the case file. During decision-making, alternative designs can be identified [13].

The ISO Standard 42010 [33] describes how the architecture of a system can be documented using architecture descriptions. The standard uses views, which are governed by viewpoints to address stakeholders' concerns and their information demands. Jugel et al. [15] introduce an annotation mechanism to add additional knowledge to an architecture description represented by an architectural model. In addition, [15] refines the viewpoint concept of [33] by dividing it into Atomic Viewpoint and Viewpoint Composition to model coherent viewpoints that can be applied simultaneously in a architecture cockpit with central and mobile environments to support stakeholders in decision-making. Architectural Issues and Decisions, were already introduced in the inspiring model of Plataniotis et al. [34]. As described in [34], architectural decisions can be decomposed, translated and substituted into other decisions.

6 Conclusion

We have discussed in this paper the need for a managed bottom-up integration of a huge amount of micro-granular systems and services, that is dynamically growing, like the Internet of Things and Microservices. Following our three mentioned research questions we have leveraged a new digital architecture approach to model a living digital enterprise architecture, which is in line with adaptive models and digital transformation mechanisms. We have investigated new architectural properties of micro-granular systems and services, like of Internet of Things and Microservices as a base for integrating them into our digital reference architecture. Strength of our research results from our novel integration of micro-granular structures and systems, while limits are still resulting from an ongoing validation of our research in practice.

We are currently working on an architectural cockpit for digital enterprise architectures and related engineering processes using extended decision support mechanisms. Both mechanisms for adaptation and flexible integration of digital enterprise architectures as well as decisional processes with rationales and explanations will be subject of future research. Similarly, it may be of interest to support the manual integration decision by automated systems, e.g. via mathematical comparisons (similarity, Euclidean distance), ontologies with semantic integration rules, or data analytics and data mining.

References

1. Porter, M.E., Heppelmann, J.E.: How smart connected products are transforming competition. Harv. Bus. Rev. **92**, 1–23 (2014)
2. Schmidt, R., Zimmermann, A., Möhring, M., Nurcan, S., Keller, B., Bär, F.: Digitization – perspectives for conceptualization. In: Celesti, A., Leitner, P. (eds.) Advances in Service-Oriented and Cloud Computing, pp. 263–275. Springer, Cham (2016)
3. Patel, P., Cassou, D.: Enabling high-level application development for the Internet of Things. CoRR abs/1501.05080, Journal of Systems and Software (2015, submitted)
4. Uckelmann, D., Harrison, M., Michahelles, F.: Architecting the Internet of Things. Springer, Heidelberg (2011)
5. Atzori, L., Iera, A., Morabito, G.: The Internet of Things: a survey. J. Comput. Netw. **54**, 2787–2805 (2010)
6. Zimmermann, A., Schmidt, R., Sandkuhl, K., Wißotzki, M., Jugel, D., Möhring, M.: Digital enterprise architecture – transformation for the Internet of Things. In: Kolb, J., Weber, B., Hall, S., Mayer, W., Ghose, A.K., Grossmann, G. (eds.) IEEE Proceedings of the EDOC 2015 with SoEA4EE, Adelaide, Australia, pp. 130–138, 21–25 September 2015
7. Zimmermann, A., Schmidt, R., Sandkuhl, K., Jugel, D., Bogner, J., Möhring, M.: Multi-perspective digitization architecture for the Internet of Things. In: Abramowicz, W., Alt, R., Franczyk, B. (eds.) Business Information Systems Workshops, BIS 2016 International Workshops, Leipzig, Germany, Revised Papers. Lecture Notes in Business Information Processing, vol. 263, pp. 289–298. Springer, Cham (2017)
8. Bogner, J., Zimmermann, A.: Towards integrating microservices with adaptable enterprise architecture. In: Dijkman, R., Pires, L.F., Rinderle-Ma, S.: IEEE – EDOC Conference Workshops EDOCW 2016, Vienna, pp. 158–163. IEEE (2016)

9. Johannesson, P., Perjons, E.: An Introduction to Design Science. Springer, Cham (2014)
10. Bente, S., Bombosch, U., Langade, S.: Collaborative Enterprise Architecture. Morgan Kaufmann, Waltham (2012)
11. Lankhorst, M.: Enterprise Architecture at Work: Modelling, Communication and Analysis. Springer, Heidelberg (2013)
12. Open Group: TOGAF Version 9.1. Van Haren Publishing (2011)
13. Open Group: ArchiMate 3.0 Specification. Van Haren Publishing (2016)
14. Zimmermann, A., Gonen, B., Schmidt, R., El-Sheikh, E., Bagui, S., Wilde, N.: Adaptable enterprise architectures for software evolution of SmartLife ecosystems. In: Proceedings of EDOCW, Ulm, Germany, pp. 316–323 (2014)
15. Jugel, D., Schweda, C.M., Zimmermann, A.: Modeling decisions for collaborative enterprise architecture engineering. In: 10th Workshop Trends in Enterprise Architecture Research (TEAR), Held on CAISE 2015, Stockholm, Sweden, pp. 351–362. Springer (2015)
16. Jugel, D., Schweda, C. M.: Interactive functions of a cockpit for enterprise architecture planning. In: International Enterprise Distributed Object Computing Conference Workshops and Demonstrations (EDOCW), Ulm, Germany, pp. 33–40. IEEE (2014)
17. Object Management Group: OMG Meta Object Facility (MOF). Core Specification, Version 2.5 (2011)
18. W3C: OWL 2 Web Ontology Language. Structural Specification and Functional-Style Syntax (2009)
19. WSO2 White Paper: Reference Architecture for the Internet of Things. Version 0.8.0 (2015). http://wso2.com
20. Trojer, T. et al.: Living modeling of IT architectures: challenges and solutions. In: Software, Services, and Systems 2015, pp. 458–474 (2015)
21. Farwick, M., Pasquazzo, W., Breu, R., Schweda, C.M., Voges, K., Hanschke, I.: A meta-model for automated enterprise architecture model maintenance. In: EDOC 2012, pp. 1–10 (2012)
22. Altunes, G., et al.: Using ontologies for enterprise architecture analysis. In: IEEE EDOCW 2013, pp. 361–368 (2013)
23. Tiwana, A.: Platform Ecosystems: Aligning Architecture, Governance, and Strategy. Morgan Kaufmann, New York (2013)
24. Schmidt, R., Möhring, M., Härting, R.-C., Reichstein, C., Neumaier, P., Jozinovic, P.: Industry 4.0 - potentials for creating smart products: empirical research results. In: 18th Conference on Business Information Systems, Poznan. Lecture Notes in Business Information Processing. Springer (2015)
25. Gubbi, J., Buyya, R., Marusic, S., Palaniswami, M.: Internet of Things (IoT): a vision, architectural elements, and future directions. Future Gener. Comput. Syst. **29**(7), 1645–1660 (2013)
26. Ganz, F., Li, R., Barunaghi, P., Harai, H.: A resource mobility scheme for service-continuity in the Internet of Things. In: GreenCom 2012, pp. 261–264 (2012)
27. Evans, E.: Domain-driven Design: Tackling Complexity in the Heart of Software. Addison Wesley, Boston (2004)
28. Ollife, G.: Microservices: building services with the guts on outside (2015). http://blogs. gartner.com/garryollife/2015/01/30/microservces-guts-on-the-outside/
29. Newman, S.: Building Microservices: Designing Fine-Grained Systems. O'Reilly, Sebastopol (2015)
30. Op't Land, M., Proper, H.A., Waage, M., Cloo, J., Steghuis, C.: Enterprise Architecture – Creating Value by Informed Governance. Springer, Heidelberg (2009)
31. Keen, P. G. W.: Decision support systems: the next decade. In: Decision Support Systems, vol. 3, no. 3, pp. 253–265. Elsevier (1987)

32. Swenson, K.D.: Mastering the Unpredictable: How Adaptive Case Management Will Revolutionize the Way That Knowledge Workers Get Things Done. Meghan-Kiffer Press, Tampa (2010)
33. Emery, D., Hilliard, R.: Every architecture description needs a framework: expressing architecture frameworks using ISO/IEC 42010. In: IEEE/IFIP WICSA/ECSA, pp. 31–39 (2009)
34. Plataniotis, G., De Kinderen, S., Proper, H.A.: EA anamnesis: an approach for decision making analysis in enterprise architecture. Int. J. Inf. Syst. Model. Des. 4(1), 75–95 (2014)

Digital Enterprise Architecture Management in Tourism – State of the Art and Future Directions

Rainer Schmidt[1], Michael Möhring[1(✉)], Barbara Keller[1], Alfred Zimmermann[2], Martina Toni[3], and Laura Di Pietro[3]

[1] Munich University of Applied Sciences, Lothstrasse 64, 80335 Munich, Germany
{Rainer.Schmidt,michael.moehring,barbara.keller}@hm.edu
[2] Reutlingen University, Alteburgstraße 150, 72762 Reutlingen, Germany
alfred.zimmermann@reutlingen-university.de
[3] University of Roma Tre, Via Silvio D'Amico 77, 00145 Rome, Italy
{Martina.Toni,Laura.Dipietro}@uniroma3.it

Abstract. The advance of information technology impacts Tourism more than many other industries, due to the service character of its products. Most offerings in tourism are immaterial in nature and challenging in coordination. Therefore, the alignment of IT and strategy and digitization is of crucial importance to enterprises in Tourism. To cope with the resulting challenges, methods for the management of enterprise architectures are necessary. Therefore, we scrutinize approaches for managing enterprise architectures based on a literature research. We found many areas for future research on the use of Enterprise Architecture in Tourism.

Keywords: Enterprise architecture management · Tourism · EAM · SOA · Analytics · Hospitality

1 Introduction

Tourism is offers its products based on complex service-systems [1] with a multitude of operand and operant resources [2]. This service-nature of the tourism industry makes it particularly susceptible to technological changes in Information Technology (IT). Therefore, it does not surprise that ICT has reshaped the value chain of the tourism industry by enhancing the ability to produce competitive advantage [3]. No other industry is so strongly impacted by the application of ICT (Information and Communication Technology) [4].

Digitization [5] triggers disruptive changes in business processes, value chains, and business models of Tourism. Tickets formerly printed on paper are now purely digital links between an individual and a service. The electronic tickets in aviation are a prominent example. Value chains are extended and shortened [6]. New intermediaries improve matching between offer and demand. Traditional intermediaries are excluded by short-circuiting provider and consumer of services [6]. E.g. many hotels offer their services directly to their customers. New business models arise, with platforms at their core, that can fulfill customers' demands without own resources [7]. In summary,

© Springer International Publishing AG 2018
I. Czarnowski et al. (eds.), *Intelligent Decision Technologies 2017*,
Smart Innovation, Systems and Technologies 73, DOI 10.1007/978-3-319-59424-8_9

digitization in tourism leads to low transaction costs, lower barriers and to the possibility to obtain more information about tourists' activities and transactions to build a stronger relationship with them [8].

Digitization creates new opportunities but also new challenges arise in Tourism. For instance, in the past, the consumer behavior and their preferences could be hardly observed. Nowadays, many customers share information (e.g. in social networks) and as a result [9], the data become available for the Tourism enterprises. However, this newly available data have to collected and processed before being useful for improving processes and decisions. For managing this new data sources appropriately, the design and implementation of an proper IT architecture becomes a key challenge for companies within the tourism sector.

In literature, the general advantages of IT-systems are investigated quite well. IT-Systems are defined as computers and communication technologies used for the acquisition, processing analysis, storage, retrieval, dissemination and application of information [10, 11]. In particular, many authors emphasize the advantages of the application of IT-Systems in terms of the competitiveness, [3, 10–12] productivity and performance of managing information as well as building closer relationship with the stakeholders [11, 13].

Derived from this research, the use of IT-Systems to digitize processes in tourism, e.g. travel, hospitality, and catering industries is advisable. IT-systems enable partnerships for developing customer-centric strategies as well as the increase of profitability [8]. However, there is only few research specific to tourism. Werthner et al. [14] highlighted the benefits and opportunities that IT brings to each layer of the tourism ecosystem composed of: (1) individual, (2) group/social, (3) corporate/enterprise, (4) network/industry, and (5) government/policy. As Berne et al. [3] state that ICT increases the vertical and horizontal relationship by enabling a large volume of information exchange among sellers and buyers. Furthermore, geographical barriers are eliminated quick and easily. This is especially advantageous in the Tourism sector, where suppliers and consumers are typically widespread.

With regards to the importance of enterprise architecture management and modern ICT in Tourism, we want to address the following research question: *"What is the state of the art of enterprise architecture management research in Tourism?"*.

To investigate this question, we first implemented a systematic literature review. Afterward, we analyzed and summarized our findings with regards to the research question. Finally, we concluded and described interesting possibilities for future research.

2   The Need for Enterprise Architecture Management in Tourism

A huge number of external factors impact enterprises in tourism. In this way tourism companies are continuously challenged to extend their capabilities [15] and adapt their Business Operating Model [16]. Enterprise Architectures (EA) as the architectural part of IT Governance [17] provides a foundation for the conceptualization and planning of initiatives.

Enterprise Architecture covers the logic for business processes and the IT infrastructure model as well as reflecting the dimensions of standardization and integration of the business operating model [16]. Enterprise Architecture Management (EAM) [18] and Services Computing [19] are approaches of choice to align strategy, organize, build, utilize, and distribute capabilities for the digital Enterprise Architecture [5, 20].

Enterprise Architecture requires changes in traditional cultures of enterprises and the development of specific architectural capabilities and roles. Typically for EA is the close interrelationship between an architectural framework [21], which defines a set of aligned architectural viewpoints by a set of visual and procedural elements, an architectural modeling language [22], and an architectural development process, like the architecture development method ADM [21]. Important EAM frameworks and reference models are [23, 24]: The Open Group Architecture Framework (TOGAF), Zachman Framework, US Federal Enterprise Architecture Framework (FEAF), Enterprise Services Architecture Reference Cube (ESARC).

Enterprise Architecture addresses issues typical for industries such as tourism, where many stakeholders cooperate to provide services. Enterprise Architecture provides an open path for stakeholder collaboration, joint analytics, and cooperative decision support [25]. By describing the participating entities in tourism industry, Enterprise Architecture Management provides leadership for business and IT with a powerful decision support environment [26]. In fact, this ranges from strategy development and support for the management to support the digital transformation of enterprises with digitized customer-oriented products and services. As a result, Enterprise Architecture enables organizations to adapt more flexibly, quickly, and effectively. EAM provides an important benefit for enterprises and organizations [27].

An example for enterprise architectures is the "ESARC" (Enterprise Services Architecture Reference Cube) [28] (Fig. 1). ESARC specializes existing architectural standards of EAM – Enterprise Architecture Management [21, 22] and extends them for service-oriented digital enterprise architectures. ESARC defines an integral classification model within eight architectural domains. These architectural domains cover specific architectural viewpoint descriptions [29] in accordance with the orthogonal

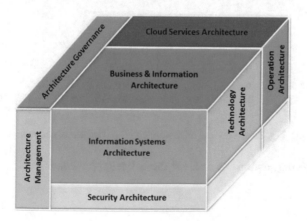

Fig. 1. Enterprise Services Architecture Reference Cube [24, 30]

dimensions of both architectural layers and architectural aspects [30]. ESARC abstracts from a concrete business scenario or technologies: ESARC is an architectural framework, which is applicable for concrete architectural instantiations to support the digital transformation of products and services. The different dimensions the ESARC contains and describes are depicted in the following figure (Fig. 1):

3   Research Method and Data Collection

For investigating the state of the art of enterprise architecture management in Tourism, we designed a systematic literature review [31]. Regarding this research approach, we looked up for research papers in this area in scientific databases like SpringerLink, AISel, ScienceDirect, IEEExplore. Therefore, we searched for keywords such as "EAM", "Enterprise Architecture Management", "Enterprise Architecture", "Tourism", "Hospitality", etc. The timeframe of our search was defined for the last 20 years regarding the development of modern ICT systems in general and particularly in Tourism.

We carefully reviewed every single one of the collected papers. Following this, we extracted only papers, which are in scope of our research question. Besides, we also found interesting research about other topics like "Easy Access Market's" or "efficiency achievement measure" in our literature review. But, because of the sparse congruence with the actual research question, we had to drop. Finally, we selected ten papers for our final review. According to Fig. 2 research with regards to our topic was published from the year of 2000 until now.

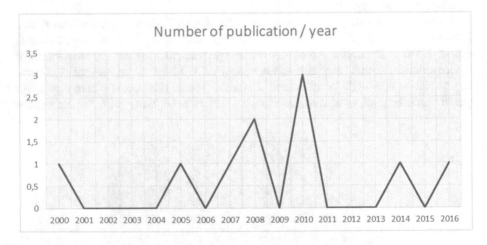

Fig. 2. Overview of the collected publications per year

In the following section, we analyse the collected publications.

4   Results

The selected publications built the basement of our final review. Finally, we analyzed the collected papers with regards to our research question. The following table shows the results we finally got from our systematic literature review (Table 1).

Table 1. Results of the literature review

Reference	Authors	Year	Short description of the aspects of EAM in Tourism
[32]	Abdi and Dominic	2010	Evaluation of an alignment of IT and Business strategy related to a service-oriented architecture (SOA) with a sample of IT managers in hospitality
[33]	Afsarmanesh and Camarinha-Mato	2000	Define the requirements for enterprise inter-operation in the tourism sector based on task sharing and federated information management
[34]	Engels et al.	2008	Development of a method for engineering of a service-oriented architecture. A Tourism sample case was used for illustration
[35]	Franke et al.	2010	Survey of Trends of Enterprise Architecture Management. Some participants of the study are from the Tourism sector
[36]	Hess et al.	2007	Approach for structuring application landscapes. The application landscape of Tourism enterprises was part of the empirical observation/project experience
[37]	Keller et al.	2016	Possibilities of data-centered platforms in Tourism based on Big Data and Predictive Analytics
[38]	Ramasubu et al.	2014	Aspects of service and enterprise architectures with one sample Tourism case "animal reserve"
[39]	Schuck	2010	Design of an enterprise architecture for a national park case
[40]	Vom Brocke et al.	2008	Tourism sample case for illustration of a method for service-oriented process controlling
[41]	Weill and Ross	2005	A hospitality and travel group as a sample case for aspects of IT Governance and related aspects of IT architecture

First of all, there is obviously sparse research in the field of EAM in Tourism in general. Franke et al. [35] show in their research about trends of enterprise architecture management that experts from Tourism sector are interested in this topic because they are participating in this study. Hess et al. [36] developed an approach for structuring application landscapes of enterprises within different domains. Tourism enterprises were observed and consist of different applications in their enterprise architecture with, e.g., different business functions, business objects as well as channels.

Other papers use Tourism cases for the illustration of their EAM method or model. For instance, vom Brocke et al. [40] provide insights into the use of service-oriented process controlling illustrated by a Tourism example. The same was conducted by Engels et al. [34]. Therefore, some research uses Tourism as an example case, but do not focus on EAM in Tourism in a deep and fundamental way. In our systematic literature review, we found only some papers national park protection by Schuck [39], which focusing mainly on aspects of EAM in Tourism. Weill and Ross [41] showed different aspects of IT Governance and related IT architectural consequences based on a Tourism and Hospitality Group case. Abdi and Dominic [32] evaluated aspects of a service-oriented architecture based on IT Business alignment with managers from the hospitality sector.

In [37] different important requirements for data-centered platforms in Tourism are defined. Enterprises in the Tourism sector consists of many different, heterogeneous information systems for different business divisions. Old (transactional) information systems should be adapted as well as integrated to a more analytical oriented infrastructure. Furthermore, the authors show how Big Data can be used in different areas in Tourism. This paper shows that a more flexible and analytical architecture is needed in the future to process a high volume of structured and unstructured data. Furthermore, more external data providers should be integrated into the architectural landscape. This transformation to a more analytical architecture can support more business goals and processes [37]. For instance, the provided information could help to deepen the understanding of the customers about a local and seasonal or even typical event. This could be the starting point for improved marketing activities with regards to new target groups or even to increase the positive experience of the participants by focusing on favorable aspects mentioned in the past. Another interesting aspect could be seen in the reorganization of existing offers and services. Due to the available data and information the menu of a restaurant could be readjusted in terms of vegan dishes or organic food, for example.

In conclusion, enterprise architecture in the field of Tourism consists of many different (heterogeneous) information systems (e.g. for marketing, financial and HR planning.) within the enterprise as well as outside of them (e.g., supplier and customer as well as governmental systems). The integration of processes, data, functions, and organizational aspects [42] are more or less automatic. Data silos are inherent. Furthermore, some research argues for flexible architectural paradigm like service-oriented architectures (SOA). Architectural research with focus to a more analytical enterprise landscape is in an early stage. The current enterprise architecture is designed to process more or less only structured data. There is a need to align architectural components to process semi- and unstructured data (e.g., from social media [37]).

In fact, there exists no big picture and structured research for the whole Tourism sector. Some research papers only uses Tourism as an example case. However, there might be more specific research in some special segments of Tourism. Regarding the importance of ICT in Tourism and the opportunities of EAM in this sector, this topic should be more discovered in the future. Especially in a more common sense with possible generalizability.

In the following, we provided a short abstract overview of the general architecture of a Tourism enterprise (Fig. 3) based on the reviewed literature:

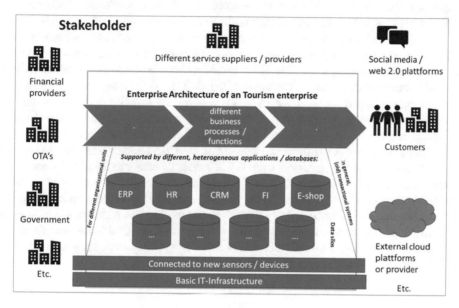

Fig. 3. Abstract overview of the general architecture of a Tourism enterprise based on previous work

5   Conclusion and Future Directions

Enterprise architecture management is important to manage the complex information systems infrastructure in Tourism needed to achieve Business and IT goals. Furthermore, it is essential for coping with the challenges of the more and more technology driven environment of the Tourism sector. Unfortunately, there is only sparse research about EAM in Tourism from a general point of view. We contribute to the current research literature by describing the state of the art of EAM research in Tourism.

Practically oriented users can be better informed about this important and challenging topic in information systems as well as Tourism. Limitations of our research can be found in the used methodology. We cannot address all possibilities of keywords and databases. We only selected the most important ones according to general recommendations [31]. Few research on the use of Enterprise Architecture management in Tourism shows that additional layers of abstraction fill the gap between these two areas. Tourism is well-suited for digitization due to its service-centered character. The first layer of abstraction could be the classification of Tourism models according to the model introduced by Weill and Woerner [43]. They identified four strategic alternatives for enterprises: omnichannel business, ecosystem driver, supplier and modular producer allowing to define the strategic alternatives in Tourism. A further layer could be digital capabilities [44].

The need for a methodological enterprise architecture management becomes more urging, as a change in the value creation mechanism of the tourism industry is about to happen. In the past, many offers were designed like physical goods. Standardized services were "produced" in large quantities with only a few or no possibilities to adapt them to individual customer wishes. Nowadays, a co-creation approach [2] is used increasingly. The preferences of the customer captured by digital means are used to tailor individual experiences. On a theoretical level, the change can be described as the move from a goods-dominant to a service-dominant perspective [2]. To be more precise, this means that the customer provides individual data and helps the supplier in term of value co-creation to improve the offered service. It is enabled on a technological level by the wide diffusion of the internet, the large use of social media and e-commerce.

Future research should follow this interesting research topic in Tourism and transfer aspects and other research results from other industry sectors to Tourism. Case study research as well as other empirical studies for evaluation different aspects of EAM in Tourism could be a good opportunity for future research. Furthermore, discovering the current offerings of consultant and IT services for the tourism sector could be a good opportunity for future research. The development of a conceptional framework for EAM in Tourism as well as the use of enterprise architecture principles and capability aspects of EAM are also possible next research steps.

References

1. Alter, S.: Does service-dominant logic provide insight about operational IT service systems? AMCIS 2010 Proceedings (2010)
2. Vargo, S., Lusch, R.: Service-dominant logic: continuing the evolution. J. Acad. Mark. Sci. **36**, 1–10 (2008)
3. Berne, C., Garcia-Gonzalez, M., Mugica, J.: How ICT shifts the power balance of tourism distribution channels. Tour. Manage. **33**, 205–214 (2012)
4. Buhalis, D., Amaranggana, A.: Smart tourism destinations enhancing tourism experience through personalisation of services. In: Information and Communication Technologies in Tourism 2015, pp. 377–389. Springer (2015)
5. Schmidt, R., Zimmermann, A., Nurcan, S., Möhring, M., Bär, F., Keller, B.: Digitization – perspectives for conceptualization. In: Celesti, A., Leitner, P. (eds) Advances in Service-Oriented and Cloud Computing, ESOCC Workshops 2015, Taormina, Italy. Springer, Cham (2016)
6. Buhalis, D., Law, R.: Progress in information technology and tourism management: 20 years on and 10 years after the internet—the state of eTourism research. Tour. Manage. **29**, 609–623 (2008)
7. Souto, J.E.: Business model innovation and business concept innovation as the context of incremental innovation and radical innovation. Tour. Manage. **51**, 142–155 (2015)
8. Buhalis, D., O'Connor, P.: Information communication technology revolutionizing tourism. Tour. Recreat. Res. **30**, 7–16 (2005)
9. Munar, A.M., Jacobsen, J.K.S.: Motivations for sharing tourism experiences through social media. Tour. Manage. **43**, 46–54 (2014)

10. Poon, A., et al.: Tourism, Technology and Competitive Strategies. CAB international, Wallingford (1993)

11. Buhalis, D.: Strategic use of information technologies in the tourism industry. Tour. Manage. **19**, 409–421 (1998)

12. O'Connor, P., et al.: Electronic information Distribution in Tourism and Hospitality. CAB International, Wallingford (1999)

13. Peppard, J.: IT strategy for business. Financial Times Management, Upper Saddle River (1993)

14. Werthner, H., Alzua-Sorzabal, A., Cantoni, L., Dickinger, A., Gretzel, U., Jannach, D., Neidhardt, J., Pröll, B., Ricci, F., Scaglione, M., et al.: Future research issues in IT and tourism. Inf. Technol. Tour. **15**, 1–15 (2015)

15. Agarwal, N., Soh, C., Sia, S.K.: IT Capabilities in Global Enterprises (2014)

16. Ross, J.W., Weill, P., Robertson, D.C.: Enterprise Architecture as Strategy. Havard Business School Press, Boston (2006)

17. Weill, P., Ross, J.W.: IT Governance: How Top Performers Manage IT Decision Rights for Superior Results. Harvard Business School Press, Boston (2004)

18. Lankhorst, M.: State of the art. In: Enterprise Architecture at Work, pp. 11–41. Springer, Heidelberg (2013)

19. Papazoglou, M.P.: Service-oriented computing: concepts, characteristics and directions. In: Proceedings of the Fourth International Conference on Web Information Systems Engineering, WISE 2003, pp. 3–12 (2003)

20. Zimmermann, A., Schmidt, R., Sandkuhl, K., Jugel, D., Bogner, J., Möhring, M.: Multi-perspective digitization architecture for the internet of things. Presented at the 2nd International Workshop on Digital Enterprise Engineering and Architecture (IDEA 2016), Vienna (2016)

21. Haren, V.: TOGAF Version 9.1 (2011)

22. Haren, V.: Archimate 2.0 Specification. Van Haren Publishing Series. Van Haren Publishing, Amersfoort (2012)

23. Schekkerman, J.: Trends in Enterprise Architecture 2005: How are Organizations Progressing. Institute for Enterprise Architecture Developments, Amersfoort (2005)

24. Zimmermann, A., Pretz, M., Zimmermann, G., Firesmith, D.G., El-Sheikh, E.: Towards service-oriented enterprise architecture for big data applications in the cloud. In: IEEE-EDOCW, Vancouver, Canada, pp. 130–135 (2013)

25. Zimmermann, A., Schmidt, R., Sandkuhl, K., El-Sheikh, E., Jugel, D., Schweda, C., Möhring, M., Wißotzki, M., Lantow, B.: Leveraging analytics for digital transformation of enterprise services and architectures. In: El-Sheikh, E., Zimmermann, A., Jain, L.C. (eds.) Emerging Trends in the Evolution of Service-Oriented and Enterprise Architectures, pp. 91–112. Springer International Publishing, Cham (2016)

26. Jugel, D., Schweda, C.M., Zimmermann, A.: Modeling decisions for collaborative enterprise architecture engineering. In: 10th Workshop Trends in Enterprise Architecture Research (TEAR), held on CAISE 2015, Stockholm, Sweden (2015)

27. Schmidt, R., Möhring, M., Härting, R.-C., Reichstein, C., Zimmermann, A., Luceri, S.: Benefits of enterprise architecture management – insights from european experts. In: Ralyté, J., España, S., Pastor, Ó. (eds.) The Practice of Enterprise Modeling, pp. 223–236. Springer International Publishing, Cham (2015)

28. Zimmermann, A., Jugel, D., Sandkuhl, K., Schmidt, R., Bogner, J., Kehrer, S.: Multi-perspective decision management for digitization architecture and governance. Presented at the Eighth Workshop on Service oriented Enterprise Architecture for Enterprise Engineering, Vienna (2016)

29. Jugel, D., Schweda, C.M.: Interactive functions of a cockpit for enterprise architecture planning. In: 2014 IEEE 18th International Enterprise Distributed Object Computing Conference Workshops and Demonstrations, pp. 33–40. IEEE (2014)
30. Zimmermann, A., Schmidt, R., Sandkuhl, K., Wissotzki, M., Jugel, D., Möhring, M.: Digital enterprise architecture - transformation for the internet of things, Adelaide 2015 (2015)
31. Cooper, H.M.: Synthesizing Research: A Guide for Literature Reviews. Sage, Thousand Oaks (1998)
32. Abdi, M., Dominic, P.D.D.: Strategic IT alignment with business strategy: service oriented architecture approach. In: 2010 International Symposium on Information Technology, pp. 1473–1478. IEEE (2010)
33. Afsarmanesh, H., Camarinha-Matos, L.M.: Future smart-organizations: a virtual tourism enterprise. In: 2000 Proceedings of the First International Conference on Web Information Systems Engineering, pp. 456–461. IEEE (2000)
34. Engels, G., Hess, A., Humm, B., Juwig, O., Lohmann, M., Richter, J.-P., Voss, M., Willkomm, J.: A method for engineering a true service-oriented architecture. In: ICEIS (3–2), pp. 272–281 (2008)
35. Franke, U., Ekstedt, M., Lagerström, R., Saat, J., Winter, R.: Trends in enterprise architecture practice–a survey. In: International Workshop on Trends in Enterprise Architecture Research, pp. 16–29. Springer (2010)
36. Hess, A., Humm, B., Voss, M., Engels, G.: Structuring software cities a multidimensional approach. In: 11th IEEE International Enterprise Distributed Object Computing Conference, EDOC 2007, pp. 122–122. IEEE (2007)
37. Keller, B., Möhring, M., Toni, M., Pietro, L.D., Schmidt, R.: Data-centered platforms in tourism: advantages and challenges for digital enterprise architecture. In: Business Information Systems Workshops, pp. 299–310. Springer International Publishing, Cham (2017)
38. Ramasubbu, N., Woodard, C.J., Mithas, S.: Orchestrating service innovation using design moves: the dynamics of fit between service and enterprise IT Architectures (2014)
39. Schuck, T.M.: An extended enterprise architecture for a network-enabled, effects-based approach for national park protection. Syst. Eng. **13**, 209–216 (2010)
40. vom Brocke, J., Thomas, O., Sonnenberg, C.: Towards an economic justification of service oriented architectures-measuring the financial impact (2008)
41. Weill, P., Ross, J.: Designing IT governance. MIT Sloan Manage. Rev. **46**(2), 26–34 (2005)
42. Scheer, A.-W., Nüttgens, M.: ARIS architecture and reference models for business process management. In: van der Aalst, W., Desel, J., Oberweis, A. (eds.) Business Process Management, pp. 376–389. Springer, Heidelberg (2000)
43. Weill, P., Woerner, S.L.: Thriving in an increasingly digital ecosystem. MIT Sloan Manage. Rev. **56**, 27 (2015)
44. Pelletier, C., Raymond, L.: The IT Strategic Alignment Process: A Dynamic Capabilities Conceptualization (2014)

Advances of Soft Computing
in Industrial and Management
Engineering: New Trends
and Applications

Building Fuzzy Variance Gamma Option Pricing Models with Jump Levy Process

Huiming Zhang[1(✉)] and Junzo Watada[2]

[1] Graduate School of Information, Production and Systems, Waseda University,
2-7 Hibikino, Wakamatsu, Kitakyushu 808-0135, Japan
huimingde@gmail.com
[2] Computer and Information Sciences Department, PETRONAS University
of Technology, 32610 Seri Iskandar, Perak Darul Ridzuan, Malaysia
junzow@osb.att.ne.jp

Abstract. Option pricing models are at core of financial area, and it includes various uncertain factors, such as the randomness and fuzziness. This paper constructs an jump Levy process by combining option pricing models with fuzzy theory, and it sets the drift, diffusion and trend terms as fuzzy random variable. Then, we adopts a Monte Carlo algorithm for numerical simulation, compares and analyses the variance gamma (VG) option pricing model through a simulation experiment, and determines the VG option pricing model and BS model pricing results. The results indicate that VG option pricing with fuzzy settings is feasible.

Keywords: Fuzzy random variable · Levy process · European-style option · Monte Carlo simulation

1 Introduction

In 1900, Louis Bachelier proposed a random model of stock price based on a random walk, which was considered as a milestone in financial mathematics. However, the stock price in the model was assumed to follow the arithmetic of Brownian motion. Consequently, the underlying stock price may be negative and inconsistent with the real market situation. In 1965, Paul Samuelson assumed that stock prices obeyed geometric Brownian motion and developed the European call option pricing formula based on partial differential equations. Later, in 1973, F. Black and M. Scholes created the famous Black–Scholes (BS) European call option pricing formula [1], which did not depend on the personal preferences of investors and could obtain analytic solutions for risk parameters, leverage effects, and so on; thus, it was a grand innovation in stock pricing theory. In the same year, R. Merton expanded the BS model to many other types of financial transactions as Black-Scholes-Merton (BSM) model.

The above theoretical models assume ideal conditions, while the real financial trading market is not like that and includes various uncertain factors. On the one hand, the rate of return of underlying assets is skewness and sharp peak and

© Springer International Publishing AG 2018
I. Czarnowski et al. (eds.), *Intelligent Decision Technologies 2017*,
Smart Innovation, Systems and Technologies 73, DOI 10.1007/978-3-319-59424-8_10

fat tail characteristics rather than a normal distribution, and this phenomenon has been widely accepted. On the other hand, all parameters of the BSM model are crisp value. However, due to the volatility of the financial market, these parameters are fuzzy and cannot be represented with crisp constant value. As for the non-normality of random variables, the Levy process can capture the sharp peaks and fat tails of the underlying assets and is widely used in the option pricing model. Fuzzy set theory is a powerful tool to address fuzziness and complexity of the social environment.

The Levy process can effectively capture jump characteristics of underlying assets, and the pricing model assumption in fuzzy settings is better close to the actual situation. Therefore, some reseachers introduce the Levy process and fuzzy set theory in the option pricing model. Xu *et al.* [6] using fuzzy set theory in the jump-diffusion model, verified the effectiveness of the model through numerical experiment. Romaniuk *et al.* [4] constructed the application of stochastic analysis and fuzzy set theory based on the option pricing method, adopted the Monte Carlo simulation method for numerical experiment. Zhang *et al.* [7] make parameters such as the risk-free interest rate, drift rate and jump intensity as fuzzy numbers, researched the double exponential jump-diffusion model pricing formula of the European option in a fuzzy environment, and explained the feasibility of the method. Nowark and Romaniuk [5] combined fuzzy set theory and a geometric Levy process into the European option pricing model.

In conclusion, with the aim of complementing the existing research, this paper considers fuzzy set theory and jump Levy process conditions based on the variance gamma option pricing model.

The remainder of this paper is structured as follows: Sect. 2 describes the option pricing formula for a VG process with fuzzy settings; Sect. 3 demonstrates its application using European call options with a Monte Carlo simulation method and analyses the sensitivity of the pricing model; Sect. 4 summarises the findings. The framework is shown in Fig. 1.

2   Building Fuzzy Variance Gamma Option Pricing Models with Levy Process

2.1   Pricing Model with the Jump Levy Process

In the probability space (Ω, F, P), the adaptive process $L = \{L_t : t \geq 0\}$ is a Levy process. If $L_0 = 0$ and L_t has an independent stationary random increment, then $\Delta L_t = L_{t+\Delta t} - L_t$, and $t \geq 0$, which are independent and have the same distribution. The characteristic function of the Levy process is as follow: $\Phi_{x_t} = (u \,|\, F_t) = E\{\exp(iux_t)\} = \exp(\phi(u))$ where $\phi(u)$ is a characteristic exponent of a characteristic function with the following structure:

$$\phi(u) = i\gamma u - \tfrac{1}{2}\sigma^2 u^2 + \int_{-\infty}^{+\infty} (e^{iux} - 1 - iux1_{|x|\leq 1})v(dx) \tag{1}$$

γ and σ are measures of drift terms and diffusion terms; v is a measure of jump terms; (γ, σ, v) are called the three Levy terms, where $v(dx)$ is the arrival

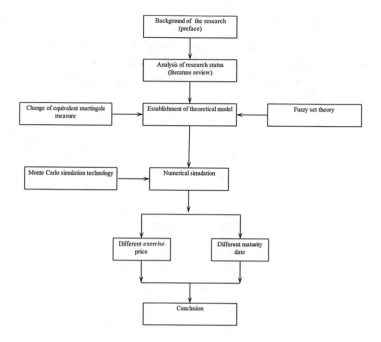

Fig. 1. Conceptual framework

rate of a jump intensity within a unit of time, and $v(R)$ is the sum of the possibilities of all jump intensities.

A gamma process is a slave process of a VG process, with the following density function:

$$f_{Gamma}(g) = \frac{b^a}{\Gamma(g)} g^{a-1} e^{-bg} \tag{2}$$

where $\Gamma(\cdot)$ is the gamma function, and $g > 0$, $a > 0$, and $b > 0$ are boundary conditions. The characteristic function of the gamma distribution is expressed as follows:

$$\Phi_{Gamma}(u; a, b) = E(e^{iux}) = \left(1 - \frac{iu}{b}\right)^{-a} \tag{3}$$

The density function of the VG process is as follows:

$$f_{x_t}(x) = \int_0^\infty \frac{g^{\frac{t}{v}-1} e^{-\frac{g}{v}}}{v^{\frac{t}{v}} \Gamma\left(\frac{t}{v}\right)} * \frac{1}{\sigma\sqrt{2\pi g}} \exp\left(-\frac{(x-\theta g)^2}{2\sigma^2 g}\right) dg \tag{4}$$

The characteristic function of jump process is written as follows:

$$E(e^{iuX_t}) = \varphi(u; \sigma, v, \theta) = (1 - iuv\theta + \tfrac{1}{2}\sigma^2 vu^2)^{-\frac{1}{v}} \tag{5}$$

2.2   Risk Neutral Model

Equivalent martingale transformation mainly transforms assets in random process X_t under measure $\mathbb{P}$ into random process $\widehat{X_t}$ under risk neutral measure $\mathbb{Q}$. The requirements of equivalent martingale in a risk neutral environment are written as follows:

$$E_t^{\mathbb{Q}}[\widehat{S_T}|\mathcal{F}_t] = \widehat{S_t} = S_t \exp(-rt) \tag{6}$$

The above formula is consistent with the characteristic function of a VG process, $\varphi_x(u) = E[\exp(iux)]$. The risk neutral return on assets $\widehat{X_t}$ for risk asset yield sequence X_t under measure $\mathbb{P}$ is corrected to:

$$\widehat{X_t} = r - \varphi(-i) + X_t \tag{7}$$

$$E(\widehat{X_t}) = r - E(\varphi(-i)) + E(X_t) = r \tag{8}$$

When the characteristic function of the VG process is substituted with $r - \varphi(-i)$, the risk neutral drift rate of the VG process is written as follows:

$$u_{VG}^* = r + \frac{\ln\left(1 - v\theta - \frac{\sigma^2 v}{2}\right)}{v} \tag{9}$$

The pricing model of the underlying assets during the VG process is expressed as follows:

$$S_t = S_0 * \exp[u_{VG}^* * t + X_{VG}]$$

$$= S_0 * \exp\left[rt + \frac{\ln\left(1 - v\theta - \frac{\sigma^2 v}{2}\right)}{v} * t + X_{VG}\right]$$

$$= S_0 * \exp\left[rt + \frac{\ln\left(1 - v\theta - \frac{\sigma^2 v}{2}\right)}{v} * t + \theta * g_t + \sigma W(g_t)\right] \tag{10}$$

where $g_t \sim gamma(a, b)$.

According to no-arbitrage pricing method, the European call option price is mentioned as follows:

$$C(S(0); K, t) = e^{-rt} E[\max(S(t) - K, 0)] \tag{11}$$

Under measure $\mathbb{Q}$, according to desirable properties:

$$C(S_0; K, t) = E\left[e^{-rt} E[\max(S(t) - K, 0) \,|\, g_t = g]\right] \tag{12}$$

Set $c(g) = e^{-rt} E[\max(S(t) - K, 0) \,|\, g_t = g]$. Madan and Milne [2] proved:

$$c(g) = S_0 \left(1 - \frac{v(\alpha + s)^2}{2}\right)^{\frac{t}{v}}$$

$$* \exp\left(\frac{(\alpha + s)^2 g}{2}\right) * N\left(\frac{d}{\sqrt{g}} + (\alpha + s)\sqrt{g}\right)$$

$$- K \exp(-rt)\left(1 - \frac{v\alpha^2}{2}\right)^{\frac{t}{v}} \exp\left(\frac{\alpha^2 g}{2}\right) * N\left(\frac{d}{\sqrt{g}} + \alpha\sqrt{g}\right) \tag{13}$$

where

$$s = \frac{\sigma}{\sqrt{1 + \left(\frac{\theta}{\sigma}\right)^2 \frac{v}{2}}}, \quad c_1 = \frac{v(\alpha + s)^2}{2}, \quad c_2 = \frac{v\alpha^2}{2},$$

$$d = \frac{1}{s}\left[\ln\frac{S_0}{K} + rt + \frac{t}{v}\ln\left(\frac{1 - c_1}{1 - c_2}\right)\right].$$

Therefore, $C(S_0; K, t) = E[c(g)|g_t = g]$:

$$C(S_0; K, t) = \int_0^\infty c(g)\frac{g^{\frac{t}{v}-1}e^{-\frac{g}{v}}}{v^{\frac{t}{v}}\Gamma\left(\frac{t}{v}\right)}dg \tag{14}$$

Set $y = g/v$, $\gamma = \frac{t}{v}$:

$$C(S_0; K, t)$$
$$= \int_0^\infty S_0(1 - c_1)^\gamma e^{c_1 y} * N\left(\frac{d/\sqrt{v}}{\sqrt{y}} + (\alpha + s)\sqrt{v}\sqrt{y}\right)$$
$$- Ke^{-rt}(1 - c_2)^\gamma e^{c_2 y} * N\left(\frac{d/\sqrt{v}}{\sqrt{y}} + \alpha\sqrt{v}\sqrt{y}\right)\frac{y^{\gamma-1}e^{-y}}{\Gamma(\gamma)}dy \tag{15}$$

Then,

$$\Psi(a, b, \gamma) = \int_0^\infty N\left(\frac{a}{\sqrt{u}} + b\sqrt{u}\right)\frac{u^{\gamma-1}e^{-u}}{\Gamma(\gamma)}du$$

Hence,

$$C(S_0; K, t)$$
$$= S_0\Psi\left(d\sqrt{\frac{1 - c_1}{v}}, (\alpha + s)\sqrt{\frac{v}{1 - c_1}}, \frac{t}{v}\right)$$
$$- K\exp(-rt)\Psi\left(d\sqrt{\frac{1 - c_2}{v}}, (\alpha + s)\sqrt{\frac{v}{1 - c_2}}, \frac{t}{v}\right) \tag{16}$$

2.3 Pricing Model of Fuzzy Random Variables Settings

Fuzzy random variable $\tilde{X}$ is a mapping of probability measure space (Ω, F, P) to fuzzy number set $\tilde{F}_c$. Mapping $\tilde{X}$ follows measurability conditions, i.e., $\forall \alpha \in [0, 1]$, $\tilde{X}_\alpha(w) = [\tilde{X}_\alpha^L(w), \tilde{X}_\alpha^U(w)]$, and $w \in \Omega$ is a random interval. Therefore, $\tilde{X}_\alpha^L(w)$ and $\tilde{X}_\alpha^U(w)$ are common random variables.

To establish the option pricing model with fuzzy settings, we would replace the yield rate, volatility and jump arrival rate with the fuzzy yield rate, fuzzy volatility and fuzzy jump arrival rate. The European call option pricing formula is written as follows:

$$\tilde{C}(S_0; K, t)$$
$$= \tilde{C}(S_0; K, t, \tilde{\theta}, \tilde{\sigma}, \tilde{v})$$
$$= S_0 \Psi\left(\tilde{d}\sqrt{\frac{1-\tilde{c}_1}{\tilde{v}}}, (\alpha + \tilde{s})\sqrt{\frac{\tilde{v}}{1-\tilde{c}_1}}, \frac{t}{\tilde{v}}\right)$$
$$- K\exp(-rt)\Psi\left(\tilde{d}\sqrt{\frac{1-\tilde{c}_2}{\tilde{v}}}, (\alpha + \tilde{s})\sqrt{\frac{\tilde{v}}{1-\tilde{c}_2}}, \frac{t}{\tilde{v}}\right) \quad (17)$$

where,

$$\tilde{s} = \frac{\tilde{\sigma}}{\sqrt{1 + \left(\frac{\tilde{\theta}}{\tilde{\sigma}}\right)^2 \frac{\tilde{v}}{2}}}, \quad \tilde{c}_1 = \frac{\tilde{v}(\alpha + \tilde{s})^2}{2}, \quad \tilde{c}_2 = \frac{\tilde{v}\alpha^2}{2},$$
$$\tilde{d} = \frac{1}{\tilde{s}}\left[\ln\frac{S_0}{K} + rt + \frac{t}{\tilde{v}}\ln\left(\frac{1-\tilde{c}_1}{1-\tilde{c}_2}\right)\right].$$

According to the extension principle, the membership function of $\tilde{C}(S_0; K, t, \tilde{\theta}, \tilde{\sigma}, \tilde{v})$ is written as follows:

$$u_{\tilde{C}_t}(c) = \sup_{\{(\theta,\sigma,v):c=C(S_0;K,t,\theta,\sigma,v)\}} \min\{u_{\tilde{\theta}}(\theta), u_{\tilde{\sigma}}(\sigma), u_{\tilde{v}}(v)\} \quad (18)$$

For value C of fuzzy price $\tilde{C}(S_0; K, t, \tilde{\theta}, \tilde{\sigma}, \tilde{v})$ at t, we need to know its membership degree λ. The membership function of $\tilde{C}$ can also be called $u_{\tilde{C}_t}(c) = \sup_{0\le\lambda\le 1}\lambda \cdot 1_{(\tilde{C}_t)_\lambda}(c)$ in which $(\tilde{C}_t)_\lambda$ is the λ-level set of $\tilde{C}_t$. The λ-level sets of $\tilde{\theta}, \tilde{\sigma}$ and $\tilde{v}$ are $\tilde{\theta}_\lambda = [\tilde{\theta}_\lambda^L, \tilde{\theta}_\lambda^U]$, $\tilde{\sigma}_\lambda = [\tilde{\sigma}_\lambda^L, \tilde{\sigma}_\lambda^U]$ and $\tilde{v}_\lambda = [\tilde{v}_\lambda^L, \tilde{v}_\lambda^U]$. Therefore, the λ-level set of $\tilde{C}_t$ is written as follows:

$$(\tilde{C}_t)_\lambda = [(\tilde{C}_t)_\lambda^L, (\tilde{C}_t)_\lambda^U]$$
$$= \left[\min_{\tilde{\theta}_\lambda^L \le \theta \le \tilde{\theta}_\lambda^U, \tilde{\sigma}_\lambda^L \le \sigma \le \tilde{\sigma}_\lambda^U, \tilde{v}_\lambda^L \le v \le \tilde{v}_\lambda^U} C(S_0; K, t, \theta, \sigma, v),\right.$$
$$\left.\max_{\tilde{\theta}_\lambda^L \le \theta \le \tilde{\theta}_\lambda^U, \tilde{\sigma}_\lambda^L \le \sigma \le \tilde{\sigma}_\lambda^U, \tilde{v}_\lambda^L \le v \le \tilde{v}_\lambda^U} C(S_0; K, t, \theta, \sigma, v)\right] \quad (19)$$

This paper selects the general defuzzification method given by Feller and Majlender [3], and obtains a fuzzy expectation with upper and lower weights of the λ-level set of $\tilde{C}_t$ with the following formula:

$$M(\tilde{C}) = \frac{M(\tilde{C})^L + M(\tilde{C})^U}{2}$$
$$= \frac{\int_0^1 f(\lambda)\tilde{C}_\lambda^L d\lambda + \int_0^1 f(\lambda)\tilde{C}_\lambda^U d\lambda}{2}$$
$$= \int_0^1 \frac{f(\lambda)}{2}(\tilde{C}_\lambda^L + \tilde{C}_\lambda^U)d\lambda \quad (20)$$

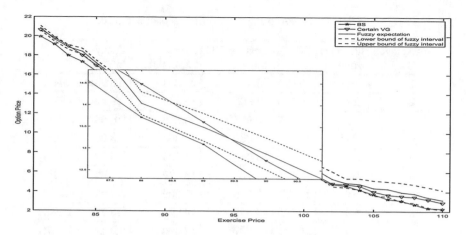

Fig. 2. Option price with different exercise price

3 Monte Carlo Numerical Simulation

3.1 European Option Pricing Under Different Exercise Price

We set the original values of the underlying assets and parameters of the model. Assume that the original value of the underlying assets is $S_0 = 100$, the annual risk free interest rate is $r = 0.02$, the exercise price at maturity is $K = 95$, and $T = 0.5$. For the parameters of the BS model, the return volatility σ is 0.2. For the parameters of the VG model in a certain environment, the drift term θ is 0.05, the extension term σ is 0.2, and the jump term v is 1.5. For the parameters of the VG model with fuzzy settings, the membership functions of fuzzy random variables $\tilde{\theta}$, $\tilde{\sigma}$ and $\tilde{v}$ are set as the most widely used trapezoidal function forms. Set $\tilde{\theta} = [0.035, 0.04, 0.05, 0.055]$, $\tilde{\sigma} = [0.14, 0.19, 0.25, 0.28]$, $\tilde{v} = [1.25, 1.5, 1.75, 2]$, with a confidence level $\lambda = 0.8$. During Monte Carlo simulation, divide the time interval into M pieces for N simulations, where $M = 10$, and $N = 100000$.

Under different exercise prices, the results of the BS model, VG model in a certain environment and VG model with fuzzy settings are shown in Table 1 and Fig. 2.

Table 2 analyses the sensitivity of the VG model to jump parameter v in certain and fuzzy environment. When the exercise price is $K = 95$ in the VG model in a certain environment, with v gradually increasing from 1 to 4, the option price drops from 11.361 to 6.437; for the VG model with fuzzy settings, with the trapezoid membership parameter interval of v increasing, the fuzzy interval of option gradually becomes narrower, and the fuzzy expectation gradually decreases.

Table 1. Option price with different exercise price

Sequence	Exercise price	BS	Certain VG model	Fuzzy expectation	Fuzzy interval
1	81	19.925	20.625	20.804	[20.561, 21.047]
2	82	19.139	19.800	19.650	[19.427, 19.873]
3	83	17.977	18.957	18.826	[18.583, 19.069]
4	84	17.325	18.006	18.393	[18.081, 18.704]
5	85	16.337	16.967	17.209	[16.921, 17.496]
6	86	15.532	16.109	16.172	[15.89, 16.454]
7	87	14.693	15.335	15.809	[15.449, 16.168]
8	88	13.713	14.480	14.035	[13.772, 14.298]
9	89	13.097	13.607	13.490	[13.173, 13.807]
10	90	12.165	12.728	12.883	[12.518, 13.247]
11	91	11.432	11.965	12.248	[11.839, 12.656]
12	92	10.656	11.179	11.505	[11.065, 11.944]
13	93	10.136	10.544	11.226	[10.697, 11.754]
14	94	8.973	9.709	9.945	[9.453, 10.437]
15	95	8.725	9.261	9.512	[8.95, 10.073]
16	96	8.036	8.333	8.748	[8.157, 9.338]
17	97	7.260	7.650	8.674	[7.967, 9.38]
18	98	6.913	7.181	7.825	[7.201, 8.449]
19	99	6.278	6.596	7.185	[6.616, 7.754]
20	100	5.702	5.962	6.381	[5.888, 6.873]
21	101	5.420	5.590	6.153	[5.562, 6.743]
22	102	4.776	4.848	5.316	[4.549, 6.083]
23	103	4.614	4.762	4.906	[4.449, 5.363]
24	104	4.202	4.593	4.792	[4.222, 5.362]
25	105	3.751	4.034	4.382	[3.607, 5.156]
26	106	3.262	3.688	4.287	[3.525, 5.049]
27	107	3.097	3.564	3.952	[2.973, 4.931]
28	108	2.723	3.479	3.818	[2.89, 4.746]
29	109	2.348	3.215	3.434	[2.392, 4.476]
30	110	2.289	2.891	3.161	[2.155, 4.166]

Table 2. Option prices with different jump intensities

Certain VG model	Jump term v value	1.000	1.250	1.500	1.750	2.000	2.250	2.500
	Option price	11.361	9.939	9.028	8.544	7.999	7.809	7.527
	Jump term v interval	2.750	3.000	3.250	3.500	3.750	4.000	
	Option price	7.184	6.956	6.959	6.717	6.640	6.437	
Fuzzy VG model	Jump term v interval	[1.2, 1.45, 1.7, 1.95]		[1.25, 1.5, 1.75, 2]		[1.3, 1.55, 1.8, 2.05]		
	Fuzzy expectation	9.999		9.986		9.976		
	Fuzzy interval	[9.303, 10.694]		[9.364, 10.608]		[9.478, 10.474]		

The membership function of the option price determined by the VG model with fuzzy settings is shown in Fig. 3. Option pricing intervals of different confidence levels λ are shown in Table 3. When $\lambda = 0.9$, the corresponding option price is in a closed interval [9.44, 10.517]. According to Fig. 4, with an increase in the confidence level λ, the fuzzy interval gradually becomes narrower, while the fuzzy expectation basically remains the same.

Table 3. Option prices with different confidence levels λ

λ	Fuzzy expectation	Fuzzy interval
0.7	9.992	[9.284, 10.699]
0.75	9.989	[9.322, 10.656]
0.8	9.986	[9.364, 10.608]
0.85	9.979	[9.397, 10.56]
0.9	9.979	[9.44, 10.517]
0.95	9.976	[9.478, 10.474]

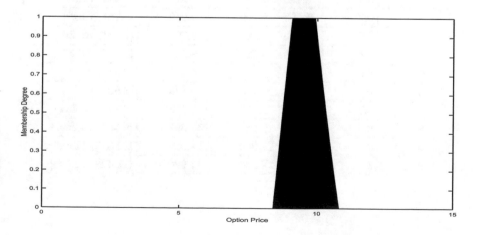

Fig. 3. Membership function of fuzzy option price

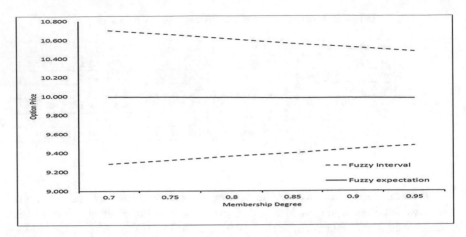

Fig. 4. Option prices with different confidence levels λ

3.2 European Option Pricing Under a Change in Maturity Dates

Under different maturity dates, the pricing results of the VG model in a certain environment and with fuzzy settings are shown in Table 4 and Fig. 5. According to the table and figure, with an increase in maturity time T, the option price increases.

Table 4. Option price with different Maturity time

Sequence	Maturity time	BS	Certain VG model	Fuzzy expectation	Fuzzy interval
1	0.1	6.436	6.894	6.915	[6.641, 7.188]
2	0.2	6.599	7.100	7.040	[6.761, 7.319]
3	0.3	7.223	7.894	7.704	[7.471, 7.936]
4	0.4	7.885	8.222	8.337	[8.005, 8.668]
5	0.5	8.454	9.039	9.077	[8.582, 9.572]
6	0.6	8.933	9.465	9.440	[9.004, 9.876]
7	0.7	9.383	9.925	10.102	[9.743, 10.461]
8	0.8	9.700	10.249	10.363	[9.964, 10.762]
9	0.9	10.173	11.066	11.074	[10.525, 11.623]
10	1	10.571	11.883	11.986	[11.428, 12.743]

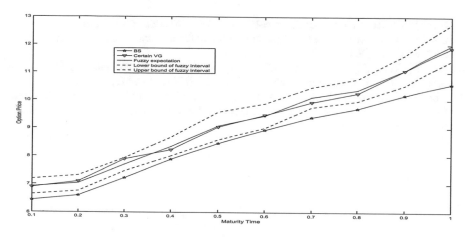

Fig. 5. The option price with different maturity time

4 Conclusions

This paper combines fuzzy set theory to construct an option pricing model with jump Levy process. The model sets the drift term, diffusion term and jump term as fuzzy random vectors, adopts the Monte Carlo simulation algorithm for simulation, then compares option pricing using the VG model, the BS model with fuzzy settings and the VG model in a certain environment. The main conclusions are below, (1) According to the Monte Carlo numerical simulation analysis results, the option pricing model with the drift term, diffusion term and jump term are assumed to be fuzzy random variables are feasible, and the results are more reasonable. (2) The option prices determined by the VG model with fuzzy settings are higher than those determined by the VG and BS models in a certain environment. (3) VG models in certain and with fuzzy settings are sensitive to changes in the jump parameters. The option price decreases when the value of the jump parameter increases.

References

1. Black, F., Scholes, M.: The pricing of options and corporate liabilities. J. Pol. Econ. **81**(3), 637–654 (1973)
2. Daal, E.A., Madan, D.B.: An empirical examination of the variance-gamma model for foreign currency options. J. Bus. **78**(6), 2121–2152 (2005)
3. Fullér, R., Majlender, P.: On weighted possibilistic mean and variance of fuzzy numbers. Fuzzy Sets Syst. **136**(3), 363–374 (2003)
4. Nowak, P., Romaniuk, M.: Computing option price for Levy process with fuzzy parameters. Eur. J. Oper. Res. **201**(1), 206–210 (2010)

5. Nowak, P., Romaniuk, M.: Application of Levy processes and Esscher transformed martingale measures for option pricing in fuzzy framework. J. Comput. Appl. Math. **263**, 129–151 (2014)
6. Xu, W., Wu, C., Xu, W., Li, H.: A jump-diffusion model for option pricing under fuzzy environments. Insur. Math. Econ. **44**(3), 337–344 (2009)
7. Zhang, L.H., Zhang, W.G., Xu, W.J., Xiao, W.L.: The double exponential jump diffusion model for pricing european options under fuzzy environments. Econ. Model. **29**(3), 780–786 (2012)

Evolutionary Regressor Selection in ARIMA Model for Stock Price Time Series Forecasting

Ruxandra Stoean[(⊠)], Catalin Stoean, and Adrian Sandita

Faculty of Sciences, University of Craiova, Craiova, Romania
{ruxandra.stoean,catalin.stoean,asandita}@inf.ucv.ro

Abstract. Stock price prediction over time is a problem of practical concern in economics and of scientific interest in financial time series forecasting. The matter also expands toward detecting the variables that play an important role in its behaviour. The current study thus appoints an ARIMA model with regressors to predict the daily return of ten companies enlisted in the Romanian stock market on the base of nine exogenous predictors. In order to additionally outline the most informative attributes for the prediction, feature selection is also considered and performed by means of genetic algorithms. The experimental results justify the benefits of the model with the evolutionary selector.

Keywords: Stock price · Financial time series forecasting · ARIMA · Regressor · Feature selection · Genetic algorithms

1 Introduction

The fluctuation of the price for the stocks of a company over time may be also influenced by the past behaviour of many exogenous variables, like the number of transactions, the number of stocks, their price range or the daily exchange rate. All these data recorded over time constitute a time series, whose modelling should be able to provide good predictions of the subsequent course.

An Auto-Regressive Integrated Moving Average (ARIMA) model with regressors is appointed within present work for the stock price time series forecasting task related to ten companies enlisted in the Romanian stock market. Nevertheless, forecasting alone is not sufficient for delivering a real decision-making support. The specialists are also interested in knowing those indicators that had decisive influence on the prediction. More, these dominant features may be different for each of the examined companies. The study continues with their heuristic determination through genetic algorithms (GA).

The paper is organized as follows. Section 2 presents the real-world problem of stock price prediction that is modelled by the ARIMA approach with evolutionary selected regressors outlined in Sect. 3. Experiments are described in the corresponding Sect. 4 and the conclusions with respect to the obtained results are given in Sect. 5.

© Springer International Publishing AG 2018
I. Czarnowski et al. (eds.), *Intelligent Decision Technologies 2017*,
Smart Innovation, Systems and Technologies 73, DOI 10.1007/978-3-319-59424-8_11

2   Materials

The data examined in the paper had been collected from the Romanian stock market - Bursa de Valori Bucuresti (BVB)[1]. Ten well-known companies enlisted on the BVB have thus been selected for stock price forecasting. Their stock symbols are BIO, BRD, BVB, EBS, EL, FP, SNP, TEL, TGN, TLV. Nine exogenous predictors are also taken into account in the process: the number of transactions, total number of stocks, total value of stocks, low price, average price, high price, opening price, closing price and the Romanian Leu-EUR exchange rate. Their values were recorded daily (obviously excluding the weekends) for 11 months of year 2016, specifically from January 4th until December 5th. The chosen indicators with their range of values are given in Table 1.

Table 1. Predictors and value ranges for the BVB stock price data set.

Variable	Range
Number of transactions	[1, 1946]
Total number of stocks	[1, 158015692]
Total value of stocks	[26.2, 36720940.64]
Low price	[0.214, 307]
Average price	[0.217, 308]
High price	[0.22, 310]
Opening price	[0.2155, 309]
Closing price	[0.2155, 310]
Romanian Leu-EUR exchange rate	[4.4444, 4.5396]

Following the argument of [1], the dependent variable is taken to be the daily return (denoted by DR in (1)) that shows the direction (increase, maintenance or decrease) of the stock closing price (denoted by CP) at time t.

$$DR_t = \frac{CP_t - CP_{t-1}}{CP_{t-1}} \qquad (1)$$

It is its logarithmic formulation (2) that is further specifically considered, as it has normal distribution [1].

$$y_t = ln(1 + DR_t) \qquad (2)$$

3   ARIMA Model with GA-Selected Regressors for Stock Price Forecasting

The nine exogenous variables together with five lags of the dependent variable are considered in the time series forecasting. An ARIMA model with regressors [10]

---
[1] www.bvb.ro.

is constructed on their base. Subsequent feature selection is performed by a GA [4] to determine the most influential predictors on the forecast.

3.1 ARIMA Model for Stock Price Prediction

ARIMA forecasting models can be interpreted in a regression-like fashion, as the output is the expression of its weighted last p values (lags) and of the weighted q prediction error values (lags of errors) (3) [10].

$$\hat{y}_t = \mu + \phi_1 y_{t-1} + ... + \phi_p y_{t-p} - \theta_1 e_{t-1} - ... - \theta_q e_{t-q}, \qquad (3)$$

where μ is the constant, ϕ_k and θ_k are the AR and MA coefficients at lag k, and y_{t-k} and $e_{t-k} = y_{t-k} - \hat{y}_{t-k}$ the values of the dependent variable and of the forecast error at lag k, respectively.

If exogenous variables are also available, the additional regressors can be also integrated into the equation, which is now reformulated into (4) [10].

$$\hat{y}_t - \phi_1 y_{t-1} - ... - \phi_p y_{t-p} = \mu - \theta_1 e_{t-1} - ... - \theta_q e_{t-q} +$$
$$+ \beta^1 (x_t^1 - \phi_1 x_{t-1}^1 - ... - \phi_p x_{t-p}^1) + ... + \beta^n (x_t^n - \phi_1 x_{t-1}^n - ... - \phi_p x_{t-p}^n), \quad (4)$$

where n is the number of exogenous variables, β^i is every constant and x_{t-k}^i is indicator i at lag k.

3.2 Genetic Algorithm for Variable Selection

Evolutionary algorithms (EA) have been used before for optimization within time series forecasting in several directions. Parametrization of the ARIMA model was achieved by means of evolution strategies in [17]. The AR and MA coefficients were evolved in [3]. An evolutionary selection of time lags in a fuzzy inference system was performed in [9]. A population of evolutionary polynomials to match a past concordant pattern with the current behaviour was appointed by [8]. Evolutionary strategies and particle swarm optimization solved the forecasting formulation in [15], in opposition to the gradient alternative.

The GA used in the present paper does not interfere with the ARIMA model, instead it aims at selecting the most important regressors with respect to a smaller error on the forecast. The approach is then typical of the traditional EA employment in feature selection for machine learning [13,14,18]. The length of an individual corresponds to the number of regressors. The encoding is binary, where a value of 1 signifies that the corresponding indicator is taken into account and a value of 0 denotes that the attribute is being skipped. The fitness evaluation is quantified by the prediction error of the ARIMA model encompassing the variables selected by the measured individual. The GA common one-point recombination and bit-flip mutation induce variation into the population.

4 Experimental Results

The current section is organized according to the recommendations in [2], which assumes the creation of several subsections that encourage the provision of the details needed to understand and also replicate the experiments.

4.1 Pre-experimental Planning

The GA is considered for finding an optimal subset of regressors for the ARIMA model and for improving the forecasting accuracy. While initially it was decided to use solely the root mean squared error (RMSE) to measure the fitness of the GA individuals, it was noticed that, when having changed the manner of evaluation, the resulted chosen regressors did not always remain the same. A different modeling of the fitness landscape helps through a better exploration of the search space and improves the problem solving performance [11]. In this sense, it was decided to use two versions of GAs with respect to their fitness function: one that uses RMSE and one that computes the mean absolute error (MAE). During pre-experimentation, it was observed that indeed some indicators were validated by both fitness measures, while others were revealed as being of high importance as well for different stock symbols.

4.2 Task

The aim is to forecast the daily return for each company. Moreover, by employing the GA for selecting which regressors to be used, another equally important task is to determine what are the decisive indicators for every company.

4.3 Experimental Setup

The total number of records for each company is 225. The first 70% are considered in the training set, while the rest represent the test part.

As mentioned in Subsect. 3.1, beside the 9 exogenous indicators, there are 5 lag attributes obtained from the daily return. This conducts to 14 indicators in total, so the GA encoding will have 14 binary genes, each corresponding to taking or not (1 or 0) into account the regressor for building the model. A population size of 50 individuals is considered, the mutation probability is taken as 0.3 and the iterative process stops after 50 steps. Each GA run is repeated 30 times in order to have statistical significance for the results.

The training set is used for building the ARIMA model. The genes from the GA individual that are equal to 1 decide the representative predictors that are engaged in constructing the model. Next, for each involved predictor, future values are estimated for the period of time corresponding to the dates from the test set and these are used in the subsequent forecasting process. The ARIMA models are found using the automatic model detection available in the R software platform. Subsequently, the two options used for measuring the errors, RMSE and MAE, are each considered in turn as fitness of the GA.

The RSME represents the sample standard deviation of the discrepancies between the predicted outputs and the actual ones on the test set [7] and is described in Eq. (5), with $\hat{y}$ and y being the forecasted and the known value for the outcome, respectively. The MAE also measures how close are the estimated values to the actual observations using (6) with the used symbols having the same meaning as those in the previous equation.

$$RSME(\hat{y}) = \sqrt{\frac{1}{n} \sum_{i=1}^{n} (\hat{y}_i - y_i)^2} \tag{5}$$

$$MAE(\hat{y}) = \frac{1}{n} \sum_{i=1}^{n} |\hat{y}_i - y_i| \tag{6}$$

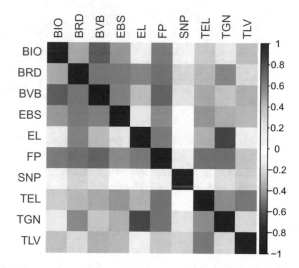

Fig. 1. Pearson correlation coefficient matrix that uses the daily return for every stock symbol.

In each of the two GA variants, the same evolutionary parameters are utilized and the best individual from the final population is reported. There are 30 repeated runs of the GA for each of the 10 stock symbols and there are two fitness functions considered in turn. In conclusion, the entire experimental setup contains 20 different runs of the GA (2 different fitness evaluations and 10 stock symbols) and every one of them is repeated 30 times, which leads to a total of 600 GA runs. The entire code lying at the base of the current research is implemented in R, i.e. ARIMA time series forecasting [5,6], and the GA [16].

4.4   Results and Visualization

Figure 1 illustrates the Pearson correlation between the $\hat{y}_t$ values for each pair of stock symbols. Table 2 shows the RMSE and MAE results for the outcomes forecasted by ARIMA when using the best predictor combinations of the GA.

Table 2. RMSE and MAE obtained from the ARIMA model with the best GA predictor solution after 30 repeated runs. The standard deviation for the $\hat{y}_t$ variable on the test set is also added along with the interval of this dependant variable on the entire data set.

Stock symbol	RSME	MAE	StDev $\hat{y}_t$ test set	$\hat{y}_t$ interval
BVB	9.27E−03	7.37E−03	9.27e−03	[−0.034, 0.056]
BIO	1.06E−02	8.02E−03	1.07e−02	[−0.088, 0.070]
BRD	1.07E−02	8.25E−03	1.079e−02	[−0.072, 0.052]
EBS	1.44E−02	1.05E−02	1.46e−02	[−0.107, 0.112]
EL	6.80E−03	4.67E−03	6.80e−03	[−0.053, 0.048]
FP	5.08E−03	3.44E−03	5.16e−03	[−0.040, 0.043]
SNP	1.22E−02	9.34E−03	1.22e−02	[−0.081, 0.058]
TEL	5.78E−03	4.08E−03	5.79e−03	[−0.073, 0.059]
TGN	8.38E−03	6.20E−03	8.56e−03	[−0.093, 0.045]
TLV	9.23E−03	7.28E−03	9.34e−03	[−0.301, 0.058]

Figure 2 considers each predictor in turn and shows what is its contribution for the daily return prediction of each stock symbol when using the RMSE as the GA fitness function. Lastly, Fig. 3 brings together the solutions as discovered by each of the two fitness options in turn and the common attributes are highlighted.

4.5 Discussion

Figure 1 shows that there are no strong correlations between the daily returns of the considered stock symbols. SNP is the least correlated with the others, BRD appears to be the one that has the highest correlation degree. The general lack of correlation represented an important factor that triggered the search for the specific indicators that affect the results for each company in turn.

Usually small values for RMSE and MAE represent good results. However, in order to reduce the overappreciation of the results in Table 2, a column with the interval of the daily return outputs is included, as the computed error is related to the length of these intervals. Nevertheless, the estimated white noise standard deviation (i.e. RMSE) represents a lower bound of the standard deviation of the forecast error [10] and, in order to consider the prediction acceptable, it should be less than or equal to the standard deviation of the dependant variable in the test data. In this respect the standard deviation is computed for each stock symbol in turn and included in the same Table 2. The obtained RSME obeys the inequality in all situations. In order to further validate the results, a linear model without regressors is fit to the time series and the obtained results were slightly worse than the ones in Table 2 both as regards RSME and MAE.

Figure 2 is generated from the results of the GA with RMSE as fitness. Each of the 14 indicators is taken in turn. The number of times in which the GA chromosome gene that corresponds to a regressor is found within the best candidate

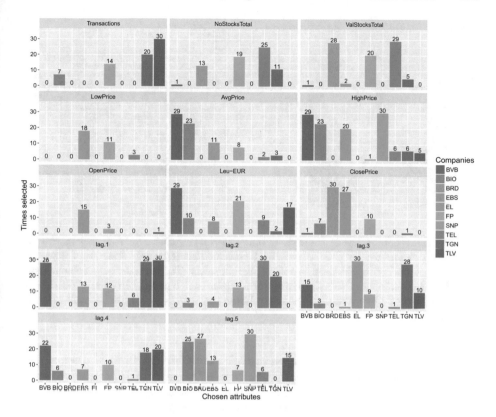

Fig. 2. Attributes involved in the daily return forecasting for each company in turn. The results are obtained after 30 repeated runs of the GA using RMSE for measuring the prediction error.

solution from the final population out of the 30 repeated runs is illustrated in the plots. While some indicators prove to be crucial for the forecasting of the daily return of certain stock symbols, like the number of transactions and lag.1 for TLV, or the high price and lag.5 for SNP etc., certain regressors prove to have a relatively low effect on the outcome forecast for any of the involved companies. Examples for the latter case are represented by the opening and low prices. In order to find the most important indicators with respect to their involvement in the overall prediction of the daily return, the number of times each attribute is chosen for all the stock symbols was summed up. Additionally, for each indicator, the number of companies for which it was involved at least once was also counted. The top 5 indicators were found to be the same in both cases, although not in the same order (meaning that those that had high values for certain stock symbols were also chosen as predictors for many companies): lag.5, high price, lag.3, lag.1 and the leu-EUR exchange rate. From the viewpoint of a stock market expert, the value of the stocks highly depends on the exchange rate. This happens because the most active entities on the Romanian stock market are

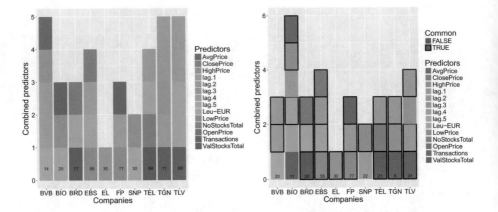

Fig. 3. The combination of indicators that are the most often associated in the best solutions determined in the 30 repeated runs of the GA with RMSE as fitness on the left plot and with MAE on the right, respectively. The number of times each such association occurs out of the 30 runs is outlined for each company in turn. Additionally, on the right plot, the attributes that are chosen for both types of fitness evaluation are highlighted as *common*.

the external funds with speculative capital. As the daily values of the exchange rates are relatively small, the market entrance of the external speculative capital conducts to the decrease of the value of the EUR with respect to the Romanian leu, because the latter are bought and invested into the market and the stock values increase. On the contrary, when the stock injected capital is redrawn, the sales affect the decrease of the stock values and the EUR appreciation over leu goes up because the sums obtained in local currency are converted into EUR. In short, the stock value increases with the appreciation of the leu and decreases when the leu depreciates.

Figure 3 puts side by side the results obtained by the GA with both types of fitness, RMSE on the left and MAE on the right. The plots illustrate for each stock symbol in turn the combination of predictors that was most often met. Both plots contain information about the number of times each such combination of attributes was achieved by the GA out of the maximum possible 30. The number of times varies from one stock symbol to another from a minimum number of 5 repeated configurations up to 30. It is important to notice that the number of selected predictors for every stock symbol in turn is not high, of only up to 5–6, while the maximum possibility would have been to take all attributes into consideration, that is 14.

For some companies, there are attributes that prove to be determinant, as they are very often chosen by the GA for the daily return forecasting. For instance, for EL and SNP there are 30 times (out of the total 30) when a certain combination of (rather limited) indicators were repeatedly found by the GA, at least in the RSME case. For EL the same single lag.3 indicator is found in all cases as being the only decisive regressor, while for SNP the high price proved to be the common and most important attribute in both GA versions.

A small number of repeated configurations in Fig. 3 signifies that there is a larger variety of attribute combinations that reach a good solution. A different successful configuration than that plotted in Fig. 3 for the same stock symbol can also contain a subset of the illustrated attributes. In order to clarify the information, the left plot from Fig. 3 should be analyzed in conjunction with Fig. 2, as the information in the former is included in the latter.

The right plot from Fig. 3 outlines the attributes that are validated as more important (with highlighted border) for each company daily return. The validation comes from the second considered type of fitness evaluation, allowing thus the GA to explore a slightly different fitness landscape. As noticed, although not all sets of attributes are similarly found, a relatively high number of them are common.

5   Conclusions and Future Work

A data set containing information about the number of transactions, total number of stocks, low, high, average, opening and closing prices of ten stock symbols is collected for 11 months. The Romanian Leu-EUR exchange rate is also added to the collected data as another exogenous attribute. In addition to these, the daily return is computed for each day and every company in turn and up to 5 lags are considered. Next, a GA is employed to find the successful combination of predictors that can be integrated in an ARIMA model that would forecast the daily return. The best candidate solutions do not only provide an acceptable forecasting, but especially reveal important information on the factors that matter most in forecasting the daily return of each company.

The future research envisages the gathering of further data and also the investigation of other time series (possibly non-linear) models. Such models proved efficient before for U.S. stocks [12] and are expected to reveal valuable insights for the currently studied time series.

References

1. Arratia, A.: Computational finance. In: An Introductory Course with R, Atlantis Studies in Computational Finance and Financial Engineering, vol. 1 (2014)
2. Bartz-Beielstein, T., Preuss, M.: The Future of Experimental Research, pp. 17–49. Springer, Heidelberg (2010)
3. Cortez, P., Rocha, M., Neves, J.: Genetic and Evolutionary Algorithms for Time Series Forecasting, pp. 393–402. Springer, Heidelberg (2001)
4. Eiben, A., Smith, J.: Introduction to Evolutionary Computing. Springer, Heidelberg (2003)
5. Hyndman, R., Athanasopoulos, G.: Forecasting: Principles and Practice. OTexts, Melbourne (2013). https://www.otexts.org/fpp. Accessed Jan 2017
6. Hyndman, R.J., Khandakar, Y.: Automatic time series forecasting: the forecast package for R. J. Stat. Softw. **26**(3), 1–22 (2008)

7. Hyndman, R.J., Koehler, A.B.: Another look at measures of forecast accuracy. Int. J. Forecast. **22**(4), 679–688 (2006)
8. Khadka, M.S., George, K.M., Park, N., Kim, J.B.: Application of Intervention Analysis on Stock Market Forecasting, pp. 891–899. Springer, Heidelberg (2012)
9. Lukoseviciute, K., Ragulskis, M.: Evolutionary algorithms for the selection of time lags for time series forecasting by fuzzy inference systems. Neurocomputing **73**(10–12), 2077–2088 (2010)
10. Nau, R.: Statistical forecasting: notes on regression and time series analysis. https://people.duke.edu/~rnau/411home.htm, Accessed Jan 2017
11. Popovici, E., De Jong, K.: Understanding cooperative co-evolutionary dynamics via simple fitness landscapes. In: Proceedings of the 7th Annual Conference on Genetic and Evolutionary Computation. pp. 507–514. ACM, NY (2005)
12. Qi, M.: Nonlinear predictability of stock returns using financial and economic variables. J. Bus. Econ. Stat. **17**(4), 419–429 (1999)
13. Stoean, C.: In search of the optimal set of indicators when classifying histopathological images. In: 2016 18th International Symposium on Symbolic and Numeric Algorithms for Scientific Computing (SYNASC), pp. 449–455, September 2016
14. Stoean, C., Stoean, R., Lupsor, M., Stefanescu, H., Badea, R.: Feature selection for a cooperative coevolutionary classifier in liver fibrosis diagnosis. Comput. Biol. Med. **41**(4), 238–246 (2011)
15. Wang, B., Tai, N.L., Zhai, H.Q., Ye, J., Zhu, J.D., Qi, L.B.: A new ARMAX model based on evolutionary algorithm and particle swarm optimization for short-term load forecasting. Electr. Power Syst. Res. **78**(10), 1679–1685 (2008)
16. Willighagen, E., Ballings, M.: Genalg: R Based Genetic Algorithm (2015)
17. Wurdinger, K.: Investigating an Evolutionary Strategy to Forecast Time Series. Universiteit Leiden Opleiding Informatica (2009)
18. Xue, B., Zhang, M., Browne, W.N., Yao, X.: A survey on evolutionary computation approaches to feature selection. IEEE Trans. Evol. Comput. **20**(4), 606–626 (2016)

Fuzzy ARTMAP with Binary Relevance for Multi-label Classification

Lik Xun Yuan[1], Shing Chiang Tan[1(✉)], Pey Yun Goh[1],
Chee Peng Lim[2], and Junzo Watada[3]

[1] Faculty of Information Science and Technology,
Multimedia University, Cyberjaya, Malaysia
lxyuan0420@gmail.com, {sctan,pygoh}@mmu.edu.my
[2] Institute for Intelligent Systems Research and Innovation,
Deakin University, Burwood, Australia
chee.lim@deakin.edu.au
[3] Department of Computer and Information Sciences,
Universiti Teknologi PETRONAS, Seri Iskandar, Malaysia
junzo.watada@utp.edu.my

Abstract. In this paper, we propose a modified supervised adaptive resonance theory neural network, namely Fuzzy ARTMAP (FAM), to undertake multi-label data classification tasks. FAM is integrated with the binary relevance (BR) technique to form BR-FAM. The effectiveness of BR-FAM is evaluated using two benchmark multi-label data classification problems. Its results are compared with those other methods in the literature. The performance of BR-FAM is encouraging, which indicate the potential of FAM-based models for handling multi-label data classification tasks.

Keywords: Fuzzy ARTMAP · Binary relevance · Multi-label classification

1 Introduction

Multi-label data classification is different from the traditional single-label classification problems. In the later, each data sample is assigned to a class from a set of predefined class labels, while in the former, each data sample could be labeled with more than one class [1]. The usefulness of multi-label data classification has been demonstrated in several research areas. As an example, in semantic scene classification [2], a home picture can be annotated with at least one conceptual class such as *sofa*, *chair* and *tv monitor* simultaneously. Similarly, in semantic video categorization, a violent video [3] can be annotated as *rope* and *bind* simultaneously. Other applications include social video [4] and music [5] classification into emotions, as well as protein function prediction [6].

Recently, multi-label data classification has attracted close attention by the machine-learning community. Conventional machine learning models can be used for classifying data samples with a single label. To perform multi-label data classification, these machine-learning models need to be modified, e.g. customized k-nearest neighbour (kNN) [7] and support vector machine (SVM) [8, 9] models.

© Springer International Publishing AG 2018
I. Czarnowski et al. (eds.), *Intelligent Decision Technologies 2017*,
Smart Innovation, Systems and Technologies 73, DOI 10.1007/978-3-319-59424-8_12

In this paper, a supervised artificial neural network based on the adaptive resonance theory (ART) is proposed to classify data samples into multiple classes. Specifically, a fuzzy adaptive resonance theory with mapping (Fuzzy ARTMAP or simply FAM [10]) is integrated with a binary relevance [11, 12] technique to form BR-FAM. The organization of this paper is as follows. In Sect. 2, the state-of-art of multi-label data classification methods is described. In Sect. 3, the methods for designing BR-FAM are explained. In Sect. 4, BR-FAM is evaluated using two benchmark multi-label data sets, with the results compared and analyzed. A summary of this research work is presented in Sect. 5.

2 Literature Review

In general, methods for learning multi-label data samples can be divided into two groups, namely *problem transformation* and *algorithm adaptation* [12]. The methods of *problem transformation* are applied to convert multi-label data samples into at least a set of one single-label data samples, either with or without considering label ranking subject to relevancy of a query of interest. On the other hand, the methods of *algorithm adaptation* are extension from single-label classifiers, and they classify multi-label data samples directly.

Four popular methods of *problem transformation* are binary relevance (BR) [11, 12], label power-set (LP) [12], ranking by pairwise comparison (RPC) [13], and calibrated label ranking (CLR) [14]. BR is a data transformation technique to decompose a multi-label data set into several single-label binary data sets. The idea of BR is described in detail in Sect. 3.1. BR and its variant [15] have been integrated with some base classifiers, which include decision tree [15, 16], Naive Bayes [15], k-nearest neighbor [15] and support vector machine [15]. LP converts each unique set of labels of a data sample into a new single label. When a new data instance is provided, the classifier assigns it to a class label that actually indicates a set of labels. The number of transformed labels in LP depends on the total number of class labels in a data set and also the combination of these class labels assigned to the data samples. RPC converts a multi-label data set into binary data sets. Each data set is based on a pair of labels, and consists of data samples of either class label but not both. Each binary data set, RPC is assigned to a classifier for training. Given a new instance, all classifiers in RPC make predictions. The final output is determined by ranking the votes of each class label. CLR is an extended version of RPC. It introduces an artificial label for multi-label ranking. The artificial label is a breaking point between relevant and irrelevant labels.

On the other hand, the learning algorithms of several single-label classification methods have been modified to perform multi-label classification. They are, for instance, multi-label variants of k-nearest neighbor [7], decision tree [18], support vector machine [19] and neural network [20, 21] models.

3   Methods

BR-FAM is an extended version of the original FAM model. It is proposed to deal with multi-label data classification tasks. The details of BR-FAM are as follows.

3.1   Binary Relevance (BR)

BR [11, 12] is one of the popular problem transformation techniques [12] dealing with a multi-label data set. The core idea of BR is to divide a multi-label data set into two groups: either relevant or irrelevant to a class label of interest. BR is algorithm independent. It transforms a multi-label data set into at least one single label data set for a classifier to perform supervised learning.

Assume $L = \{\lambda_j : j = 1, \cdots, c\}$ is a set of labels in a multi-label data set; $D = \{(x_i, Y_i), i = 1, \cdots, m\}$ is a set of original multi-label data samples, where x_i denotes a feature vector, $Y_i \subseteq L$ represents the corresponding multi labels of the i-th sample. BR processes the original data set D into c data sets with two classes D_{λ_j}, $j = 1, \cdots, c$ where all data samples from D having λ_j are labeled positively, otherwise labeled negatively.

3.2   Fuzzy ARTMAP (FAM)

FAM [10] consists of two fuzzy ART modules that are connected through a map field, F^{ab}. One of these two fuzzy ART modules is the input module that processes the input vectors, whereas another is the output module that processes the output labels. Each fuzzy ART model contains nodes interconnected in three layers: (i) a normalization layer, F_0, that normalizes an M-dimensional input vector a or an N-dimensional output label b through a complement-coding process [10] to a 2 M- dimensional input vector A or 2 N-dimensional output vector B (i.e., $A = (a, 1 - a)$ or $B = (b, 1 - b)$); (ii) an input layer that receives A (or B); (iii) a recognition layer that contains a group of prototype nodes whereby each prototype node represents a cluster of information elicited from training samples. The map field is an associative memory that links the prototype nodes from the F_2 layer of the input and output fuzzy modules during training. FAM undergoes an incremental learning process wherein new prototype nodes can be added to F_2 to store new information.

Both the input and output modules perform the same information processing operation. After the input vector a is complement-coded to A, it is forwarded to F_2^a, where a choice function [10] is utilized to compute the activation of each prototype node with respect to A, as follows:

$$T_j = \frac{\left| A \wedge w_j^a \right|}{\alpha + \left| w_j^a \right|} \tag{1}$$

where α is the choice parameter, which is set to a small positive value close to 0 [10]; w_j^a denotes the connection weight of the j-th prototype node; $\wedge$ represents the fuzzy AND operator that performs element-wise minimum of two vectors. The prototype node with the highest activation, namely node J, is identified as the winning node. A vigilance test is applied to compute the similarity between w_j^a and A against a vigilance parameter [10] $\rho_a \in [0, 1]$.

$$\frac{\left|A \wedge w_J^a\right|}{|A|} \geq \rho_a \tag{2}$$

If the vigilance test is not passed, a new cycle of search for the next winning prototype node is undergone. This search process for a new winning prototype node is only terminated once the winning node succeeds to pass in the vigilance test. Nevertheless, when none of the existing prototype nodes can satisfy the vigilance test, a new prototype node is introduced in F_2^a to encode A.

After each fuzzy ART module has identified a winning node, a map-field vigilance test [10] is executed to evaluate prediction accuracy, as follows:

$$\frac{\left|y^b \wedge w_J^{ab}\right|}{|y^b|} \geq \rho_{ab} \tag{3}$$

where y^b denotes the output vector; w_J^{ab} represents the connection weight of the winning node from F_2^a to F^{ab}; and $\rho_{ab} \in [0, 1]$ represents the map-field vigilance parameter.

If the map-field vigilance test fails, it indicates an incorrect prediction of the output class. Consequently, a match-tracking process [10] is triggered, where ρ_a is raised slightly higher from its baseline setting of $\bar{\rho}_a$ as follows:

$$\rho_a = \frac{\left|A \wedge w_J^a\right|}{|A|} + \delta \tag{4}$$

where δ is set as a positive value close to 0. The adjustment of ρ_a causes the vigilance test in the input fuzzy module to fail. As such, a new search cycle in the input fuzzy module is initiated again with the updated ρ_a setting. The effort for searching a winning node is continuously made until a correct prediction of the output class is made.

When the map-field vigilance test is satisfied, a learning process ensues where w_J^a is updated [10] as follows:

$$w_J^{a(new)} = \beta_a \left(A \wedge w_J^{a(old)}\right) + (1 - \beta_a)w_J^{a(old)} \tag{5}$$

where $\beta_a \in [0, 1]$ denotes the learning parameter of the input fuzzy module. The output fuzzy module undergoes the same operation for pattern matching and learning as in the input fuzzy module from Eqs. (1)–(5) by replacing a with b.

3.3 Fuzzy ARTMAP with Binary Relevance (BR-FAM)

BR-FAM is a modified version of FAM for tackling multi-label data classification tasks. In BR-FAM, an L-label data set D ($L = \{\lambda_j : j = 1, \cdots, c\}$) is converted to c datasets. Each D_{λ_j} contains data samples with binary classes subject to a class λ_j of interest. In this case, a total of c FAM models are created. Each FAM is trained with D_{λ_j}. The outputs are the union prediction of λ_j made by all FAMs.

Two performance metrics are used to measure classification performance of BR-FAM. They are from the harmonic mean of precision and recall, namely the F measure [22]:

$$F1 = \frac{2 * tp}{2 * tp + fp + fn} \qquad (6)$$

where tp denotes the number of true positive correctly classified; fp denotes the number of false positive; fn denotes the number of false negative. These two performance metrics are micro-averaged and macro-averaged versions of $F1$, i.e., micro $F1$ (B_{micro}) and macro $F1$ (B_{macro}) [23, 24]. For clarity, consider a binary classification task of D_k,

$$B(tp_k, fp_k, tn_k, fn_k) \text{ for } k = 1, \cdots, c \qquad (7)$$

where fp_k, fp_k, tn_k, fn_k are respectively the number of true positive, false positive, true negative and false negative after classifying samples from D_k, then

$$B_{micro} = B\left(\sum_{k=1}^{c} tp_k, \sum_{k=1}^{c} fp_k, \sum_{k=1}^{c} tn_k, \sum_{k=1}^{c} fn_k \right) \qquad (8)$$

$$B_{macro} = \frac{1}{c} \sum_{k=1}^{c} B(tp_k, fp_k, tn_k, fn_k) \qquad (9)$$

4 Evaluation

4.1 Benchmark Data

Two multi-label data sets that are available from Mulan [25] are used in the experiment to evaluate the classification performance of BR-FAM. The *scene* data set comprises numerical records of 2407 images that are labeled up to 6 concepts, for example, *beach*, *field*, and *mountain*. The *yeast* data set contains numerical records of 2417 micro-array expressions and phylogenetic profiles that are labeled with at least one of 14 functional categories such as *metabolism*, *energy*. Table 1 lists the statistics of both data sets in terms of number of instances, input features, and labels.

Table 1. Information of two multi-label data sets

Dataset	Number of instances (#Training: #Test)	Number of input features	Number of labels
Scene	2407 (1211:1196)	294	6
Yeast	2417 (1500:917)	103	14

4.2 Experimental Setup

We refer to the experimental setup as in [23] to execute BR-FAM for ten times with different sequences of training data samples. Upon completion of a training session with all training samples, the classification performance of BR-FAM is evaluated with all test data samples. Each FAM is trained using $\bar{\rho}_a = 0.5$ and $\beta_a = \beta_b = 1$ within ten epochs. The numbers of training and test samples from the two data sets are listed in Table 1, which follows the original quantity of data samples in the training and test sets in [25]. The classification results of BR-FAM are averaged.

4.3 Results and Analysis

The classification performance of BR-FAM is compared with C4.5 integrated with: (i) different problem transformation methods [23], which include BR, LP, Calibrated Label Ranking (CLR) [14] and two efficient versions of LP (i.e., Random k-Labelsets of a disjoint version, namely RAkEL$_d$, and Random k-Labelsets of an overlapping version, namely RAkEL$_o$); (ii) two modified methods for multi-label data classification, which include a multi-label version of the backpropagation algorithm for perceptrons (BPMLL) [20] and a multi-label version of k-nearest neighbor algorithm (MLkNN) [7]. Notably, except for BR-FAM, all the aforementioned classification methods used in this benchmark study had been trained with 66% of the samples from the entire data set and the rest as the test samples [23]. For clarity, BR-FAM has been trained using fewer number of data samples, i.e., approximately 50% of *scene* and 62% of *yeast* data sets. The rationale is to compare rigorously the classification performance between BR-FAM and those of existing multi-label classification methods.

Tables 2 and 3 present the classification results in terms of micro $F1$ (based on B_{micro}) and macro $F1$ (based on B_{macro}) among BR-FAM, the four versions of multi-label C4.5 (with BR, LP, RAkEL$_d$, and RAkEL$_o$), CLR, MLkNN, and BPMLL. From these results, BR-FAM achieves the highest rates of micro $F1$ and macro $F1$ when classifying the *scene* data set. The classification performances of BR-FAM are moderate in *yeast* where its micro $F1$ is ranked at the sixth position and its macro $F1$ is the second highest among the eight classifiers. Based on these results, BR-FAM appears to be a moderate model for multi-label data classification. However, a further analysis of the results of BR-FAM and a group of five multi-label classifiers developed using different problem transformation methods (i.e., CLR and the four C4.5 versions with BR, LP, RAkEL$_d$ and RAkEL$_o$) in the *yeast* classification task is made. BR-FAM could achieve micro $F1$ (i.e., 55.15%) that is within the performance range of these five classifiers (53.04%–61.89%). On the other hand, BR-FAM is inferior to MLkNN and BPMLL. These two multi-label classifiers have been developed by an algorithm

Table 2. The results of micro $F1$ (standard deviation is typed in round brackets)

Classifier	Classification task (%)	
	Scene	Yeast
BR	62.36 (1.01)	57.67 (1.89)
LP	60.05 (1.14)	53.04 (1.03)
MLkNN	72.29 (1.08)	**63.93 (1.06)**
RAkEL$_d$	59.87 (0.82)	54.26 (0.58)
RAkEL$_o$	69.58 (1.53)	61.89 (0.74)
CLR	62.82 (0.92)	61.69 (1.29)
BPMLL	48.18 (5.19)	63.11 (1.47)
BR-FAM	**77.43** (3.24)	55.15 (0.69)

Table 3. The results of macro $F1$ (standard deviation is typed in round brackets)

Classifier	Classification task	
	Scene	Yeast
BR	63.41 (0.91)	38.29 (0.59)
LP	61.04 (1.16)	37.26 (1.09)
MLkNN	72.63 (1.37)	36.34 (0.79)
RAkEL$_d$	60.90 (0.88)	38.84 (0.50)
RAkEL$_o$	70.26 (1.64)	40.66 (0.77)
CLR	64.23 (0.89)	38.52 (0.96)
BPMLL	51.29 (5.26)	**42.85 (1.02)**
BR-FAM	**78.58 (4.35)**	41.46 (0.76)

adaptation approach achieving micro $F1$ within between 63% and 64%. In other words, the performance of BR-FAM in *yeast* is competitive with those classifiers developed using the same approach, i.e., the problem transformation methods.

5 Summary

In this paper, the FAM model is integrate with a binary relevant technique to handle multi-label data classification problems. The effectiveness of BR-FAM is evaluated using two benchmark data sets. The empirical results show that BR-FAM is comparable with other multi-label classifiers, especially those developed with problem transformation approach.

As part of future work, additional experiment will be carried out to evaluate the classification capability of BR-FAM using additional multi-label data sets available in different application areas. We will also develop a multi-label FAM model using the algorithm adaptation approach.

References

1. Boutell, M.R., Luo, J., Shen, X., Brown, C.M.: Learning multi-label scene classification. Pattern Recogn. **37**, 1757–1771 (2004)
2. Tamaazousti, Y., Le Borgne, H., Popescu, A.: Constrained local enhancement of semantic features by content-based sparsity. In: Proceedings of the 2016 ACM on International Conference on Multimedia Retrieval, ICME 2016, pp. 119–126. ACM, New York (2016)
3. Li, X., Huo, Y., Jin, Q., Xu, J.: Detecting violence in video using subclasses. In: Proceedings of the 2016 ACM on Multimedia Conference, MM 2016, pp. 586–590. ACM, Amsterdam (2016)
4. Chávez-Martínez, G., Ruiz-Correa, S., Gatica-Perez, D.: Happy and agreeable?: multi-label classification of impressions in social video. In: Proceedings of the 14th International Conference on Mobile and Ubiquitous Multimedia, MUM 2015, pp. 109–120. ACM, Austria (2015)
5. Lin, Y.-C., Yang, Y.-H., Chen, Homer H.: Exploiting online music tags for music emotion classification. ACM Trans. Multimedia Comput. Commun. Appl. **7 s**, Article 26 (2011)
6. Yu, G., Rangwala, H., Domeniconi, C., Zhang, G., Yu, Z.: Protein Function Prediction with Incomplete Annotations. IEEE/ACM Trans. Comput. Biol. Bioinf. **11**, 579–591 (2014)
7. Zhang, M.-L., Zhou, Z.-H.: ML-kNN: a lazy learning approach to multi-label learning. Pattern Recogn. **40**, 2038–2048 (2007)
8. Xu, J.: An extended one-versus-rest support vector machine for multi-label classification. Neurocomputing **74**, 3114–3124 (2011)
9. Xu, J.: Multi-label core vector machine with a zero label. Pattern Recogn. **47**, 2542–2557 (2014)
10. Carpenter, G.A., Grossberg, S., Markuzon, N., Reynolds, J.H., Rosen, D.B.: Fuzzy artmap: a neural network architecture for incremental supervised learning of analog multidimensional maps. IEEE Trans. Neural Netw. **3**, 698–713 (1992)
11. Godbole, S., Sarawagi, S.: Discriminative methods for multi-labeled classification. Lecture Notes in Artificial Intelligence **3056**, 22–30 (2004)
12. Tsoumakas, G., Katakis, I.: Multi-label classification: an overview. Int. J. Data Warehous. Min. **3**, 1–13 (2007)
13. Hüllermeier, E., Fürnkranz, J., Cheng, W., Bringer, K.: Label ranking by learning pairwise preferences. Artif. Intell. **172**, 1897–1916 (2008)
14. Fürnkranz, J., Hüllermeier, E., Mencia, L., Brinker, K.: Multi-label classification via calibrated label ranking. Mach. Learn. **73**, 133–153 (2008)
15. Cherman, E.A., Monard, M.C., Metz, J.: Multi-label problem transformation methods: a case study. CLEI Electron. J. **14**, 4 (2011)
16. Tanaka, E.A., Nozawa, S.R., Macedo, A.A., Baranauskas, J.A.: A multi-label approach using binary relevance and decision tree applied to functional genomics. J. Biomed. Inf. **53**, 85–95 (2015)
17. Chou, S., Hsu, C.-L.: MMDT: a multi-valued and multi-labeled decision tree classifier for data mining. Expert Syst. Appl. **28**, 799–812 (2005)
18. Wu, Q., Ye, Y., Zhang, H. Chow, Tommy W.S., Ho, S.-S.: ML-TREE: a tree-structure-based approach to multilabel learning. IEEE Trans. Neural Netw. Learn. Syst. **26**, 430–443 (2015)
19. Chen, W.-J., Shao, Y.-H., Li, C.-N., Deng, N.-Y.: MLTSVM: a novel twin support vector machine to multi-label. Learning **52**, 61–74 (2016)
20. Zhang, M.-L., Zhou, Z.-H.: Multi-label neural networks with applications to functional genomics and text categorization. IEEE Trans. Knowl. Data Eng. **18**, 1338–1351 (2006)

21. Chen, Z., Chi, Z., Fu, H., Feng, D.: Multi-instance multi-label image classification: a neural approach. Neurocomputing **99**, 298–306 (2013)
22. van Rijsbergen, C.J.: Information Retrieval. Butterworths, London (1979)
23. Tsoumakas, G., Katakis, I., Vlahavas, I.: Random k-labelsets for multilabel classification. IEEE Trans. Knowl. Data Eng. **23**, 1079–1089 (2011)
24. Yang, Y.: An evaluation of statistical approaches to text categorization. J. Inf. Retrieval **1**, 78–88 (1999)
25. Mulan: a Java library for multi-label learning. http://mulan.sourceforge.net/datasets-mlc.html

Rough Set-Based Text Mining from a Large Data Repository of Experts' Diagnoses for Power Systems

Junzo Watada[1(✉)], Shing Chiang Tan[2], Yoshiyuki Matsumoto[3], and Pandian Vasant[4]

[1] Department of Computer and Information Sciences, Universiti Teknologi PETRONAS,
Seri Iskandar, Perak, Malaysia
junzo.watada@gmail.com
[2] Faculty of Information Science and Technology, Multimedia University, Melaka, Malaysia
sctan@mmu.edu.my
[3] Faculty of Economics, Shimonoseki City University, Gakuencho, Shimonoseki, Japan
matsumoto@shimonoseki-cu.ac.jp
[4] Department of Fundamental Mathematics and Sciences, Universiti Teknologi PETRONAS,
Seri Iskandar, Perak, Malaysia
pvasant@gmail.com

Abstract. Usually it is hard to classify the situation where uncertainty of randomness and fuzziness exists simultaneously. This paper presents a rough set approach applying fuzzy random variable and statistical t-test to text-mine a large data repository of experts' diagnoses provided by a Japanese power company. The algorithms of rough set and statistical t-test are used to distinguish whether a subset can be classified in the object set or not. The expected-value-approach is also applied to calculate the fuzzy value with probability into a scalar value.

Keywords: Fuzzy statistical test · Rough set · Expected-value-approach · Randomness and fuzziness

1 Introduction

We often have problems in classifying data under hybrid uncertainty of both randomness and fuzziness. For example, linguistic data always have these features. However, as the meaning of each linguistic datum can be interpreted by a fuzzy set and the variability of the individual meaning may be understood as a random event, the fuzzy random variable is a concept which can be applied to such a situation. In this research linguistic data are obtained randomly as a fuzzy random variable and after the fuzzy random variables are defined, the expected-value-approach will be applied to calculate them into some scalar values. Finally the subset, using its expectation values of random samples, will be distinguished whether to be classified into the object set or not by applying the method of rough set [1] and statistical t-test.

Japan consists of the nine power companies which provide electricity to Japan industries and living people. The companies aggregate their past diagnoses as a text repository. This research study was passed to obtain the latent structure of experts' diagnoses in the past.

© Springer International Publishing AG 2018
I. Czarnowski et al. (eds.), *Intelligent Decision Technologies 2017*,
Smart Innovation, Systems and Technologies 73, DOI 10.1007/978-3-319-59424-8_13

McGil et al. [2] overviewed fuzzy random variables, recently. The concept of fuzzy random variables was developed by Kwakernaak [3, 4]. Puri and Ralescu [5, 6] established the mathematical basis of fuzzy random variables. Other authors also discussed fuzzy random variables, related works can be found in [7] and so on. In the fields of applying statistical test with concepts of fuzzy theory, Pei-Chun Lin et al. [8, 9] also defined a new function, weight function, which can be used to deal with continuous fuzzy data. Comparing with these researches, the novelty of this research is, to propose a rough set approach to the classification field based on both fuzzy random variable and statistical test which is few researched in the similar field.

Other researches also performed a confidence-internal-based fuzzy random regression models to characterize the real data (Junzo Watada et al.) [10–12]. In their research, fuzzy random variables are introduced to address regression problems in presence of such hybrid uncertain data, serving as an integral component of regression models.

The remainder of this paper is organized as follows. We give an overview of fuzzy random variables in Sect. 2. The expected-value-approach, which can calculate the fuzzy random variables into scalar values, will also be explained in Sect. 2. In Sect. 3, the statistical t-test model which is built according to the rough sets theory will be explained including its principles and features. In Sect. 4, a data analysis process applying above methods will be explained. Finally, we will summarize this paper in conclusions.

2 Fuzzy Random Variables

2.1 Fuzzy Variable

In this section we recall some basic concepts on fuzzy variable and fuzzy random variable which make it easier to follow further discussions on the models. Assume that $(\Gamma, P(\Gamma), Pos)$ is a possibility space, where $P(\Gamma)$ is the power set of Γ, X is fuzzy variable defined on $(\Gamma, P(\Gamma), Pos)$, with membership function μ_x, and r is a real number. As a well-known fuzzy measure, possibility measure of a fuzzy event $X \leq r$ is defined as:

$$Pos\{X \leq r\} = \sup_{t \leq r} \mu_x(t) \tag{1}$$

Lacking the self-duality, the possibility measure is not always the optimal approach to characterizing the fuzziness or vagueness in decision making problems. As a simple example, we consider an event $X > 3$ induced by a triangular fuzzy variable $X = (1, 2, 10)$, through possibility, we can calculate the credibility level of $X > 3$ is 0.875. This fact makes decision-makers confused. To overcome the above drawback, a self-dual set function, named credibility measure, is formed as follows:

$$Cr\{X \leq r\} = (1 + \sup_{t \leq r} \mu_x(t) - \sup_{t > r} \mu_x(t)) \tag{2}$$

In the above example, we can calculate by credibility, condense of $X > 3$ is 0.4345, and the credibility level of $X <= 3$ based on credibility is $1 - 0.4345 = 0.5655$ [11].

2.2 Fuzzy Random Variables and the Expected-Value Approach

Given some universe Γ, let Pos is a possibility measure defined on the power set $P(\Gamma)$ of Γ. Let $\mathfrak{R}$ be the set of real numbers. A function $Y:\Gamma \to \mathfrak{R}$ is said to be a fuzzy variable defined on Γ [12]. The possibility distribution μ_Y of Y is defined by $\mu_Y(t) = Pos\{Y = t\}$. For fuzzy variable Y with possibility distribution μ_Y, the possibility, necessity and credibility of event $\{Y \leq r\}$ are given, as follows:

$$Pos\{Y \leq r\} = \sup_{t \leq r} \mu_Y(t),$$

$$Nec\left\{Y \leq r\right\} = 1 - \sup_{t > r} \mu_Y(t), \tag{3}$$

$$Cr\left\{Y \leq r\right\} = \frac{1}{2}(1 + \sup_{t \leq r} \mu_Y(t) - \sup_{t > r} \mu_Y(t)).$$

It should be noted that the credibility measure is an average of the possibility and the necessity measure, i.e. $Cr\{\cdot\} = (Pos\{\cdot\} + Nec\{\cdot\})/2$, and it is a self-dual set function for any set in $P(\Gamma)$. The motivation behind the introduction of the credibility measure is to develop a certain measure which is a sound aggregate of the two extreme cases such as the possibility (expressing a level of overlap and being highly optimistic in this sense) and necessity. Based on credibility measure, the expected value of a fuzzy variable is presented as follows.

Definition 1 [13]: Let Y be a fuzzy variable. The expected value of Y is defined as:

$$E[Y] = \int_0^\infty Cr\{Y \geq r\}dr - \int_{-\infty}^0 Cr\{Y \leq r\}dr \tag{4}$$

provided that the two integrals are finite.

Example 1: Assume that $Y = (c, a^l, a^r)_T$ is a triangular fuzzy variable whose possibility distribution is

$$\mu_Y(x) = \begin{cases} \dfrac{x - a^l}{c - a^l}, a^l \leq x \leq c \\ \dfrac{a^r - x}{a^r - c}.c \leq x \leq a^r \\ 0, \qquad otherwise. \end{cases}$$

Making use of (2), we determine the expected value of Y to be

$$E[Y] = \frac{a^l + 2c + a^r}{4}. \tag{5}$$

Next the definitions of fuzzy random variable and its expected value and variance operators will be explained. For more theoretical results on fuzzy random variables, one may refer to Gil et al. [1], Liu and Liu [14], and Wang and Watada [#10&8], [15].

Definition 2 [13]: Suppose that $(\Omega, \Sigma, \Pr)$ is a probability space, F_v is a collection of fuzzy variables defined a possibility space $(\Gamma, P(T), Pos)$. A fuzzy random variable is a mapping $X: \Omega \rightarrow F_v$ such that for any Borel subset B of $\mathfrak{R}$, $Pos\{X(\omega) \in B\}$ is a measurable function of ω.

Let X be fuzzy random variable on Ω. From the above definition, we can know for each $\omega \in \Omega$, $X(\omega)$ is a fuzzy variable. Furthermore, a fuzzy random variable X is said to be positive if for almost every ω, fuzzy variable $X(\omega)$ is positive almost surely.

Definition 3 [14]: Let X be fuzzy random variable defined on a probability space $(\Omega, \Sigma, \Pr)$. The expected value of X is defined as:

$$E[\xi] = \int_{\Omega} \left[\int_0^{\infty} Cr\{\xi(\omega) \geq r\} dr - \int_{-\infty}^{0} Cr\{\xi(\omega) \leq r\} dr \right] \Pr(d\omega). \qquad (6)$$

3 T-Testing Processing

3.1 Overview of the Student's T-Test

The t-statistic was introduced in 1908 by William Sealy Gosset, a chemist working for the Guinness brewery in Dublin, Ireland. The t-test is the most commonly used method to evaluate the differences in means between two groups. The groups can be independent (e.g., blood pressure of patients who were given a drug vs. a control group who received a placebo) or dependent (e.g., blood pressure of patients "before" vs. "after" they received a drug, see below). Theoretically, the t-test can be used even if the sample sizes are very small (e.g., as small as 10; some researchers claim that even smaller n's are possible), as long as the variables are approximately normally distributed and the variation of scores in the two groups is not reliably different.

3.2 T-Test Procedure

After we obtained the expected value of the fuzzy random variables through the mathematical approach, a new algorithm applying statistical t-test is designed to identify the inclusion efficiency of the upper approximation in rough set theory.

Suppose 30 subsets are included in the collectivity set and each subset has 50 sample values. A discrimination ratio for the collectivity set also exists which means when the classification result of each subset is larger than the discrimination ratio, the subset will be included into the object set. The statistical t-test is applied to verify the result of each subset in order to make out whether the result of samples accords with the result of the whole subset.

In this study, one-side testing is applied and the testing equation is shown as follows:

$$T = \frac{\bar{x} - \mu_0}{\frac{S_n^*}{\sqrt{n}}}$$

where $\bar{x}$ is the average of each subset. μ_0 is the threshold set for each subset. S_n^* is the modified sample variance of each subset, and n is the number of the subset values. The significance level used in this study is 5%.

The sequence of the test is shown as follows:

1. Calculate the modified sample variance and the value T.

2. Comparing the value T and $t_{1-\frac{n}{2}}(n-1)$

3. If the value $T < t_{1-\frac{\alpha}{2}}(n-1)$, then accept the result of sampling.

4. If the value $T > = t_{1-\frac{\alpha}{2}}(n-1)$, then reject the result of sampling.

According to Pei-Chun Lin et al. [8], we can calculate the test based on Kolmogorov-Smirnov two sample test with continuous data. But in this case, we approximately apply the fuzzy mean and fuzzy numbers to the above t-test based on t distribution.

4   Data Analysis Process

4.1   Data Description

The data studied in this paper came from a certain electrical power companies in Japan. Totally 2416 samples were given and each sample has 12 attributes and one expenses value which can be regarded as the decision attribute. The examples of attributes are shown in Figs. 1 and 2.

Attribute 1: The value of the electronic circuit voltage of the power system
Attribute 2: The kind of the transformer of the power system
Attribute 3: The maker of the transformer of the power system.
Attribute 9: The detail of the operations conducted on the power system.

Fig. 1. Contents of attributes

A1=110: The value of the electronic circuit voltage of the power system is 110 KV.
A1=220: The value of the electronic circuit voltage of the power system is 220 KV.
A3=3: The maker of the transformer of the power system is company C.
A9=1: The detail of the operations conducted on the power system is "gas check."

Fig. 2. Examples of attributes

These data have to be referred to as raw data because some Japanese structure appeared to be confusing. In order to get meaningful analyzed result from the raw data, pre-processing analysis must be performed. Also when the data appears to contain fuzziness and randomness, we will format it into a scalar value by applying the functions below explained in Sect. 4. After pre-processing of the raw data, the attributes of each sample (power system) are defined. According to these coordinated data, the analysis result is conducted by using Eqs. (5) and (6).

4.2 Data Analysis Result

The rough set analysis is applied in this research. After improving the above coordinated data into the rough set model, the diversified results of analysis are conducted (Table 1).

Table 1. The result of approximation

Class	Object number	Lower approximation	Upper approximation	Accuracy
1	649	632	658	0.8929
2	1248	1186	1327	0.8059
3	519	475	586	0.6883
Quality of classification	0.8893			

Table 2. The result of core

Core	Quality of classification
For all condition attributes	0.8893
For condition attributes in core	0.8356
Attributes in core	CoreA3, CoreA4, CoreA7, CoreA9, CoreA10

Table 3. The result of reduction

Number	Reduction	Length
1	A1, A3, A4, A6, A7, A9, A12, A13	8
2	A1, A2, A3, A4, A6, A7, A8, A10	8
3	A1, A2, A3, A4, A8, A9, A10, A11, A12, A13	10
4	A1, A2, A3, A4, A8, A10, A11, A12	8
5	A1, A2, A3, A6, A7, A9, A11, A12, A13	9
6	A1, A3, A4, A6, A9, A11, A12, A13	8
7	A1, A2, A3, A4, A6, A7, A9, A10	8
8	A1, A3, A4, A5, A6, A7, A8, A9, A10	9
9	A1, A3, A4, A6, A7, A8, A12	7
10	A1, A3, A4, A6, A7, A9, A12	7
11	A1, A3, A4, A6, A9, A11, A12	7
12	A1, A3, A4, A6, A8, A11, A12	7
13	A1, A2, A3, A6, A7, A8, A11, A12	8

Table 4. Examples of meaningful rules

Rule	Rule content	Rule accuracy
Rule1	$(A7 = 4)\&(A10 = 1)\&(A11 = 1966) \to (Dec = 1)$	62.5%
Rule2	$(A5 = 1965)\&(A9 = 8) \to (Dec = 1)$	73.75%
Rule7	$(A3 = 6000)\&(A7 = 2)\&(A9 = 9)\&(A10 = 1) \to (Dec = 3)$	60.00%
Rule218	$(A5 = 1956)\&(A7 = 2) \to (Dec = 3)$	52.00%

(1) Approximations

From the above result we can get to know over half (1248) of the samples (the expenses of power system) have been classified into class 2, that means over half of the operation expenses are controlled from 0.002 million yen to 0.01 million yen as shown in Table 5. And both the lower and upper approximations of Decision Class 2 are more than the ones of others.

Table 5. Explanations about some meaningful rules

Rule	Rule explanation
Rule1	If the method of protecting the power system from degradation is "the method of diaphragm", the kind of the power system is "present", and the power system is produced in 1966, then the expenses of operation of this power system is more than 0.01 million yen
Rule2	If the power system is produced in 1965, and the operation conducted is"exterior check", then the expenses of operation of this power system is more than 0.01 million yen
Rule7	If the standard capacity of the power system is "6000KVA", the method of protecting the power system from degradation is "the method of aluminum", the detail of the operation conducted is "repair operation" and the kind of the power system is "child", then the expenses of operation of this power system is less than 0.002 million yen
Rule218	If the power system is produced in 1956 and the method of protecting the power system from degradation is "the method of aluminum", then the expenses of operation of this power system is less than 0.002 million yen

According to this approximation result we can conclude that over half of the operation expenses are controlled from 0.002 million yen to 0.01 million yen as mentioned above. Besides, when any other unknown operation is taken, its expense is also most likely to be inferred from 0.002 million yen to 0.01 million yen.

Besides, the above analysis result is also believable because the accuracies of the three approximation results (0.8929, 0.8059, and 0.6883) are all more than 50%.

(2) Core

Next step of the analysis is to generate core of attributes in order to find out the most meaningful feature appeared in the power systems. The result is shown in Table 2.

As we can see from the above result, attributes A3, A4, A7, A9 and A10 belong to the core of attributes. That means, in deciding the operation expenses of a power system, the attribute A3 (The maker of the transformer of the power system), attribute A4 (The standard capacity of the transformer of the power system), attribute A7 (The method to protect the power system from degradation), attribute A9 (The detail of the operation conducted on the power system) and A10 (The kind of the operation conducted on the power system.) are important.

(3) Reduction

The concept of Reductions in Rough set helps us to find out the most discriminable attributes of all features (Table 3).

From the above result we can find there are total 13 reductions at all. By using each group of the reduction we can easily distinguish all the 2416 samples.

(4) Rule generation

One of the most important elements of Rough set data analysis is rule generation. Totally 318 rules including 258 exact rules and 60 approximate rules are extracted by applying the rule generation method of rough set analysis. From the rules we can know the composition of the operation expenses. Some meaningful rules and their explanations are shown below (Table 4).

From the analysis result we can conclude that the Rough set approach with statistical test to dealing with data with randomness and fuzziness is particularly useful when the data are not convenient for traditional, ordinary statistical methods. The result obtained by the Rough set approach is easy to understand by the analyst and allow to building the justification or explanation for derived data analysis conclusions. In this case if an operation conducted on the power system has one or more of twelve attributes described above, the detail expenses of the operation can be soon inferred.

5 Conclusions

This research aims to solve the problem of classification when the object contains vagueness, randomness and fuzziness. At first, we proposed a rough set approach because rough set deals well with the vagueness. Secondly we apply the concepts of fuzzy random variable as well as the method of expected-value-approach to handle the problem of randomness and fuzziness. In data analysis process we demonstrate one situation briefly. Suppose that we want to clarify the difference between the meaning of some word. In this case we have to distinguish the fuzziness included in the word from the fluctuation caused by individuality, and then the approach of fuzzy random variable is adopted. After we obtained the expected value of fuzzy random variables through the above method, the algorithm of statistical t-test is adopted to reach the goal of classification.

In the data process section, we analyzed 2416 data and obtained 318 rules finally. From these rules we can infer the total expenses from each operation conducted on the power systems. Further studies are expected for improving the accuracy of the data analysis process.

Acknowledgement. This work was supported partially by Petroleum Research Fund (PRF) No. 0153AB-A33 through Universiti Teknologi PETRONAS.

References

1. Pawlak, Z.: Rough sets. Int. J. Comput. Inf. Sci. **11**(5), 341–356 (1982)
2. Gil, M.A., Miguel, L.D., Ralescu, D.A.: Overview on the development of fuzzy random variables. Fuzzy Sets Syst. **157**(19), 2546–2557 (2006)
3. Kwakernaak, H.: Fuzzy random variables-1. definitions and theorems. Inf. Sci. **15**(1), 1–29 (1978)
4. Kwakernaak, H.: Fuzzy random variables-2. algorithms and examples for discrete case. Inf. Sci. **17**(3), 253–278 (1979)
5. Puri, M.L., Ralescu, D.A.: The concept of normality for fuzzy random variables. Ann. Probab. **13**(4), 1373–1379 (1985)
6. Puri, M.L., Ralescu, D.A.: Fuzzy random variables. J. Math. Anal. Appl. **114**(2), 409–422 (1986)
7. Nahmias, S.: Fuzzy variables. Fuzzy Sets Syst. **1**(2), 97–111 (1978)
8. Lin, P.-C., Wu, B., Watada, J.: Kolmogorov-Smirnov two sample test with continuous fuzzy data. Working paper ISME20090401001, pp. 1–21 (2009)
9. Watada, J., Lin, L.-C., Qian, M., Lin, P.-C.: A fuzzy random variable approach to restructuring of rough sets through statistical test. In: Proceedings of RSFGr2009, pp. 2–3 (2009)
10. Watada, J., Wang, S., Pedrycz, W.: Building confidence-interval-based fuzzy random regression models. IEEE Trans. Fuzzy Syst. **17**(6), 1273–1283 (2009). doi:10.1109/TFUZZ.2009.2028331
11. Wang, S., Watada, J.: T-independence condition for fuzzy random vector based on continuous triangular norms. J. Uncertain Syst. **2**(2), 155–160 (2008)
12. Wang, S., Watada, J.: Studying distribution functions of fuzzy random variables and its applications to critical value functions. Int. J. Innov. Comput. Inf. Control **5**(2), 279–292 (2009)
13. Wang, S., Watada, J.: Fuzzy Stochastic Optimization: Theory, Models and Applications. Springer, New York (2012). ISBN 978-1-4419-9559-9
14. Liu, B., Liu, Y.K.: Expected value of fuzzy variable and fuzzy expected valued models. IEEE Trans. Fuzzy Syst. **10**(4), 445–450 (2002)
15. Liu, Y.-K., Liu, B.: Fuzzy random variable: a scalar expected value operator. Fuzzy Optim. Decis. Making **2**(2), 143–160 (2003)
16. Liu, Y., Liu, B.: Fuzzy random programming with equilibrium chance constraints. Inf. Sci. **170**(2–4), 363–395 (2005)

Interdisciplinary Approaches
in Business Intelligence Research
and Practice

Applying Domain Knowledge for Data Quality Assessment in Dermatology

Nemanja Igić[1]([✉]), Branko Terzić[1], Milan Matić[2], Vladimir Ivančević[1], and Ivan Luković[1]

[1] Faculty of Technical Sciences, University of Novi Sad, Novi Sad, Serbia
{nemanjaigic,branko.terzic,dragoman,ivan}@uns.ac.rs
[2] Faculty of Medicine, University of Novi Sad, Novi Sad, Serbia
milan.matic@mf.uns.ac.rs

Abstract. The Dermatology Clinic at the Clinical Center of Vojvodina, Novi Sad, Serbia, has actively collected data regarding patients' treatment, health insurance and examinations. These data were stored in documents in the comma-separated values (CSV) format. Since many fields in these documents were presented as free form text or allow null values, there are many data records that are inconsistent with the real-world system. Currently, there is a large need for an analytic system that can analyze these data and find relevant patterns. Since such an analytic system would require clean and accurate data, there is a need to assess data quality. Therefore, a data quality system should be designed and built with a goal of identifying inaccurate records so that they can be aligned with the real-world state. In our approach to data quality assessment, the domain knowledge about data is used to define rules which are then used to evaluate the quality of the data. In this paper, we present the architecture of a data quality system that is used to define and apply these rules. The rules are first defined by a domain expert and then applied to data in order to determine the number of records that do not match the defined rules and identify the exact anomalies in the given records. Also, we present a case study in which we applied this data quality system to the data collected by the Dermatology Clinic.

Keywords: Dermatology · Data quality assessment · Domain knowledge application

1 Introduction

Collection and analysis of multi-year data are important activities in many large systems. Finding relevant patterns, analyzing trends and building good predictive models are the key elements of good decision making. In medicine, analytic systems are very important in identifying key indications in disease diagnosis [1] and the right prevention is essential both financially and in terms of patient health. There are many examples of advanced analysis of medical data and extraction of meaningful information, such as using neural networks in diagnosis [2], using big data analysis to find business value in health care [3], and using data mining in the treatment of heart diseases [4].

© Springer International Publishing AG 2018
I. Czarnowski et al. (eds.), *Intelligent Decision Technologies 2017*,
Smart Innovation, Systems and Technologies 73, DOI 10.1007/978-3-319-59424-8_14

To get maximum benefits from using an analytic system, the analyzed data must be correct and trustful. Therefore, each anomaly in data records needs to be detected and fixed before any analysis is performed. There must be a well-defined process that aims to evaluate data and measure their quality by using the provided metrics. Failing to define such a process may result in poor and inaccurate results, which can lead to making bad and, in the medical domain, potentially dangerous conclusions.

Data quality assessment generally includes evaluation of data quality based on some predefined metrics. By using a generic model of data quality assessment that relies on well-defined objective and subjective metrics based on questionnaires [5], we can perform unbiased data quality assessment. However, in an analysis of data gathered from the Dermatology Clinic at the Clinical Center of Vojvodina (CCV), Novi Sad, Serbia, a more specific model of data quality assessment is required. For this purpose, we focus more on the domain experts' view of data because the domain experts will be interpreting the results of the analytic system that will be built for these data. Thus, this kind of data quality assessment gives emphasis to the domain experts' interpretation of data rather than a general one.

Our main goal is to construct a data quality system that will allow domain experts to easily present their knowledge about the data by offering them a wide range of easily understandable concepts for defining rules. After applying the defined rules to the provided data records, our system will generate the results of data quality assessment in the form of a report. Domain experts can then use the results from the report to conclude which data records are inconsistent with the defined rules. After that, domain experts should be able to investigate the causes of the discovered inconsistencies and suggest the preferable data values.

The system presented in this paper relies on the domain knowledge about the data. Since the Dermatology Clinic is the holder of data, we presume that the experts at the clinic may offer the most relevant interpretation of data. By letting these experts present their view about data as a set of rules, we can easily analyze how much inconsistency there is and which concrete rules trigger negative results for the analyzed data records. In this way, we can determine exact differences between the real world and the data gathered by the Dermatology Clinic. Our hope is that we can get better results by using domain specific metrics that are rule based rather than the generic metrics, which are predefined for any case study, as presented in [5].

Besides Introduction and Conclusion, this paper has four more sections. In Sect. 2, we describe various approaches to data quality assessment, with emphasis on data quality assessment in medicine. In Sect. 3, we describe data gathered from the Dermatology Clinic at CCV. In Sect. 4, we present the architecture of the system that we used to assess data quality. In Sect. 5, we present a case study in which we applied our data quality system to the data gathered by the Dermatology Clinic. In this case study, domain experts defined a set of rules, which were applied to the provided data. Also, we present results of the report generated by our data quality system and the domain experts' conclusions.

2   Related Work

In this section, we first describe various approaches in data quality assessment. Afterwards, we present examples of data quality assessment in medicine.

2.1   Data Quality Assessment Approaches

In [6], there is a comparison between many different data quality methodologies that can be applied in various types of information systems. All those methodologies have a goal to define a set of objective metrics and activities in data quality assessment for the given type of information system. For example, in [7], a system for managing data quality for cooperative information systems is presented. In [8], an example of a modeling data quality process for multi-input and multi-output information system is presented. There are cases where data quality is treated as product quality and product quality procedures are applied [9]. There are also methodologies that are based on questionnaires to gather data on the status of the organizational information quality [10]. In [11, 12], there are procedures on how to manage data quality in big data and large multi-organization systems, which are very popular nowadays. As discussed in [13], there are also approaches where organizational structure is taken into account. All these approaches are using predefined metrics which are easy to apply and interpret, but give more generic results. In our approach, we emphasize domain experts' knowledge in order to get in-depth results of data quality assessment.

Besides defining metrics and processes for data quality assessment, there are approaches in which data mining techniques are used to accomplish the same task. As presented in [14, 15], association rules can be used to find data anomalies based on very high or very low values of confidence. This rule-based approach gives some hidden insights about the data that might not be obvious, but there is also possibility not to emphasize other very important conclusions about the data. On the other hand, our approach gives a wider set of concepts to define different rules, rather than just using association rules. Also, we are more focused on important aspects of data quality assessment from the domain experts' viewpoint, rather than finding hidden knowledge in data.

During our research, within the available literature, we did not find any approaches that are based directly on the domain knowledge from persons who will interpret data after the data are cleaned and transformed.

2.2   Data Quality in Medicine

There are also research studies concerning data quality assessment in the field of medicine. In [16], there is a description of a framework that enables clinical institutions to plan and control the data quality assessment process based on dimensions that were predefined for the clinical domain. In [17], there is a discussion of problems in data quality assessment that are related to integration of various sources in a single warehouse when different naming conventions are used, e.g., for gender and race. The proposed solutions are the standardization of the values and strict database constraints. In [18], a complete process of data quality assessment is proposed with defined metrics that were proven in practice. Even though these approaches are used in medicine, they are more focused on defining various metrics for this field rather than applying domain knowledge in data quality assessment.

3 Data Sources

In this section, we describe the data collected at the Dermatology Clinic at CCV. The data were gathered during the period from 2010 to 2015. Those data contain information about patients treated at the Dermatology Clinic and they were pulled from the CCV information system. The data are split into several comma-separated values (CSV) documents, based on database tables. More details about these tables are presented in the following subsections.

3.1 Outpatients and Hospital Treated Patients

There are two documents describing the patient's relationship with the clinic. The first document contains data about patients treated at the outpatient department of the clinic, i.e., patients who are not treated at the hospital and are present at the clinic only for the examination. The second document contains data about inpatient treatment. The document that contains data about the outpatient treated patients has 60516 data records and the document that contains data about hospital treated patients has 4507 data records. Both documents have the same data structure. There are 51 attributes, including the examination time, patient's health insurance, basic patient information and information about the doctor performing the examination. Based on the fields from these two documents, the domain experts have defined a set of rules that we used in the data quality assessment case study, described in Sect. 5.

3.2 Examination Reports

There are 14960 examination reports for the given period of time. Each examination report contains data about the patient, doctor, and examination time, as well as a text report, which is created as an unstructured text description. These documents were not used in the presented case study, due to simplicity and ease of its interpretation.

3.3 Medicines and Supplies

This document contains data about medicines and supplies that were prescribed to patients or used during treatment. There are 344676 data records. The attributes from this document describe the medicine or supply, date, patient and quantity. This document was not used in the presented case study, for the same reason as the previous one.

4   Architecture

As presented in Fig. 1, the architecture of our system comprises six components: a rule builder, a tokenizer, a parser, a rule assembler, a data quality assessment (DQA) engine, and a report engine. These components constitute a pipeline and run in the sequential order depicted in Fig. 1. They were implemented in Java using the Spring framework [19] and the Java Data Mining Package [20]. In the remainder of this section, we describe each component, as well as input files and produced report files.

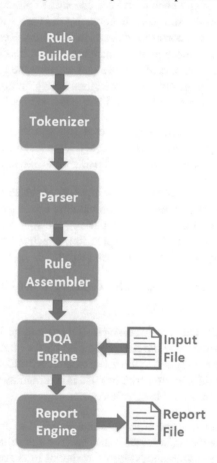

Fig. 1. A presentation of the system architecture

Rule Builder. A domain expert may use the rule builder through its user interface to define a set of rule specifications for the given business rule. In a guided process, the domain expert may use one of the provided rule patterns, select attributes of interest and create rule specification as a relationship between the attributes based on the given patterns. Rule patterns, which are defined separately by the developer, represent meta-rules through which concrete rules may be defined. Based on the created set of rule specifications, a domain expert may conduct data quality assessment.

Tokenizer. The tokenizer is responsible for extracting tokens from the rule specifications created by the domain expert. Tokens are used to determine to which rule pattern the provided rule specification belongs. Tokens represent predefined labels that describe one concrete concept and are used as a part of rule patterns. Tokens are defined as part of a data quality grammar and each token corresponds to one regular expression [21]. Tokens are recognized in the rule specifications by the corresponding regular expressions. Each rule specification corresponds to a list of tokens. Regular expressions must be ordered from specific to more general since regular expressions are applied to rule specifications sequentially. A part of the rule that satisfies the given rule expression is marked as a token. After that, regular expressions are again applied to the rest of the rule. This procedure continues until the whole rule is tokenized.

Parser. The parser takes a list of tokens for each rule specification as input and checks which rule pattern corresponds to the given list of tokens. Each list of tokens corresponds to one rule pattern. Rule patterns are defined using EBNF notation [22].

Rule Assembler. This component gets, as input from the Parser component, a map where the key is a rule specification and the value is a rule pattern to which the rule specification corresponds. Based on a rule pattern, the rule assembler transforms the concrete rule specification to a Boolean expression, which is later applied to each data record to check if the rule is true for the given record.

Input File. The input file represents a CSV document based on which the data quality assessment is made.

DQA Engine. The Data Quality Assessment (DQA) engine is used to check if the Boolean expressions provided by the rule assembler are true for the given data records. Each Boolean expression is applied to the given data record and, for each pair consisting of a rule specification a data record, true or false is returned as a result, i.e., if the rule specification for the given record is satisfied or not.

Report Engine. For each rule specification - data record pair for which the DQA engine returns false, a template message is generated based on the rule specification and the data record values. Each template message corresponds to the rule pattern that satisfies the given rule specification.

Report File. Compiled template messages from the report engine are appended to the report file, which is presented to the user as the final result. The report file also presents statistics about the data quality assessment process, such as the number of data records that yielded false for the given rule.

5 Case Study

In this case study, we present an example of using the proposed data quality system. We show how to define a rule specification based on the given set of rule patterns. This case study is based on a subset of real business rules that are defined by domain experts from the Dermatology Clinic in CCV and are applied on the data set from that clinic. The goal of this case study is to determine how many, and which, data records are not in accordance with the provided rules.

To define rule specifications, based on which the data quality assessment will be executed, rule patterns need to be defined. Rule patterns are used to define in what way a concrete rule will be specified and to define the rule semantics. Table 1 presents supported rule patterns for this case study. Rule patterns should be defined in the EBNF notation and there are no other limitations regarding their definition. Rule pattern are defined, as presented in Table 1, in the parser component and the pattern semantics, i.e., how the rules defined by the pattern will be presented as Boolean expressions by which data records will be evaluated, are implemented in the DQA engine.

Table 1. Supported rule patterns in the case study

Rule pattern	Description
IS_DATE : = DATE	Checks if the token is date
IS_NULL : = FIELD NOT NULL	Checks if the field is not null
REL_EXP : = NUM_FIELD REL_EXP NUM_FIELD REL_EXPS : = REL_EXP I REL_EXPS AND REL_EXP I REL_EXPS OR REL_EXP	Checks if the relational expression is true. This rule can be applied recursively by using the operators AND and OR
FUN_REL : = FIELDS => FIELDS I FIELDS : = FIELD FIELDS FIELD	Checks if the left hand side attributes are functionally related with the right hand side attributes

To present concrete business rules, the domain expert needs to write rule specifications using the given rule patterns. For example, we can define that medical insurance number is mandatory and write a rule specification as: INS_NUM NOT NULL, where INS_NUM represents the name of the insurance number column. We can also define a rule specification stating that the number of days spent in hospital for outpatients is one: STATUS = A AND NDAYS = 1, where STATUS represents a value defining if the patient has the outpatient status and NDAYS represents the number of days the patient has been hospitalized. We can assume that the information about the patient's clinic substation functionally determines the information about the patient's home town since patients have to be registered at the clinic substation nearest to their place of living. Therefore, we can get a rule specification which is presented as SUBSTA-TION => HOME_TOWN. After we define the set of rule specifications, we can apply them to our current data set. After running rules over our documents regarding outpatients' data, we get the results presented in Tables 2 and 3.

Table 2. The resulting statistics of applying the given rules over the data set.

Rule	Number of anomalies	Percentage of records with anomaly
INS_NUM NOT NULL	674	1.13%
STATUS = A AND NDAYS = 1	17	0.028%
SUBSTATION => HOME_TOWN	121	0.12%

Table 3. A part of the resulting report file

Row	Rule	Explanation
4	STATUS = A AND NDAYS = 1	When STATUS is A, NDAYS is 1
15	INS_NUM NOT NULL	INS_NUM must not have a null value
28	STATUS = A AND NDAYS = 1	When STATUS is A, NDAYS is 1
455	INS_NUM NOT NULL	INS_NUM must not have a null value
460	INS_NUM NOT NULL	INS_NUM must not have a null value
1499	SUBSTATION => HOME_TOWN	HOME_TOWN must not have different values for the same value of the field SUBSTATION

In Table 2, for each rule specification that has been defined, we present both the number of anomalies and the percentage of records with the corresponding anomaly. For the first rule specification in Table 2, there are 674 anomalous data records in which the insurance number is not specified. This may be the case of an emergency situation when only basic information is entered while the rest of the record is never updated afterwards. There are only 17 cases that do not satisfy the second rule specification in Table 2 and this might be due to a typing error. The third rule specification in Table 2 is not matched within 121 data records. This might happen when patients change their place of living but they do not change their substation yet. Table 1. Supported rule patterns in the case study.

Based on the report segment presented in Table 3, it is possible to see the exact row number where an anomaly occurred and the rule specification that was not satisfied. For example, one rule specification was not satisfied for row 4 since the number of days an outpatient stays in hospital should be 1 but it was 4. In this case, the number of days needs to be set to 1 or the patient's status needs to be set to hospitalized.

6   Conclusion

Using domain knowledge represented as a set of rules can be viewed as a flexible approach to data quality assessment in the domain of dermatology. One of the important advantages of this approach is the absence of fixed and generic metrics in the process of data quality assessment. Also, this approach ensures that the domain expert who will interpret data can make a set of rules through which all anomalies may be found, even if these rules are simple not null rules or complex, domain specific rules. One might argue that using generic metrics might be easier to implement and use in practice. In certain situations, it may be complex to define rule patterns that can appear, tokens for the rule pattern corpus, and concrete rules. Analyzing system complexity and making a pilot solution is necessary before fully applying this approach. The presented system yielded satisfactory results in our case study about the quality of dermatologic data since we did not have many problems in detecting the needed rule patterns and tokens. The domain expert has easily detected which concrete rules should be applied, but this might not always be the case. This approach also worked very well in our scenario since we needed data interpreted from the domain expert's point of view, so we can build a more accurate and reliable analytic system based on that data. Our future research in this domain will be focused on implementing an autocorrection mechanism of the data records that are marked as false for the given rule, where applicable. Also, we plan to work on a mechanism which will resolve conflicts between rules, if they occur.

Acknowledgements. The research presented in this paper was supported by the Ministry of Education, Science, and Technological Development of the Republic of Serbia under Grant III-44010. The authors are most grateful to Clinical Center of Vojvodina for the provided data set and valuable support throughout the study.

References

1. Kwetishe, D., Osofisan, A.O.: Evaluation of predictive data mining algorithms in Erythemato-Squamous disease diagnosis. IJCSI Int. J. Comput. Sci. Issues **11**(6), 85–94 (2014)
2. Brause, R.W.: Medical analysis and diagnosis by neural networks. Med. Data Anal. **2199**, 1–13 (2001)
3. Ji, Z.: Applications analysis of big data analysis in the medical industry. Int. J. Database Theor. Appl. **8**(4), 107–116 (2015)
4. Shouman M., Turner T., Stocke R.: Using data mining techniques in heart disease diagnosis and treatment. In: Proceedings of the 2012 Japan-Egypt Conference on Electronics, Communications and Computers, pp. 173–177 (2012)
5. Pipino, L.L., Lee, Y.W., Wang, R.Y.: Data quality assessment. Commun. ACM Support. Commun. Build. Soc. Capital **45**(4), 211–218 (2002)
6. Batini, C., Cappiello, C., Francalanci, C., Maurino, A.: Methodologies for data quality assessment and improvement. ACM Comput. Surv. **41**(3), 16 (2009). Article No. 16
7. Scannapieco, M., Virgillito, A., Marchetti, C., Mecella, M., Baldoni, R.: The DaQuinCIS architecture: a platform for exchanging and improving data quality in cooperative information systems. Inf. Syst. **29**(7), 551–582 (2004)
8. Ballou, D.P., Pazer, H.L.: Modeling data and process quality in multi-input, multi-output information systems. Manage. Sci. **31**(2), 150–162 (1985)
9. Ballou, D.P., Wang, R.Y., Pazer, H., Tayi, G.K.: Modeling information manufacturing systems to determine information product quality. Manage. Sci. **44**(4), 462–484 (1998)
10. Lee, Y.W., Strong, D.M., Kahn, B.K., Wang, Y.W.: AIMQ: a methodology for information quality assessment. J. Inf. Manage. **40**(2), 133–146 (2002)
11. Laudon, K.C.: Data quality and due process in large interorganizational record systems. Commun. ACM **29**(1), 4–11 (1986)
12. Cai, L., Zhu, Y.: The challenges of data quality and data quality assessment in the big data era. Data Sci. J. **14**, 2 (2015)

13. Silvola, R., Harkonen, J., Vilppola, O., Kropsu-Vehkapera, H., Haapasalo, H.: Data quality assessment and improvement. Int. J. Bus. Inf. Syst. **22**(1), 62–81 (2016)
14. Hipp, J., Guntzer, U., Grimmer, U.: Data quality mining. In: Proceedings of the 6th ACM Sigmod Workshop on Research Issues in Data Mining and Knowledge Discovery (2001)
15. Farzi, S., Baraani, D.A.: Data quality measurement using data mining. Int. J. Comput. Theor. Eng. **2**(1), 1793–8201 (2010)
16. Nahm, M.: Data quality in clinical research. In: Richesson, R.L., Andrews, J.E. (eds.) Clinical Research Informatics, pp. 175–201. Springer, London (2012)
17. Bae, C.J., Griffith, S., Fan, Y., Dunphy, C., Thompson, N., Urchek, J., Parchman, A., Katzan, I.L.: Challenges of data quality in medical informatics data warehouses. EGEMS (Wash DC) **3**(1), 1125 (2015)
18. Zozus M.N., Ed Hammond, W., Green, B.B., Kahn, M.G., Richesson, R.L., Rusincovitch, R.A., Simon, G.E., Smerek, M.M.: Assessing data quality for healthcare systems data used in clinical research. NIH Health Care Systems Research Collaboratory
19. Spring – Spring Framework. http://spring.io/
20. JDMP – Java Data Mining Package. http://jdmp.org/
21. Karttunen, L., Chanod, J.P., Grefenstette, G., Schiller, A.: Regular expressions for language engineering. Nat. Lang. Eng. 1–24 (1997)
22. Scowen, R.S.: Extended BNF — a generic base standard. In: Proceedings of the Software Engineering Standards Symposium (1993)

A Comparison of Predictive Analytics Solutions on Hadoop

Ramin Norousi[1], Jan Bauer[1], Ralf-Christian Härting[2], and Christopher Reichstein[2(✉)]

[1] Business Field Advanced Analytics, MHP – A Porsche Company, Ludwigsburg, Germany
ramin@norousi.de
[2] Business Administration, Aalen University of Applied Sciences, Aalen, Germany
ralf.haerting@kmu-aalen.de,
christopher.reichstein@hs-aalen.de

Abstract. New approaches regarding data streaming, data storage and data analysis have been developed facing the huge volume and velocity of generated data. Enterprises are convinced that one of their key success factor is to consider available data searching for patterns and predicting the future in order to gain more insights about their business, to optimize processes and to save costs. Hence, predictive analytics has never been considered more important than it is now. Hadoop as a popular open-source framework was introduced to store and process extremely large data sets. The paper shows various ways of carrying out predictive analytics based on a Hadoop ecosystem. We investigated different solutions of both commercial vendors and open-source communities interoperating with Hadoop. Each scenario is described by its technical implementation, features and restrictions. A comparison sums up the most important issues to get a deeper insight in order to optimize Predictive Analytics Solutions based on Hadoop.

Keywords: Hadoop · Predictive analytics solutions · Big data · Spark · IBM SPSS Modeler · RapidMiner · Radoop

1 Introduction

The amount of data being collected and analyzed has been increased rapidly in the past few years. This fact caused an enormous interest in large-scale data storage and data processing. One of the initial successful approach to meet these challenges is MapReduce which was introduced in 2004 by Google [1]. MapReduce is a framework for processing parallelizable problems across huge datasets using a large number of computers (nodes), collectively referred to as a cluster. Each fragment may be executed on any node in a distributed cluster and finally the results are aggregated.

The idea of MapReduce was implemented in Hadoop [2] as an open-source framework with its underlying structure HDFS (Hadoop Distributed File System) which has become the standard for data processing of large-scale data [3, 4]. HDFS is

© Springer International Publishing AG 2018
I. Czarnowski et al. (eds.), *Intelligent Decision Technologies 2017*,
Smart Innovation, Systems and Technologies 73, DOI 10.1007/978-3-319-59424-8_15

highly fault-tolerant and is designed to be deployed on low-cost hardware. Hadoop can be considered as a trigger that led to lot of developments in the big data community which kicked off an ecosystem of parallel data analysis tool for large clusters [5, 6]. MapReduce and its variants have been successfully applied in large-scale data-intensive applications on commodity clusters. However, these techniques were not suitable for all popular applications due to the fact that they are optimized for one-pass batch processing which make them slow for interactive data exploration. Furthermore, it was impossible for more complex, multi-pass algorithms, such as the algorithms that are common in machine learning [7]. Therefore, Apache Spark framework was proposed in 2012 by researchers at the University of Berkley [8] which overcomes these problems while allowing programmers to perform in-memory computation on large clusters. Spark as a fast and open-source engine for large-scale data processing can be considered as the next generation data processing alternative to MapReduce in the big data community [6, 9]. Furthermore Spark can outperform Hadoop up to 40 × faster than MapReduce applications, which translates directly into faster applications [10].

Based on highly successful introduced frameworks for storage of large-scale data sets, exploring and analyzing of the data is becoming more important. Thus, predictive analytics algorithms gain more and more attention in order to get insights from the large-scale data sets. Initially open-source solutions like Apache Spark were also used for analytics purposes in Hadoop by its machine learning library MLlib [11, 12]. The machine learning library of Spark as an open-source solution can be applied directly on data sets which are stored in a distributed file system like HDFS. However, it has the characteristic that it is based on scripting and coding. Hence, it is difficult to use for people with a non-programming background. Hand-coded implementation analytics can be laborious, time consuming and error-prone. Further enterprise solutions based on graphical interface can be considered as further suitable approaches which are successfully established in the analytics world. Among these, the IBM SPSS application "Modeler" and the RapidMiner application "Radoop" as leaders are investigated. Both software applications provide the possibility to build analytic models in a non-programming environment based on the data sets stored inside a Hadoop cluster [12, 13].

The paper is structured as follows: In Sect. 2, all three solutions considered in this work are described by their architecture and the process of applying analytics on Hadoop. In Sect. 3, the results based on our practical experiences with real-world data sets are summarized.

2 Distributed Analytics with Hadoop

Hadoop, as generally known, is the most suitable open-source software framework for storing and running applications on clusters of commodity hardware used by companies such as Google, Yahoo and Facebook. It provides as a massive storage for any kind of data, an enormous processing power and the ability to handle virtually concurrent jobs. Nevertheless, in order to explore the data sets on Hadoop and get accurate insights from them, there are various tools available which can be either connected to or integrated in the Hadoop ecosystem. Among others Apache Spark (with extensions to Python), IBM SPSS Modeler and RapidMiner Radoop are considered and evaluated in this paper, which are described in more details below.

2.1 Analytics on Hadoop Using Apache Spark

Apache Spark is the most applied open-source engine for large-scale data processing and analyzing from Apache Hadoop project. Spark provides a simple and expressive programming model that supports a wide range of applications, including ETL, machine learning, stream processing, and graph computation [10]. The idea distinguishing Spark is its in-memory computation, allowing data to be cached in memory across iterations. Spark overcomes MapReduce by providing a new storage called Resilient Distributed Datasets (RDD) [7]. RDDs let users store data in memory across queries and provides seamless support by two types of applications: iterative algorithms which are common in machine learning (e.g. kernel support vector machines [13]), and interactive data mining tools that are hard to express using acyclic data flow model pioneered by MapReduce. RDDs are collections of elements partitioned across several nodes in a cluster [14–17]. Initially, it lacked a suite of robust and scalable learning algorithms until the creation of MLlib. Development of MLlib as part of the MLbase project [11] was introduced as an open-source program in 2013. Spark MLlib provides a wide range of data preprocessing, data modeling and evaluation steps on distributed data.

Architecture
The Spark engine can be integrated within the Hadoop ecosystem and consists of following four components (Fig. 1): Spark SQL as a package for working with structured data, Spark Streaming component that enables processing of live streams of data, GraphX as a library for manipulating graphs, and last but not least the Spark MLlib for machine learning approaches which is considered in this paper in more details [15]. In order to be able to manage all tasks in the Hadoop ecosystem and be efficient while maximizing the flexibility, Hadoop YARN as cluster manager is introduced.

Data Access
Apache Spark is written in Scala Programming language [18] and it runs either directly on Hadoop ecosystem or in a standalone mode. Furthermore it can read flat files and access to diverse data sources including following databases [15]:

- Distributed file systems such as HDFS, Cassandra and S3 (Amazon Simple Storage System)
- NoSQL database such as HBase
- Relational database management system such as MySQL

Additional API
The Spark machine learning library MLlib is scalable and interoperates with further programming languages. It provides a high level API to Scala, Java, Python and R. Hence, it eases the use for users which are familiar with other languages to write Spark

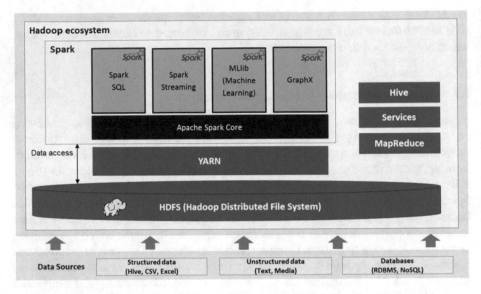

Fig. 1. Higher level architecture of Hadoop ecosystem including Spark (own source)

operations in other languages. It also offers the opportunity to access to further well-used machine learning libraries such as Numpy, Scikit-learn from Python or from R. It should be noted that in case of interoperating of Spark with further programming languages, a distributed execution is not possible.

Deployment

Due to the fact that Spark is integrated in Hadoop ecosystem, it enables to save all data preparation and predictive analytics steps in a Spark format which is able to be streamed and applied real time to Hadoop data sets. Furthermore, the MLlib supports partially model exports to the Predictive Model Markup Language (PMML), which is an XML-based interchange format to exchange models between different platform and tools [19].

Summarizing

Apache Spark is an open-source and widely-used programming model. It is integrated within the Hadoop ecosystem and it enables streaming Spark jobs in order to perform a real time execution. Furthermore, the Apache Spark community makes possible to remain constantly up-to-date regarding the analytics algorithms. Spark can also enable a distributed execution regardless of Hadoop by installing on another distributed system.

2.2   Analytics on Hadoop Using IBM SPSS Modeler and Analytic Server

IBM SPSS Modeler is a strong predictive analytics platform and one of two leaders in predictive analytics space according to a recent report from Gartner [20]. It provides a range of predictive algorithms based on a user-friendly graphical interface to support all major phases of the predictive analytics process. It has a large user base that continues

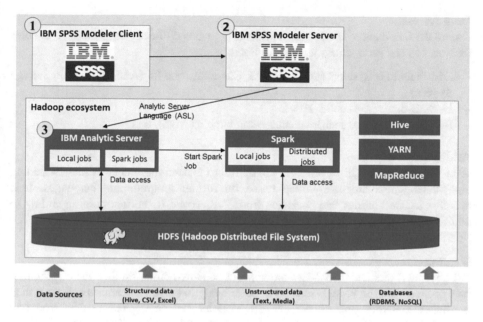

Fig. 2. Interoperating IBM SPSS and Hadoop – high level architecture (own source)

to keep up with innovation required by the market for example, integrating open-source R, Python and now Spark, to maintain high flexibility while making coding optional.

Architecture

As depicted in Fig. 2, in order to interact with data sets stored in Hadoop following three IBM components are required: IBM SPSS Modeler Client, IBM SPSS Modeler Server and IBM Analytic Server which should be initially installed as a part of the Hadoop platform and enables analysts to apply predictive analytics operations in SPSS Modeler to data stored in Hadoop. The process of performing predictive analytics jobs on Hadoop from the IBM SPSS Modeler is as follow (IBM):

1. The user develops a predictive analytics routine (called stream) on IBM SPSS Modeler Client which will be transferred into IBM SPSS Modeler Server.
2. The IBM SPSS Modeler Server receives the generated stream and translates the Stream into an IBM specific script language called Analytic Server Language (ASL) and sends it to the IBM Analytic Server which is installed on Hadoop ecosystem.
3. The IBM Analytic Server determines if the analysis should be distributed with Spark over the cluster or if it should run on the local Analytic Server JVM (Java Virtual Machine). This depends on the amount of data used for the analysis. Default settings are 128 Megabyte. Everything greater is translated into Spark Jobs and distributed over the Cluster. If spark is not available the analytic Server translates the Job into MapReduce Programs, which gets as well distributed over the cluster [21].

Data Access
Due to the fact that the IBM Analytic Server is installed directly on Hadoop, it enables to access to the same data sources like Spark:

- Distributed file systems such as HDFS, Cassandra and S3 (Amazon Simple Storage System)
- NoSQL database such as HBase
- Relational database management system such as MySQL

Additional API
IBM SPSS Modeler Client provides the option to expand the functionalities by adding algorithms which are user-written based on further programming languages. It is possible to use libraries and packages from Python and R. Furthermore, regarding to additional Hadoop services it can interoperate with Spark and use the full potential of Spark based on Analytic Server.

Deployment
All generated predictive analytics models can be stored either on the local machine or be exported with the PMML schema to other analytic tools. Furthermore, all generated predictive analytics operations in SPSS Modeler can be applied to data stored in Hadoop. Hence, IBM Analytic Server supports in-Hadoop execution of the majority of data preparation and modeling operations.

Summarizing
Providing a wide range of operations to support all major steps of predictive analytics process is an important strength of the architecture based on IBM SPSS. It can be applied on data stored in Hadoop without deep knowledge of Hadoop or Spark programming due to the graphical interface of IBM SPSS Modeler. It can access to various data bases and the major of generated operations can be applied in-Hadoop. A special feature of this architecture is that all RapidMiner operations are translated into Spark jobs first and are built on top of the MLlib in Spark. IBM SPSS Modeler is not free to use and in order to run this scenario it requires a further server for installing the IBM SPSS Server. During our practical analysis we noticed that some algorithms which are implemented are not fully distributed over the cluster, they partially run local on the Analytic Server, this may cause by the optimization engine of the Analytic Server itself.

2.3   Analytics on Hadoop Using RapidMiner Radoop

RapidMiner Radoop is a code-free analytics solution for Hadoop which has no separate service on Hadoop. RapidMiner Radoop is a fully graphical tool supporting the whole range of data analytics from ETL and ad-hoc reporting to predictive analytics [22].

Architecture
As it is depicted, RapidMiner Radoop does not require any installation on Hadoop. It offers two possibilities for analyzing and visualizing large-scale data sets stored in Hadoop:

1. RapidMiner client with Radoop as a client application that simplifies creating, maintaining and running analytics jobs over Hadoop directly.
2. RapidMiner server with Radoop as collaboration, scheduling, web reporting and web service integration to make it easier deploying big data analytics processes into an existing enterprise environment. It is a great choice for companies with a large Hadoop cluster and many users who wish to analyze and visualize big data.

In order to access the data on Hadoop, all operations developed on RapidMiner are translated into Spark or Hive jobs and transmitted to the Hadoop ecosystem. Hive is structuring the data for further analysis. It uses concepts like tables or columns to present the data. It translates SQL familiar Queries into MapReduce and HDFS tasks [23] (Fig. 3).

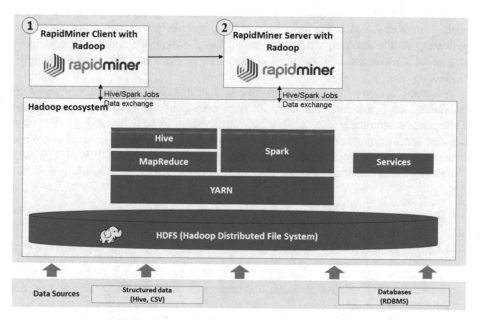

Fig. 3. Interoperating RadipMiner and Hadoop – high level architecture (own source)

Data Access
As RapidMiner Radoop runs all data preprocessing steps in Hive, data must be converted in Hive to use it with RapidMiner Radoop. It is possible to convert CSV-Files directly into Hive Tables and to access different Data sources which are listed below:

– Distributed file systems such as HDFS, Cassandra and Amazon Simple Storage (S3)
– Relational database management system such as MySQL

Additional API
It is possible to expand the RapidMiner functionalities by adding user-written operations based on further programming languages like Python or R.

Regarding to additional Hadoop services, it can interoperate with Spark and use the full potential of Spark based on Analytic Server. It can also interact with Hive and Pig.

Deployment
Predictive analytics operations which are built with a Spark script can be saved within spark itself. Furthermore, it is possible to use the generated models either outside of Hadoop with RapidMiner itself on local machine or on other analytic tools based on PMML-export. It supports in-Hadoop execution the essential of data preparation and modeling operations.

Summarizing
The architecture based on RapidMiner has also the advantage that it provides a wide range of operations to support all steps of a predictive analytics process. It can be applied on data stored in Hadoop without deep knowledge of Hadoop or Spark programming due to the graphical interface of RapidMiner. Furthermore, it can access various data bases and the major of general data analysis operations can be applied in-Hadoop.

A special feature of this architecture is the fact that all RapidMiner operations are translated into Spark jobs first and are built on top of the MLlib in Spark. It can interoperate additionally with further Hadoop services like Hive and Pig. The native RapidMiner is a free of use software. The version including Radoop in order to connect a Predictive Analytic application with Hadoop is covered by licence fees. Furthermore, this scenario can run without an additional server component for RapidMiner Server.

3   Results of Applying Analytics on Hadoop

Following approaches were considered based on a large real-world data set:

- Spark as an open-source cluster computing framework from Apache. It performs in-memory computation on large data sets comprising MLlib (Machine Learning library).
- IBM as a leader vendor with a high visibility in the advanced analytics space providing its strong product IBM SPSS Modeler with a graphical user interface and a wide range of analytics functionalities.
- RapidMiner as a further leader of analytics vendor. It provides a basic and community editions which is free and open-source and a commercial professional edition Radoop that has the ability to work with Hadoop.

According to our practical implementation and test based on three mentioned scenarios, findings are summarized in this section. The table below shows different aspects of the

Table 1. Solutions for data analytics on Hadoop

Criterion	Spark	IBM SPSS Modeler & Analytic Server	RapidMiner Radoop
Data Access	Structured (CSV) Unstructured NoSQL RDBMS Hive	Structured (CSV) Unstructured NoSQL RDBMS Hive	Structured (CSV) RDBMS Hive
Additional Hadoop services interaction	Hive	Spark written in Python	Spark written in Python or R Hive Pig
Deployment	In-Hadoop execution, export with PMML	In-Hadoop execution, export with PMML	In-Hadoop execution, export with PMML
Open-source	Yes	No	No
Independent of Hadoop	Yes	No	No
Community support	Yes	No	No
Hand Coded/GUI	Hand Coded	GUI/Hand Coded	GUI/Hand Coded
Data export	PMML Data Export to local or HDFS	PMML Data Export to local or HDFS	PMML Data Export to local or HDFS

solutions compared to each other. The main parts are highlighted. It should help to gain more knowledge about the different solutions for data analytics on Hadoop (Table 1).

Furthermore, available predictive analytics algorithms in each scenario can be evaluated (Table 2).

Each predictive analytics solution interoperating with Hadoop has its strengths and weaknesses. The choice of the appropriate solution depends on each specific application. It should be determined on how many data sources should be read, how complex the data preparation steps are, which predictive analytics algorithms should be applied and finally the generated models be deployed and integrated in-Hadoop or on other databases.

To sum up, based on five selected criteria a recommendation can be made according to the three solutions as shown in Table 3. The selected criteria can be considered as main steps of a data mining process. Based on a real data set, we tested the performance and number of available operations of all three analytics solutions according to these criteria. All three scenarios are assessed based on defined facts to make recommendations from our tests.

Table 2. Predictive analytics algorithms in each scenario

Methods	Algorithms	Spark	SPSS Analytic Server	RapidMiner Radoop
Classification	Decision Tree based on Gini-Impurity	X	X	X
	Decision Tree based on Entropy	X	X	X
	Decision Tree based on Chi-square independence		X	
	SVM (Linear)	X	X	X
	Naive Bayes	X		X
	KNN			
	Neural Network	X	X	
Regression	Linear Regression	X	X	X
	Logistic Regression	X	X	X
Clustering	K-Means	X		X
	TwoStep		X	
Association Rules	FP-Gwoth	X		
	Association Rules	X	X	
	Apriori			
Others	PCA	X		X
	Time Series Analysis		X	
Ensemble	Random Forest	X	X	X

Table 3. Recommendation based on five selected criteria

Criterion	Recommendation	Facts
Data Access	Spark, IBM SPSS	Accessing to five different data sources
Data Preparation	RapidMiner, IBM SPSS	Number of available operations
Modeling	All three solution	Number of available algorithms
Evaluation	IBM SPSS Modeler, RapidMiner	Number of available statistics for error measurements
Deployment	Spark	Additionally to in-Hadoop execution, it provides streaming of generated models

4   Conclusion

The number of unexpended data is growing at a breathtaking pace year by year. Scientists as well as managers have to deal with an exponential growing number of data in the next years [24]. In order to provide a remedy, there are different solutions in the market place both commercial and open-source. This paper compared three different and widely used Predictive Analytics Solutions on Hadoop based on a large real-world data set: Spark, IBM and RapidMiner. First, we investigated each solution by itself describing its technical implementation, features and restrictions. Afterwards, we considered Predictive Analytics Algorithms in different scenarios and compared all three solutions with each other in terms of data access, data preparation, modeling, evaluation and deployment. As a result, the choice of an appropriate Predictive Analytics Solution mainly depends on its specific application, the volume of data that should be read and the complexity of the data preparation. For firms with less complex data structures, classical data management concepts (i.e. on-premises data warehouse solutions) within the company might be sufficient.

The results contribute to the current research. Using Predictive Analytics Algorithms helps to understand how these solutions work within different scenarios. A comparison shows the strengths and weaknesses of each Predictive Analytics solution to reveal potential opportunities and risks when operating with Big Data [24].

Beyond, there are also managerial implications regarding Predictive Analytics Solutions. Managers can use our study as a recommendation when and which Predictive Analytics Solution is to use. An appropriate usage with the most convenient Predictive Analytics Solution might help to better understand the mass of data. Thus, analyzed data about customers might be useful to get essential insights about customer needs which in consequence lead to higher customer satisfactions and higher profits.

There are also some limitations regarding our examination. First, there are different aspects that were not taken into consideration like company size, different usage in different industries as well as different technical know-how within a company and/or possible missing IT infrastructures. Especially the last aspect is an important issue for managing Big Data in "the age of cloud computing" [5, 6], which is connected with the choice between Cloud- or On-Premises-Hadoop-Technologies. Second, we only considered three different Predictive Analytics Solutions. In fact, there are many more services available to analyze Big Data. Some of these services are based on new concepts. So called Data Lakes are integrating classical Data Warehouses and Hadoop-Clusters. Examples are HDInsight, a Microsoft Hadoop-Platform, based on Hortonworks Data Platform (HDP) or Analytics Platform System (APS).

Hence, future research should focus on investigation of various Predictive Analytics Solutions within firms depending on different industries, levels of know-how and IT infrastructures to gain deeper knowledge about an optimized use of data analytics.

References

1. Dean, J., Ghemawat, S.: MapReduce: simplified data processing on large clusters. In: Proceedings of the 6th Conference on Operating Systems Design and Implementation (OSDI), p. 10. USENIX Association, Berkeley (2004)
2. White, T.E.: Hadoop: The Definitive Guide, 3rd edn. O'Reilly, Sebastopol (2012)
3. Shvachko, K., Kuang, H., Radia, S., Chansler, R.: The Hadoop distributed file system. In: Shvachko, K., Kuang, H., Radia, S. (eds.) 26th Symposium on Mass Storage Systems and Technologies (MSST), pp. 1–2. IEEE, Incline Village (2010)
4. Zhao, J., Wang, L., Tao, J., Chen, J., Sun, W., Ranjan, R., Georgakopoulos, D.: A security framework in G-Hadoop for big data computing across distributed Cloud data centres. J. Comput. Syst. Sci. **80**(5), 994–1007 (2014)

5. McAfee, A., Brynjolfsson, E., Davenport, T.H., Patil, D.J., Barton, D.: Big data. The management revolution. Harv. Bus. Rev. **90**(10), 61–67 (2012)
6. Hashem, I.A.T., Yaqoob, I., Anuar, N.B., Mokhtar, S., Gani, A., Khan, S.U.: The rise of "big data" on cloud computing: review and open research issues. Inf. Syst. **47**, 98–115 (2015)
7. Zaharia, M., Chowdhury, M., Franklin, M.J., Shenker, S., Stoica, I.S.: Cluster computing with working sets. In: Proceedings of the 2nd USENIX Conference on Hot Topics in Cloud Computing (HotCloud), p. 10. USIENIX Association, Berkeley (2010)
8. Srirama, S.N., Jakovits, P., Vainikko, E.: Adapting scientific computing problems to clouds using MapReduce. Future Gener. Comput. Syst. **28**(1), 184–192 (2012)
9. Sagiroglu, S., Sinanc, D.: Big data: a review. In: International Conference on Collaboration Technologies and Systems (CTS), pp. 42–47. IEEE, San Diego (2013)
10. Zaharia, M., Chowdhury, M., Das, T., Dave, A., Ma, J., McCauley, M.: Resilient Distributed Datasets: A Fault-Tolerant Abstraction for In-Memory Cluster Computing. EECS Department, University of California, Berkeley (2011)
11. Meng, X., Bradley, J., Yavuz, B., Sparks, E., Venkataraman, S., Liu, D.: MLlib: machine learning in apache spark. J. Mach. Learn. Res. **17**(34), 1–7 (2016)
12. Zikopoulos, P., Eaton, C.: Understanding Big Data: Analytics for Enterprise Class Hadoop and Streaming Data. McGraw-Hill Osborne Media, New York (2011)
13. Patel, A.B., Birla, M., Nair, U.: Addressing big data problem using Hadoop and MapReduce. In: Nirma University International Conference on Engineering (NUiCONE), pp. 1–5. IEEE, Ahmedabad (2012)
14. Duda, R.O., Hart, P.E., Stork, D.G.: Pattern Classification, 2nd edn. Wiley-Interscience, New York (2012)
15. Apache Spark: Apache Spark™ - Lightning-Fast Cluster Computing. https://spark.apache.org/. Accessed 11 Jan 2017
16. Wu, X., Zhu, X., Wu, G.Q., Ding, W.: Data mining with big data. IEEE Trans. Knowl. Data Eng. **26**(1), 97–107 (2014)
17. Witten, I.H., Frank, E., Hall, M.A., Pal, C.J.: Data Mining: Practical Machine Learning Tools and Techniques. Morgan Kaufmann, Burlington (2016)
18. Odersky, M., Venners, B., Spoon, L.: Programming in Scala, 2nd edn. Artima Press, Walnut Creek (2011)
19. DMG: Data Mining Group. http://dmg.org/. Accessed 17 Jan 2017
20. Kart, L., Herschel, G., Linden, A., Hare, J.: Magic quadrant for advanced analytics platforms. Gartner report 9 (2016)
21. IBM: IBM SPSS Analytic Server Version 3.0: Overview. ftp://public.dhe.ibm.com/software/analytics/spss/documentation/analyticserver/3.0/English/IBM_SPSS_Analytic_Server_3.0_Overview.pdf. Accessed 19 Jan 2017
22. RapidMiner Radoop: RapidMiner Radoop - RapidMiner Documentation. http://docs.rapidminer.com/radoop/. Accessed 19 Jan 2017
23. Thusoo, A., Sarma, J.S., Jain, N., Shao, Z., Chakka, P., Zhang, N.: Hive - a petabyte scale data warehouse using Hadoop. In: IEEE 26th International Conference on Data Engineering (ICDE), pp. 996–1005. IEEE, Piscataway (2010)
24. Fan, W., Bifet, A.: Mining big data: current status, and forecast to the future. ACM SIGKDD Explor. Newsl. **14**(2), 1–5 (2013)

A Data Analytics Framework for Business in Small and Medium-Sized Organizations

Michael Dittert[1], Ralf-Christian Härting[2(✉)], Christopher Reichstein[2(✉)], and Christian Bayer[2]

[1] Competence Center for Information Systems, Aalen University of Applied Sciences, Aalen, Germany
[2] Business Administration, Aalen University of Applied Sciences, Aalen, Germany
{ralf.haerting,christopher.reichstein,
christian.bayer}@hs-aalen.de

Abstract. Data Analytics and derived Data Mining are powerful approaches for the analysis of Big Data. There are a lot of commercial Data Analytics applications enterprises can take advantage of. In the past, many firms were still critical of Data Analytics. Through efforts made in the field of the establishment of process standards, managers might be convinced of Data Analytics advantages. Many small and medium-sized organizations are still exempt from this development. The main reasons are a lack of business prioritization, a lack of (IT) knowledge, and a lack of overview of Data Analytics issues. To reduce that problem, we developed a useful process framework. It resembles with existing frameworks, but is highly simplified and easy to use. To exemplify, how this framework can be put into action by the means of a retail site location analysis, we set up a case study as best practice. There we are focusing on Data Mining because it is the most important domain of Data Analytics.

Keywords: Data analytics · Data mining · Location analysis · Process framework · SME

1 Introduction

At present, behind digitization concepts such as internet of things, cloud computing and social software, lies an essential technology called Big Data. We are facing a volume of data, which formerly was unknown. The sheer volume of data as well as the growth of this volume is extraordinary [1]. The volume of data surpassed 4.4 zettabytes in 2013 [2]. It is expected to double its size every two years [3] and to continue its rise to a volume of 44 zettabytes in 2020 [4]. This data does not only comprise structured data. There are different forms of data, e.g. audio, text or video files [5]. This development yields big opportunities for enterprises. Recent studies show that companies with a systematic approach towards Big Data observed a remarkable growth in efficiency and profitability [6]. This enables them to outperform their non-data driven competitors remarkably [7]. This leads to the assumption that Big Data is a decisive competitive factor, whose importance is still going to increase in future [6]. Big Data technologies

© Springer International Publishing AG 2018
I. Czarnowski et al. (eds.), *Intelligent Decision Technologies 2017*,
Smart Innovation, Systems and Technologies 73, DOI 10.1007/978-3-319-59424-8_16

and methods are not equivalent to concepts such as Business Intelligence and Business Analytics [8]. But special methods of Business Analytics are required to gain useful knowledge out of the large volume of data [9]. One of these methods is Data Mining [10]. Even though it has proven as a valuable tool in many business cases, a quote of Karen Watterson made in the year 1999 summarizes a problem many enterprises are still facing today: "There's always been an aura of mystery, even magic, associated with Data Mining. It was a science practiced on powerful UNIX systems overseen by unsmiling statisticians and brilliant mathematicians [11]." While in the past 17 years this aura may have become more permeable – for big companies at least – many small and medium-sized enterprises (SME) still take a critical stance on Data Mining [12]. For them, expected costs and a lack of knowledge are main obstacles for the implementation of Data Mining activities [13]. Therefore, the European Commission outlined in a recent publication in order to convince SMEs "that it needn't to be such an expensive and complex process [14]." An important step to achieve this goal is to increase their awareness through case studies [3] and a conceptual framework.

Among small and medium-sized organizations, two basic types can be distinguished. Depending on relevant stakeholder groups, there are profit orientated organizations and nonprofit organizations (NPO). Profit organizations, particularly small and medium-sized enterprises, are the most important driver of the European economy. Nine out of 10 enterprises are SMEs. SMEs create 85% of new jobs [15]. That indicates a relevant potential for Data Analytics exists in SMEs. The European Union defines SME by the means of three criteria: Staff headcount, annual turnover and balance sheet total [15]. In order to be classified as a medium-sized enterprise, a company's staff headcount should be beneath 250 employees [15]. In addition, a medium-sized enterprise should feature either an annual turnover of less than 50 million Euro or a balance sheet total of less than 43 million Euro [15]. These numbers already indicate that SMEs have special characteristics. They also affect their approach towards Big Data and Data Mining.

In this paper we will develop an appropriate Data Analytics process framework for SMEs. For this purpose we will first give a short overview of Data Mining and its applications. Afterwards we will describe the most common process frameworks. Alongside the Cross Industrial Standard-Data Mining (CRISP-DM) we will develop a process framework for SMEs. This framework will finally be exemplified by a case study.

1.1 Data Mining

Data mining, also known as "knowledge discovery in databases" (KDD), is a relatively new term. The most common definition was coined in the year 1996 by Usama Fayyad [16]. He describes KDD as "the nontrivial process of identifying valid, novel, potentially useful, and ultimately understandable patterns in data [16]". The key elements of this definition are further clarified. KDD is a process consisting of various steps. The aim is to find patterns in data by adapting a model to it, which describes structures hidden in the data [9]. These structures must be valid, i.e. they must also be applicable to unseen data [16]. For the user the patterns must be novel, useful and understandable [17]. Finally, the whole process is nontrivial in a way that it differs from simple calculations like the mean or the standard deviation of the data [18].

There are several applications by which enterprises can take advantage of Data Mining. Possible applications can vary according to the industry the enterprise is in [16]. The pure business applications are one half each located in marketing and finance [18]. Applications can be classified either as descriptive or as predictive Data Mining [19]. Descriptive models describe hidden patterns in data [16]. A frequently used application is the clustering algorithm [18]. Clustering detects relations between the data [20]. On the basis of the identified relations it establishes clusters of data with similar characteristics [16]. A common business application of clustering is the identification of homogenous groups among customers, who share similar characteristics [21]. Predictive models forecast an unknown output by learning relevant patterns in the input data [18]. They can be used to predict future sales [22]. Classification models are often used in the finance sector to judge a customer's credit rating [21]. In the context of Predictive and Descriptive modeling, Data Mining is the most important domain of Data Analytics [3]. Hereinafter the comprehensive term "Data Mining" will be the main focus of this paper.

1.2 Use of Data Analytics in Small and Medium-Sized Enterprises

When it comes to the usage of Big Data analytics, SMEs are often left far behind. According to a study conducted in the United Kingdom in 2012 [3] SMEs have an adoption rate of 0.2% while the same rate was about 25% for businesses with more than thousand employees [3]. Studies conducted in Germany and Austria revealed similar results [13]. In Germany, 46% of all interviewed SMEs indicated that they are not even going to use Big Data in the future [13].

There are various reasons for this rather negative attitude towards Big Data. Most of them can be assigned to one of the four categories: Security considerations, financial restrictions, lack of knowledge, and lack of prioritization for business issues. Especially the categories lack of knowledge and financial restrictions are closely interrelated. Many of the SMEs are sheer domain specialists [3], so there is no awareness of new concepts like Big Data [23]. As a result, there is no deeper understanding of Big Data techniques. There is a lack of support that could help developing skills in Data Analytics [3]. In addition, there still is demand for case studies that can help to exemplify the benefit of Big Data to them [23]. On the labor market there is a shortage of data analysts [24]. Hence, most of the SMEs are not able to recruit experts who can help to use Big Data. As there is no in-house expertise SMEs must rely on external expertise [3]. Both, recruitment and external consulting are very expensive. Due to their financial restriction SMEs often cannot afford to take advantage of Big Data [3].

In the recent years one barrier between SMEs and Data Analytics could be removed: To perform Data Analytics activities they now can profit from open source initiatives like Apache Mahout, Moa or Vowpal Wabbit [25]. Some software solutions like R or RapidMiner offer a facilitated usage through graphical interfaces [18]. These software environments offer the advantage that programming skills are no longer prerequisites to Data Mining [18]. Modeling of Data Mining processes, previously a complex and time consuming problem, now is reduced to a drag-and-drop-activity [18]. Furthermore, the user is supported in many ways, like in-software- or video-tutorials.

2   Relevant Process Frameworks

When the first approaches of Data Analytics have been launched, bigger enterprises were also facing similar problems as described above. The authors of the Cross Industry Standard Process-Data Mining (CRISP-DM) mentioned that their aim was to "demonstrate to prospective customers that Data Mining was sufficiently mature to be adopted as a key part of their business processes" [26]. This quote indicates the existing reservations about Data Mining there were in the industry. To break down these reservations, efforts were undertaken to establish Data Mining standards. Therefore, both academic and industrial research were done [27]. While academic research especially aimed to standardize the Data Mining language, industrial research concentrated on the establishment of process standards [28]. The most famous results of the industrial research were the CRISP-DM and the Sample, Explore, Modify, Model, Assess (SEMMA) Process [18]. Both can be seen as practical implementations of the older KDD-Process [28]. All three standards bear strong similarities.

2.1   The KDD-Process

The KDD-Process consists of nine steps that should be iteratively performed. The idea of iteration does not exclude a back loop. Sometimes it is necessary to step for- and backwards between two steps for several times. The steps of KDD include [16]:

1. Understand the application domain and gain relevant prior domain knowledge. Define the goal of the process.
2. Create a target data set.
3. Clear and preprocess the data. Detect and handle noisy data and missing values.
4. Reduce the data by removing features that are not necessary to achieve the goal.
5. Look out for a concrete Data Mining method that is suitable to achieve the goal.
6. In an explanatory analysis, the model is selected and its parameters must be fitted.
7. In the actual Data Mining step patterns of interest are identified.
8. The mined patterns must be interpreted.
9. The discovered knowledge is put into action. This can be done by the incorporation into systems, by documentation and reporting to interested persons or by looking out for contradiction with previous knowledge.

An advantageous feature of this process is the independence from a particular software tool. Thus, the KDD-Process is yielding a common basis for every other process standard [28].

2.2   SEMMA

In contrast, SEMMA is closely linked to the SAS software [16]. It was developed as a guide to carry out Data Mining tasks with the SAS Enterprise Miner [28]. The authors mentioned that it is crucial to define a goal of the Data Mining activities in the beginning and to implement the results in the end in their approach. But they focused on the model development primarily [29]. The name "SEMMA" is an acronym. Every letter represents one process step. The process steps are [29]:

- **Sample:** A representative data sample is extracted. It should be large enough to contain all necessary information. By mining only a sample of the data, costs and processing power can be saved.
- **Explore:** The sampled data set is explored by visual or statistical techniques. The aim is to identify anomalies and to get a first idea of previously unknown patterns.
- **Modify:** Now the data set must be prepared for modeling. This includes the selection of necessary variables, the creation of new variables and their transformation into a form the model can work with in the next step. In this phase, the decision of how to handle outliers and missing values must also be made.
- **Model:** A model that can fulfill the demanded task is selected and executed on the data set.
- **Assess:** Finally, the usefulness and reliability of the model's findings must be evaluated. To evaluate, it is proposed to apply the trained model on an unseen data set in order to confirm results.

With the findings of the process new questions can occur. This leads back to the step exploration and an additional refinement of the data [29].

It is neither intended by the authors nor expedient to build a Data Analytics process standard for SMEs on the basis of SEMMA as important steps like defining a goal and model deployment are only implicitly included in the SEMMA-process. As mentioned above especially for SMEs it is important to show why they should conduct Data Analytics activities and how they can take advantage out of its implementation. An additional obstacle is its close linkage to the SAS software that won't be found in many SMEs. Finally, the first phase, the sampling, would be unnecessary in many SMEs, because they normally won't have problems with a too large volume of data.

Nevertheless, there are some interesting aspects touched. In particular, if there is only a relative small data set on which the Data Analytics activities can be performed, the emphasis made on detecting and handling outliers is important. Because this is often the case in SMEs, outlier detection will be an important issue in our process model, too.

2.3   CRISP-DM

The CRISP-DM was developed in collaboration between Data Mining experts of various enterprises including SPSS and Daimler Chrysler [18]. Although there is no such strong linkage to a software as between SAS and SEMMA, the authors originally thought of the SPSS Modeler software when setting up CRISP-DM [27]. Nevertheless, this process standard can also be used independent from a specific software [28].

The CRISP-DM consists of the six steps Business Understanding, Data Understanding, Data Preparation, Modeling, Evaluation and Deployment. Like in the two previously presented standards it is not necessary to perform the steps in a rigid order [26]:

1. **Business Understanding:** The aim is to establish and define a concrete business objective, i.e. a goal that should be achieved through the Data Mining project. Therefore it is necessary to gain some expertise about the business, in whose context the Data Mining project will be conducted. At the end of this step, there is a plan how the business objective can be achieved.
2. **Data Understanding:** This step comprises the collection and analysis of the data. The analysis can be supported through visualization. The quality of the data must be judged by means of integrity and accuracy.
3. **Data Preparation:** The dataset has to be prepared for modeling. This comprises the handling of missing values, a first selection of variables and their transformation into a format the model can work with.
4. **Modeling:** A model is applied to solve the Data Mining task. The best model and its parameters are chosen in a trial and error procedure.
5. **Evaluation:** The model must be evaluated in quantitative and qualitative ways. Therefore it must be asked, how accurate the results are and if they are suitable to fulfill the defined business objective.
6. **Deployment:** The results must be used in a profitable manner. The easiest way of deployment could be, simply to report the results to the decision makers, so they can decide, how to take advantage of them. A more complex form of deployment could be the implementation of the model into the decision process of the enterprise.

The success of this process standard is reflected by the fact that it is often used as a basis for practical Data Mining user guides [18]. A possible reason is that it is a very comprehensive yet also flexible standard. It can easily be adapted to every Data Analytics task in terms of Data Mining processes.

3   Process Framework for SME

The process framework for SMEs in Fig. 1 is strongly related to the CRISP-DM.

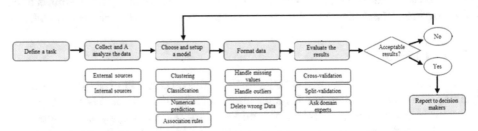

Fig. 1. Process framework for SMEs

Many SMEs, as mentioned above, are still critical towards concepts like Big Data and Data Analytics. Therefore, the process framework should be easily accessible to provide low barriers to entry into Data Analytics activities. This could be achieved by reducing the complexity of the process. The main reduction pertains to the loops between the process steps. In each of the presented process models, it is possible to step forward and back-ward in so-called loops between the process steps. As a result, the process is more flexible, but becomes unclear. Regarding the SME model, there is only one loop included. A second big change was made in the order, in which the process steps are to be performed. Before the data is prepared for modeling, in consequence to our approach, a model must be chosen and set up. Only with that arrangement, the user knows in which way he has to prepare the data. The step modeling is only included implicitly. SMEs can rely on a Data Mining software with a graphical interface. This reduces the effort of modeling to a simple drag-and-drop procedure.

To add further clarity, the steps of the process are named as instructions. They already give a first impression of the task to be performed. In the graphical representation, the most important alternatives of each step are pointed out. The content of the steps strongly resembles CRISP-DM [26]. In detail, the framework is constructed as follows:

- **Define a task:** Ideas for an appropriate Data Mining application should be generated by the SME's top management. This can be done internal based on workshops or external based on consulting support. The collected ideas have to be evaluated in terms of cost-benefit considerations. Finally, SMEs can choose between a multitude of standard Data Mining tasks. These tasks include, grouping customers, predicting customers' behavior, sales predictions, or market basket analysis [16]. SMEs often have effective services and support to solve standard tasks. Therefore it is recommended to rely on such standard tasks at the beginning of Data Mining activities. Normally the task will belong to one of the following categories: Statistical learning, classification, clustering, association analysis, and link mining [30]. The tasks will depend on the industry, where the SME is operating in [27]. The SME's domain knowledge of their own industry is very distinct normally [3]. This will help them to define an appropriate task [31]. Finally, it should be predefined, at which point the Data Mining task is fulfilled.
- **Collect and analyze the data:** The activity of collecting the data is stressed by putting it in the first place. Unlike big enterprises SMEs often do not have well organized data bases or even data warehouses themselves. This makes it necessary to set up a data basis for each Data Mining task. The data collected by SMEs may not be sufficient. Therefore it is recommended for them to think about taking advantage of external data. Data e.g. supplied by governmental statistical offices feature a high quality and are often available for free [23]. The collected data can be analyzed first by checking, if there are any unrealistic records, then by visualizing it for example by scatterplots. The visualization especially reveals if there are any outliers, which will influence the results of the Data Mining project in a negative way [18].
- **Choose and set up a model:** Depending on the task defined in the first step, a model must be chosen. For some tasks, like for class prediction, there are a lot of models available. Each of these models can be customized by setting its parameters. Here it must be stressed, that it is not necessary to choose and set up the right model at the first try. It often takes a series of trial and errors in order to achieve the most optimal model and parameters. If a Data Mining software with a graphical interface is used, the trial an error procedure can easily be done and is not very time consuming [26].

- **Format data:** First of all, it must be checked what types of data the chosen model can work with. While models like a decision tree can work with nearly every kind of values, a neural net for example only accepts numerical values. If the values can't be changed into a numerical scale, they have to be excluded. Handling missing values, wrong values or outliers is always a case to case decision. Especially in small data sets it may be necessary to exclude outliers from the modeling. They have a relatively big influence and so distort the results. So it may be often advisable, simply to exclude records with missing or wrong values as well as outlier records [26].
- **Evaluate results:** There are several possibilities to evaluate the model's quality. They depend on the kind of task that was performed. A classification model can be evaluated via split-validation, where the dataset is randomly divided into two parts [18]. One part is used to build the model, the other part to evaluate the model's performance. Numerical prediction can be evaluated by several statistical techniques [18]. In every case it is advisable to discuss the results with experts and let them judge, whether the results are useful [10]. If the results match the predefined demands, it can be passed to the next step. Otherwise the loop back to the model selection and setup should be performed.
- **Report to decision makers:** The easiest way of deployment is to support the decision makers with the knowledge gained from the Data Mining activities [26].

4   Case Study

To exemplify the practical insert of the proposed process framework for SMEs, a case study was conducted. Within the scope of that study, we used the RapidMiner software. It is an open source software with a high usability, as described above. We performed the described loop between the steps "choose and set up the model" and "evaluate results" several times. For the purpose of clarity, only the model and data formatting which lead to the best results will be described. As starting basis for the case study, we received access to the sales data of a middle-sized oil-trader located in southern Germany. It supplies about 70 petrol stations with fuels. In addition, the oil-trader operates a few of its own stations.

4.1   Define a Task

The petrol station sector is characterized by fierce competition [32]. That is comparable to the whole retail sector. Because of recent trends, e.g. more and more fuel-efficient engines and the steep ascent of the electro-mobility, the competition will continue to grow over the next decade [32]. Subsequently the number of petrol stations, which already dropped in the past years, will decline further [32].

As the establishment of own stations as well the as acquisition of dealer owned stations come along with high investments for the oil-trader, both must be carefully evaluated in advance [32]. The investment decision of course depends on the sales potential, which can be expected on a particular site. Hence the Data Mining task is to judge a given location by its sales potential. The aim is to build a model, which predicts with an accuracy of at least 85%, if at a particular site high or low sales can be expected.

4.2 Collect and Analyze the Data

There are many factors that influence the profitability of a retail site. There are various systematizations for these location factors. We decided to choose the classification of Roig-Tierno [33], which fits best to the particular situation of fuel stations. He distinguishes between the four classes' establishment, location, demographics and competition [33]. The data was obtained from internal and external sources. The extern sources include statistical offices and commercial websites. The demographic data could be either retrieved on the level of municipality or district the site is located in. All the data were compiled in an Excel-sheet. The following Table 1 provides an overview, which data were finally used, where we got it from and to which class it belongs.

Table 1. Selected Variables

Class	Factor	Level	Source
Establishment	Sales 2015	Station	Company's ERP-system
	Accepted credit cards	Station	Company's website
	Opening times	Station	Company's website
Location and demographics	Population	Municipality	Statistical office
	Motor vehicles	Municipality	Statistical office
	Disposable income per capita	District	Statistical office
	Unemployment rate	Job center district	Statistical office
	Population density	Municipality	Statistical office
Competition	Cars per station	5 km radius	Commercial website
	Capita per station	5 km radius	Commercial website
	Number of fuel stations	5 km radius	Commercial website

The advantage of the external sources is, that they offer high quality data for free, hence there were no noisy data at all. However, a closer look at the scatterplots revealed that there were some outliers. Even though these outliers are legitimate data, they occur very rarely, but exert a big influence on the results. Therefore it was decided to exclude them from modeling. This could be achieved via an automatic outlier detection in the RapidMiner Software.

4.3 Choose and Set Up a Model

RapidMiner offers several models for classification tasks, which were tested here. Finally the best results were gained by using an artificial neural net (ANN). With this model complex non-linear relations between in- and output-data can be represented [18].

It assumes weights between the independent input variables, i.e. the location factors, and the dependent output variables, i.e. the sales [20]. The model learns through a function called back propagation [2]. Here, there are quite a few training cycles performed [34]. In each cycle the model adapts the weights between input and output [2]. The model's parameters were set up by trial and error.

To make the model even more resistible, the ensemble learner technique of bagging was used. Here, the idea is to build several independent models by sampling different training data sets [34]. By joining the prediction power of the independent models, a higher accuracy can be reached as by using a single model only [20].

4.4 Format Data

The numerical target variable, i.e. the sales, had to be discretized into two classes: High sales potential and low sales potential. After a discussion with a domain expert, the dividing line was drawn by 3 million sold liters of fuel a year. Every value above was considered a high, every value beneath a low sales potential. As the ANN can only work with numerical inputs, all non-numerical inputs were filtered out. Furthermore, the detected outliers were excluded from modeling. Finally, some variables were combined by building for example the ratio between cars and petrol stations in the radius of 5 km around the locations. Table 1 already shows the variables as they were used for modeling in the end.

4.5 Evaluate Results

For evaluating the results the split-validation was used. Here, the data set is divided into a training data set, on which the model is built, and a testing data set, on which the built model is tested [18]. So the ability of the model to make right predictions can be judged [18]. After this, the process framework provides the possibility to go two steps back, if the model's quality does not match the previously defined expectations. In the case study this loop had to be performed often. Only then, the demanded accuracy of 85% could be reached. The best model could even reach a prediction accuracy of 86.67%.

4.6 Report to Decision Makers

The results can be used for several purposes. For example, the model could be used to support an investment decision. It could be also applied to judge the sales of fuel stations of the competition. A prerequisite to make decisions on the basis of these findings is decision-making authority of course. So it is necessary to report the gained findings to the decision makers.

5 Results

SMEs must leverage their data with Data Analytics methods to ensure their competitiveness over the long term. Due to special features, only a few SMEs take full advantage of their data. A significant number of obstacles which prevented Data Analytics have been removed. In that context it is imperative to emphasize the development of handy and intuitive software solutions. Previously, the development of process frameworks helped in the success of big enterprises to leverage Data Analytics. We introduced the best-known of these frameworks and checked them for their practicability in SMEs. Finally on the basis of these frameworks, we developed a new process framework which fits to the particular demands of SMEs.

6 Conclusion

The process framework we developed helps to provide small and medium-sized organizations with a low threshold entrance into the field of Data Analytics. Neither high expenses for external experts nor a big internal expertise are required to use it. Through its easy accessibility and usability, it can take part in the efforts to break down the reservations of SMEs against Big Data analysis. It enables SMEs to leverage their data and to participate in analyzing Big Data.

With the case study the usefulness of the developed process standard for SMEs could be proven. Particularly, it could be stressed this way how to handle the special features of SMEs like small data sets or the lack of internal data bases. The case study further emphasized the fact, that even with the limited capacities of SMEs, Data Analytics can proof as a powerful tool. Still, there is the need for more case studies. They should deal with some other Data Mining problems, which can be solved on a data basis available for SMEs.

Further case studies should take cost-benefit considerations into account. This should be done at the beginning of a Data Mining task. This might even lead to an extension of the process framework. In front of the step "define a task" further process steps can be included. They deal with the questions how to generate ideas for data analysis activities, how to evaluate these ideas, and how to prioritize the ideas in terms of cost-benefit considerations (see Fig. 2). In other words it is about including the SME's top management into the process of idea generation and enabling a transparent design of the evaluation and of the selection process. At the end of the pre-steps the best tasks can be chosen (out of a set of tasks) effectively. The business prioritization of Data Analytics could be even more enhanced.

Fig. 2. Proposed extension of the process framework

This process framework is for internal SME use only. Enterprises which offer Big Data analysis services do not need simplified process frameworks. It might be a challenge for further research to develop a more comprising process framework for small and medium-sized organizations.

References

1. Mohanty, H.: Big data: an introduction. In: Bhuyan, P., Chenthati, D., Mohanty, H. (eds.) Big Data: A Primer, pp. 1–28. Springer, New Delhi (2015)
2. Giudici, P.: Applied Data Mining: Statistical Methods for Business and Industry. Wiley, Chichester (2005)
3. Coleman, S., et al.: How Can SMEs Benefit from Big Data? Challenges and a Path Forward. Qual. Reliab. Eng. Int. **32**, 2151–2164 (2016)
4. IDC Digital Universe. https://www.emc.com/leadership/digital-universe/2014iview/executive-summary.htm. Accessed 6 Dec 2016
5. Hashem, I.A.T., Yaqoob, I., Anuar, N.B., Mokhtar, S., Gani, A., Khan, S.U.: The rise of "Big Data" on cloud computing: review and open research issues. Inf. Syst. **47**, 98–115 (2015)
6. Härting, R., Schmidt, R., Möhring, M.: Business intelligence & big data: Eine strategische Waffe für KMU? In: Härting, R. (ed.) Big Data – Daten strategisch nutzen, pp. 11–29. Books on Demand, Norderstedt (2014)
7. McAfee, A., Brynjolfsson, E., Davenport, T.H., Patil, D.J., Barton, D.: Big data. The management revolution. Harvard Bus. Rev. **90**(10), 61–67 (2012)
8. Schmidt, R., Möhring M., Maier, S., Pietsch, J., Härting, R.: Big data as strategic enabler – insights from central European enterprises. In: Abramowicz, W., Kokkinaki, A. (eds.) 17th International Conference on Business Information Systems, Lecture Notes in Business Information Processing, pp. 50–60. Springer, Cham (2014)
9. Hui, S.C., Jha, G.: Data mining for customer service support. Inf. Manage. **38**, 1–13 (2000)
10. Chen, L.-F., Tsai, C.-T.: Data mining framework based on rough set theory to improve location selection decisions: a case study of a restaurant chain. Tourism Manage. **53**, 197–206 (2016)
11. Watterson, K.: Datamining poised to go mainstream. http://www.datamation.com/datbus/article.php/616511/Datamining-poised-to-go-mainstream.htm. Accessed 6 Dec 2016
12. Ghaderi, H., Fei, J., Shakeizadeh, M.H.: Data mining practice in SMEs: a customer relationship management perspective. In: ANZAM, pp. 1–12 (2013)
13. Seufert, A.: Entwicklungsstand, Potentiale und zukünftige Herausforderungen von Big Data – Ergebnisse einer empirischen Studie. HMD – Praxis der Wirtschaftsinformatik **51**, 412–423 (2014)
14. European Commission. http://cordis.europa.eu/result/rcn/93077_en.html. Accessed 6 Dec 2016
15. Commission, E.: User guide to the SME Definition. Publications Office of the European Union, Luxemburg (2015)
16. Fayyad, U.M., Piatetsky-Shapiro, G., Smyth, P.: From data mining to knowledge discovery in databases. Am. Assoc. Artif. Intell. **17**(3), 37–54 (1996)
17. Fayyad, U.M., Piatetsky-Shapiro, G., Smyth, P.: The KDD process for extracting useful knowledge from volumes of data. Commun. ACM **39**(11), 27–34 (1996)
18. Deshpande, B., Kotu, V.: Predictive Analytics and Data Mining. Concepts and Practice with RapidMiner. Morgan Kaufmann, Amsterdam (2015)
19. Mitra, S., Pal, S.K., Mitra, P.: Data mining in soft computing framework: a survey. IEEE Trans. Neural Networks **13**(1), 3–14 (2002)
20. Han, J., Kamber, M.: Data Mining. Concepts and Techniques. Morgan Kaufmann, Amsterdam (2006)
21. Seng, J.-L., Chen, T.C.: An analytic approach to select data mining for business decision. Expert Syst. Appl. **37**, 8042–8057 (2010)
22. Ahmed, S.: Applications of data mining in retail business. In: International Conference on Information Technology: Coding and Computing (ITCC 2004), vol. 2, pp. 455–459 (2004)

23. Coleman, S.Y.: Data-mining opportunities for small and medium enterprises with official statistics in the UK. J. Official Stat. **32**(4), 849–865 (2016)
24. Davenport, T.H., Patil, D.J.: Data scientist: the sexiest job of the 21st century. Harvard Bus. Rev. **90**(10), 70–76 (2012)
25. Fan, W., Bifet, A.: Mining big data: current status, and forecast to the future. ACM SIGKDD Explor. Newslett. **14**(2), 1–5 (2013)
26. Chapman, P. et al.: ftp://ftp.software.ibm.com/software/analytics/spss/support/Modeler/Documentation/14/UserManual/CRISP-DM.pdf. Accessed 15 Dec 2016
27. Kurgan, L.A., Musilek, P.: A survey of knowledge discovery and data mining process models. Knowl. Eng. Rev. **21**(1), 1–21 (2006)
28. Azevedo, A., Santos, M.F.: KDD, SEMMA and CRISP-DM: a parallel overview. In: Proceedings of the IADIS European Conference on Data Mining, pp. 182–185 (2008)
29. SAS Institute Inc. https://web.archive.org/web/20120308165638/http://www.sas.com/offices/europe/uk/technologies/analytics/datamining/miner/semma.html/. Accessed 17 Dec 2016
30. Wu, X., Kumar, V., Quinlan, J.R., Ghosh, J., Yang, Q., Motoda, H., Zhou, Z.H.: Top 10 algorithms in data mining. Knowl. Inf. Syst. **14**(1), 1–37 (2008)
31. Wu, X., Zhu, X., Wu, G.Q., Ding, W.: Data mining with big data. IEEE Trans. Knowl. Data Eng. **26**(1), 97–107 (2014)
32. Gürsel, M., Tölke, O., von dem Bussche, G.: Branchenstudie Tankstellenmarkt Deutschland 2015. Scope Investor Services, Berlin (2016)
33. Roig-Tierno, N., et al.: The retail site location decision process using GIS and the analytical hierarchy process. Appl. Geogr. **40**, 191–198 (2013)
34. Witten, I.H., Frank, E., Hall, M.A., Pal, C.J.: Data Mining: Practical Machine Learning Tools and Techniques. Morgan Kaufmann, Amsterdam (2016)

Reasoning-Based Intelligent Systems

The Deep Learning Random Neural Network with a Management Cluster

Will Serrano[✉] and Erol Gelenbe

Intelligent Systems and Networks Group, Electrical and Electronic Engineering,
Imperial College London, London, UK
{g.serrano11,e.gelenbe}@imperial.ac.uk

Abstract. We propose the Random Neural Network with Deep Learning Clusters that emulates the way the brain takes decisions. We apply our model to measure and evaluate Web result relevance by associating each Deep Learning Cluster with a different Web Search Engine; in addition, we include a Deep Learning Cluster to perform as a Management Cluster that decides the final result relevance based on the inputs from the Deep Learning clusters. Our algorithm reorders the Web results obtained from different Web Search Engines after a user introduces a query. We evaluate the performance of the Management Cluster when included as an additional layer to the Deep Learning Clusters. On average; our Management Cluster improves result relevance; its inclusion increments the overall result relevance quality.

Keywords: Intelligent Search Assistant · World wide web · Random Neural Network · Web search · Clusters · Deep Learning · Management Clusters

1 Introduction

Our brain is formed of clusters of same neurons that perform different functions; they adapt and specialize when they learn tasks. Our brain decides the different actions to be taken based on a weighted decision from different sensorial inputs; an example is driving. A parallel scenario is the data available on the Web as it is formed of a large amount of information; users select relevant data obtained from different sources such as Web Search Engines or Recommender Systems. The brain retrieves a large amount of data obtained from the senses; analyses the material and finally selects the relevant information. This decision can be erroneous due different external factors such as light flashes or background noise, likewise, a user needs to select relevant Web results from a search outcome that may be influenced or manipulated by a commercial interest as well as by the users' own ambiguity in formulating their requests or queries. This paper proposes to associate the most complex biological system; our brain with the most complex artificial system represented by the Web. The link between them is the Random Neural Network.

We present an Intelligent Internet Search Assistant (ISA) that acts as an interface between an individual user's query and the different search engines. Our ISA acquires a query from the user and retrieves results from various search engines assigning one

© Springer International Publishing AG 2018
I. Czarnowski et al. (eds.), *Intelligent Decision Technologies 2017*,
Smart Innovation, Systems and Technologies 73, DOI 10.1007/978-3-319-59424-8_17

Deep Learning Cluster per each Web Search Engine; in addition we assign a Deep Learning cluster to perform as a Management Cluster. The result relevance is calculated by applying our innovative cost function based on the division of a query into a multi-dimensional vector weighting its dimension terms with different relevance parameters. Our ISA adapts and learns the perceived user's interest and reorders the retrieved snippets based in our dimension relevant centre point. Our ISA learns result relevance on an iterative process where the user evaluates directly the listed results as a supervised learning. We evaluate our Management Cluster unsupervised learning and compare its performance against other search engines with a new proposed quality definition, which combines both relevance and rank.

We describe the related work in Sect. 2. We define our Deep Learning Clusters in Sect. 3 and we present our Intelligent Search Assistant in Sect. 4. We have validated our model in Sect. 5 showing our experimental results in Sect. 6. Finally, we present our conclusions in Sect. 7 followed by the Appendix, where we list the queries used by our validators, and References.

2   Related Work

This paper is an extension of a prior model presented by Gelenbe, E. and Serrano, W. [1]. The Intelligent Search Assistant we design is based on the Random Neural Network (RNN) [2–4]. This is a spiking recurrent stochastic model for neural networks. Its main analytical properties are the "product form" and the existence of the unique network steady state solution. The RNN represents more closely how signals are transmitted in many biological neural networks where they actual travel as spikes or impulses, rather than as analogue signal levels. It has been used in different applications including network routing with cognitive packet networks, using reinforcement learning, which requires the search for paths that meet certain pre-specified quality of service requirements [5], search for exit routes for evacuees in emergency situations [6, 7], pattern based search for specific objects [8], video compression [9], and image texture learning and generation [10].

Deep Learning with Random Neural Networks is described by Gelenbe, E. and Yin, Y. [11, 12]. This model is based on the generalized queuing networks with triggered customer movement (G-networks) where customers are either "positive" or "negative" and customers can be moved from queues or leave the network. G-Networks are introduced by Gelenbe, E. [13, 14]; an extension to this model is developed by Gelenbe, E. et al. [15] where synchronised interactions of two queues could add a customer in a third queue.

Advanced learning techniques applied to the cloud are described in several publications; Wang, L et al. [16] present different experiments that compare three on-line real time techniques for task allocation to different cloud servers; they [17] also propose an experimental system that can exploit a variety of on-line QoS aware adaptive task allocation schemes. Brun, O. et al. [18] address the use of Big Data and machine learning based analytics to the real-time management of Internet scale Quality of Service route optimization with the help of an overlay network. Different search models were also proposed; Gelenbe, E. and Abdelrahman, O.H. [19] propose a Search in unknown random environments, [20] describe a search in the Universe of big networks and data and [21] present time and energy in team-based search. Schmidhuber, J. [22] reviews Deep Learning in neural networks.

3   The Cluster Random Neural Network

3.1   Mathematical Model

The Random Neural Network is composed of M neurons each of which receives excitatory (positive) and inhibitory (negative) spike signals from external sources which may be sensory sources or neurons. These spike signals occur following independent Poisson processes of rates $\lambda^+(m)$ for the excitatory spike signal and $\lambda^-(m)$ for the inhibitory spike signal respectively, to cell $m \in \{1, \ldots M\}$. In this model, each neuron is represented at time $t \geq 0$ by its internal state $k_m(t)$ which is a non-negative integer. If $k_m(t) \geq 0$, then the arrival of a negative spike to neuron m at time t results in the reduction of the internal state by one unit: $k_m(t^+) = k_m(t) - 1$. The arrival of a negative spike to a cell has no effect if $k_m(t) = 0$. On the other hand, the arrival of an excitatory spike always increases the neuron's internal state by 1; $k_m(t^+) = k_m(t) + 1$.

3.2   Clusters of Neurons

We propose the construction of special clusters of identical densely interconnected neurons. We consider a special network M(n) that contains n identically connected neurons, each which has a firing rate r and external inhibitory and excitatory signals λ^- and λ^+ respectively. The state of each cell is denoted by q, and it receives an inhibitory input from the state of some cell u which does not belong to M(n). Thus for any cell $i \in M(n)$ we have an inhibitory weight $w^-(u) \equiv w^-(u, i) > 0$ from u to i (Fig. 1).

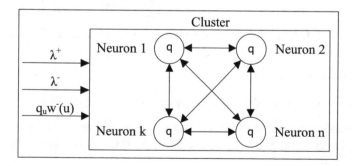

Fig. 1.   Clusters of neurons

3.3   The Random Neural Network with Multiple Clusters

We build a Deep Learning Architecture (DLA) based on multiple clusters, each of which is made up of a M(n) cluster. The DLA is composed of C clusters M(n) each with n hidden neurons. For the c-th such cluster, $c = 1, \dots, C$, the state of each of its identical cells is denoted by q_c. In addition, there are U input cells which do not belong to these C clusters, and the state of the u-th cell $u = 1, \dots, U$ is denoted by $\overline{q_u}$. The cluster network has U input cells and C clusters (Fig. 2).

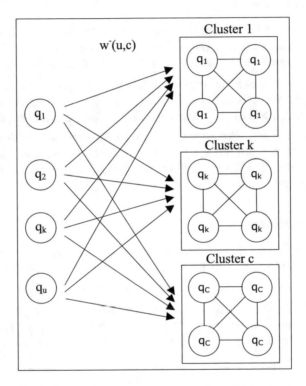

Fig. 2. The Random Neural Network with multiple clusters

3.4   Deep Learning Clusters

We propose the learning model of the Deep Learning Clusters in this section. We define:

- I, a U-dimensional vector $I \in [0, 1]^U$ that represents the input state $\overline{q_u}$ for the cell u;
- $w^-(u, c)$ is the U x C matrix of weights from the U input cells to the cells in each of the C clusters;
- Y, a C-dimensional vector $Y \in [0, 1]^C$ that represents the cell state q_c for the cluster c.

The network learns the U x C weight matrix $w^-(u, c)$ by calculating new values of the network parameters for the input X and output Y using Gradient Descent Descent learning algorithm which optimizes the network weight parameters $w^-(u, c)$ from a set of input-output pairs (i_u, y_c):

- the input vector $I = (i_1, i_2, \ldots, i_u)$ where i_u is the input state $\overline{q_u}$ for cell u;
- the output vector $Y = (y_1, y_2, \ldots, y_c)$ where y_c is the cell state q_c for the cluster c.

3.5 Management Cluster in the Random

We define the Deep Learning management Cluster model in this section. We consider:

- I_{mc}, a C-dimensional vector $I_{mc} \in [0, 1]^C$ that represents the input state $\overline{q_c}$ for the cluster c;
- $w^-(c)$ is the C-dimensional vector of weights from the C input clusters to the cells in the Management Cluster mc;
- Y_{mc}, a scalar $Y_{mc} \in [0, 1]$, the cell state q_{mc} for the Management Cluster mc.

Let us now define the activation function of the Management Cluster mc as:

$$\zeta(x_{mc}) = \frac{[np(\lambda_{mc}^+ + r) + n(\lambda_{mc}^- - x_{mc}) - p(\lambda_{mc}^+ + r) + r]}{2p_{mc}(n-1)[\lambda_{mc}^- + x_{mc}]}$$
$$- \frac{\sqrt{[np(\lambda_{mc}^+ + r) + n(\lambda_{mc}^- - x_{mc}) - p(\lambda_{mc}^+ + r) + r]^2 - 4p_{mc}(n-1)[\lambda_{mc}^- + x]n\lambda_{mc}^+}}{2p_{mc}(n-1)[\lambda_{mc}^- + x_{mc}]} \tag{1}$$

where:

$$x_{mc} = \sum_{c=1}^{C} \overline{q_c} w^-(c) \tag{2}$$

we have:

$$y_{mc} = \zeta(x_{mc})$$

The input state $\overline{q_c}$ for cell c represents the result relevance from each learning cluster; $w^-(c)$ is the C-dimensional vector of weights that represents the learning quality of each learning cluster c; y_{mc} is the final result relevance assigned by the Management Cluster (Fig. 3).

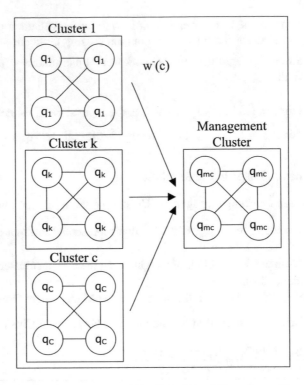

Fig. 3. The Random Neural Network with a Management Cluster

4   Implementation

The Intelligent Search Assistant we have proposed emulates how Web search engines work by using a very similar interface to introduce queries and display results. Our ISA acquires up to eight different dimensions values from the user however the Web search engine and number of result options are fixed.

Our ISA has been programmed to retrieve snippets from five Web search engines (Google, Yahoo, Ask, Lycos, Bing). Our process is transparent; our ISA gets the query from the user and sends it to the different search engines selected without altering it. The ISA provides to the user with a reordered list of results clustered by Web Search Engine. The results are reordered based on our predetermined cost function defined [1]; the user then selects relevant results and then our ISA provide with the final reordered list clustered by the different Web Search Engines (Fig. 4).

Intelligent Search Assistant

| Dimension 1 | Dimension 2 | | Dimension 8 |

⊠ Google ⊠ Yahoo ⊠ Ask ⊠ Lycos ⊠ Bing

Number of results: 10

Search!

Fig. 4. Intelligent Search Assistant user interface

5 Validation

A user in the experiment introduces a query. Our ISA assigns a learning cluster per Web Search Engine, acquires the first N results of each Web Search Engine and reorders them applying our cost function independently for each cluster. Finally our ISA shows a reordered list clustered by Web Search Engine. In our validation we have asked users to select Y relevant results per Web Search Engine, not to rank them, as they normally do using a Web search engine therefore we consider a result is either relevant or irrelevant. The results are shown to the user on random order to avoid a biased result rank evaluation where the users indirectly follow the order shown by the selected algorithm.

Each learning cluster learns each Web Search Engine Relevant Centre Point as defined in [1] where the inputs i_u are the same values as the outputs y_c. Our ISA reorders

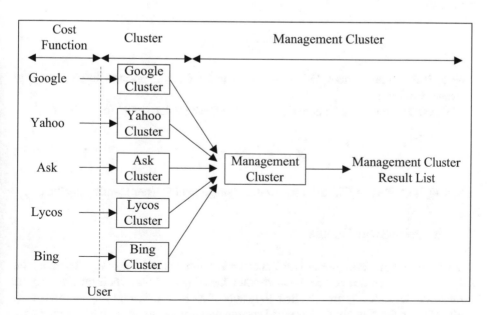

Fig. 5. Management Cluster validation

each Web Search Engine cluster list according to the minimum error to the cluster Relevant Centre Point. Once our ISA has established the best performing cluster; it uses its neural network weights to reorder the other cluster's result lists (Fig. 5).

In order to measure search quality we can affirm the better learning cluster provides with a list of more relevant results on top positions. We propose the following quality description where within a list of N results we score N to the first result and 1 to the last result, the value of the quality proposed is then the summation of the position score based of each of the selected results. Our definition of Quality, Q, can be defined as:

$$Q = \sum_{i=1}^{Y} RSE_i \qquad (3)$$

where RSE_i is the rank of the result i in a particular search engine with a value of N if the result is in the first position and 1 if the result is the last one. Y is the total number of results selected by the user. The best Deep Learning cluster or Web search engine would have the largest Quality value. We define normalized quality,.., as the division of the quality, Q, by the optimum figure which it is when the user consider relevant all the results provided by the Web search engine. On this situation Y and N have the same value:

$$\overline{Q} = \frac{Q}{\frac{N(N+1)}{2}} \qquad (4)$$

We define I as the quality improvement between a Deep Learning Cluster and a reference:

$$I = \frac{QC - QR}{QR} \qquad (5)$$

where I is the Improvement, QC is the quality of the Deep Learning Cluster and QR is the quality reference.

We validate our management cluster with different values of $w^-(c)$:

$$w^-(c) = \left(\frac{1}{Q_c}\right)^{\theta} \qquad (6)$$

where Q_c is Quality of Cluster c and Φ the Management Cluster learning coefficient.

6   Experimental Results

We have asked our validators to introduce up to 8 dimensions per query. There are no rules in what users can search, however they have been advised their queries may be published. Our ISA acquires the first 10 results of each Web Search Engine, independently reorders them applying our cost function and finally shows a 50 result randomly reordered list joining the 10 result Web Search Engine cluster list. In our validation we have asked users to select 2 relevant results per Web Search Engine cluster list; this enables us to better compare performance between learning clusters (Fig. 6).

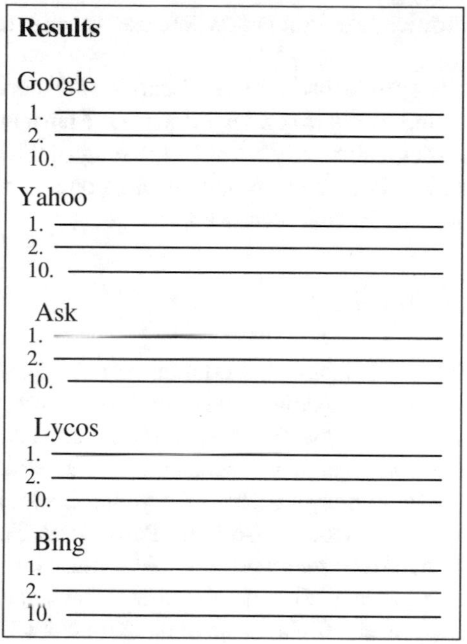

Results

Google
  1. _____
  2. _____
  10. _____

Yahoo
  1. _____
  2. _____
  10. _____

  Ask
  1. _____
  2. _____
  10. _____

  Lycos
  1. _____
  2. _____
  10. _____

  Bing
  1. _____
  2. _____
  10. _____

Fig. 6. Intelligent Search Assistant result list

We have validated our proposed Management Cluster model with 45 different user queries. The below Table 1 shows the Quality values at the different stages: before the user validation (Cost Function), after the user validation (Cluster) and with the Management Cluster (Management Cluster) for five different Management Cluster learning coefficients. We also show the Improvement between the quality of our Management cluster after and before the user validation.

Table 1. Management Cluster validation

Learning coefficient	Cost function Q	Cluster Q	Management Custer Q	MC vs. cluster I	MC vs. cost function I
$\Phi = 1/4$	0.239	0.2767	0.279	0.81%	16.75%
$\Phi = 1/2$	0.239	0.2767	0.278	0.34%	16.20%
$\Phi = 1$	0.239	0.2767	0.276	−0.30%	15.46%
$\Phi = 2$	0.239	0.2767	0.271	−1.97%	13.53%
$\Phi = 4$	0.239	0.2767	0.263	−5.05%	9.97%

7   Conclusions

We have presented a biological inspired learning algorithm: the Random Neural Network in a Deep Learning structure with a Management Cluster. We have validated it in a similar Big Data artificial environment where users need to take relevant decisions: the Web.

On average; our results prove that our Deep Learning Clusters increase significantly the search Quality after the user's first interaction. Our Management cluster improves the Deep Learning Cluster Quality only when its learning coefficient is less than one; it has a detrimental effect if it is equal to or greater than one. Our Management Cluster needs to be the tuned to take the right decisions.

Appendix – List of Queries

Random Neural Network - Art Exhibitions London - Art Galleries Berlin - Night Clubs London - Night Clubs Berlin - Vegetarian restaurant Stoke Newington - Techno Festivals Europe - Indie Rock London - Flights Tokyo - Gentrification World -Best documentaries 2016 - Book shops soho - Boutique hotels Mexico city - Brownie cafe London - Haute couture catwalks 2016 - Mid Century Patterns - Recycling Center London - Vintage Arm Chair - Wedding Design Gifts Shops - Weekend Paris Break Tory Party Conference - restaurants Forest Gate - compare energy prices - mid century furniture - reclaimed wood London - Best dinner date London - Best holiday destinations - Holiday deals packages - Best rooftop bars London - cheap flights Shanghai - Film clubs Edinburgh - fundraising jobs arts heritage Edinburgh - Holiday cottages rent Stoke on Trent 10 people - Home remedies sore throats - Volunteering opportunities Edinburgh - Brexit investments - Bristol breweries - Harmonica shops London - Narrow boats sale London - How get Irish passport - Best Pizza New York - Islington History - Southern rail overground - US presidential election forecast - Yorkshire dales walks.

References

1. Serrano, W., Gelenbe, E.: An intelligent internet search assistant based on the random neural network. Artif. Intell. Appl. Innov. **475**, 141–153 (2016)
2. Gelenbe, E.: Random neural networks with negative and positive signals and product form solution. Neural Comput. **1**, 502–510 (1989)
3. Gelenbe, E.: Product-form queueing networks with negative and positive customers. J. Appl. Probab. **28**, 656–663 (1991)
4. Gelenbe, E.: Learning in the recurrent random neural network. Neural Comput. **5**, 154–164 (1993)
5. Gelenbe, E., Lent, R., Zhiguang, X.: Networks with cognitive packets. In: Modeling, Analysis, and Simulation on Computer and Telecommunication Systems, pp. 3–10 (2000)
6. Gelenbe, E., Wu, F.: Large scale simulation for human evacuation and rescue. Comput. Math Appl. **64**, 3869–3880 (2012)
7. Filippoupolitis, A., Hey, L., Lucas, G., Gelenbe, E., Timotheou, S.: Emergency response simulation using wireless sensor networks. In: Ambient Media and Systems, p. 21 (2008)

8. Gelenbe, E., Kocak, T.: Area-based results for mine detection. IEEE Trans. Geosci. Remote Sens. **38**, 12–24 (2000)
9. Gelenbe, E., Sungur, M., Cramer, C., Gelenbe, P.: Traffic and video quality with adaptive neural compression. Multimedia Syst. **4**, 357–369 (1996)
10. Atalay, V., Gelenbe, E., Yalabik, N.: The random neural network model for texture generation. Int. J. Pattern Recogn. Artif. Intell. **6**, 131–141 (1992)
11. Gelenbe, E., Yin, Y.: Deep learning with random neural networks. In: International Joint Conference on Neural Networks, pp. 1633–1638 (2016)
12. Yin, Y., Gelenbe, E.: Deep learning in multi-layer architectures of dense nuclei. CoRR abs/1609.07160 (2016)
13. Gelenbe, E.: G-networks with triggered customer movement. J. Appl. Probab. **30**, 742 (1993)
14. Gelenbe, E.: G-networks with signals and batch removal. Probab. Eng. Inf. Sci. **7**, 335 (1993)
15. Gelenbe, E., Timotheou, S.: Random neural networks with synchronized interactions. Neural Comput. **20**, 2308–2324 (2008)
16. Wang, L., Gelenbe, E.: Experiments with smart workload allocation to cloud servers. In: IEEE 4th Symposium on Network Cloud Computing and Applications, pp. 31–35 (2015)
17. Wang, L., Gelenbe, E.: Adaptive dispatching of tasks in the cloud. IEEE Trans. Cloud Comput. 1 (2015). ISSN 2168-7161
18. Brun, O., Wang, L., Gelenbe, E.: Data driven smart intercontinental overlay networks. CoRR abs/1512.08314 (2015)
19. Gelenbe, E.: Search in unknown random environments. Phys. Rev. E **82**(6), (2010)
20. Gelenbe, E., Abdelrahman, O.: Search in the universe of big networks and data. IEEE Netw. **28**, 20–25 (2014)
21. Abdelrahman, O., Gelenbe, E.: Time and energy in team-based search. Phys. Rev. E **87** (2013)
22. Schmidhuber, J.: Deep learning in neural networks: an overview. Neural Netw. **61**, 85–117 (2015)

The Importance of Paraconsistency and Paracompleteness in Intelligent Systems

Jair M. Abe[1,2]([email]), Kazumi Nakamatsu[3], Seiki Akama[4], and João I.S. Filho[5]

[1] Graduate Program in Production Engineering, ICET - Paulista University, R. Dr. Bacelar, 1212, São Paulo, SP CEP 04026-002, Brazil
jairabe@uol.com.br
[2] Institute for Advanced Studies, University of São Paulo, São Paulo, Brazil
[3] School of Human Science and Environment/H.S.E., University of Hyogo, Kobe, Japan
nakamatu@shse.u-hyogo.ac.jp
[4] C-Republic, Tokyo, Japan
akama@jcom.home.ne.jp
[5] Santa Cecília University, Santos, Brazil
jinacsf@yahoo.com.br

Abstract. Nowadays, paraconsistent logic and paracomplete logic have established a distinctive position in a variety of fields of knowledge. The last systems are among the most original and imaginative systems of non-classical logic developed in the last and present century. Annotated logics are a kind of paraconsistent, paracomplete, and non-alethic logic.

Keywords: Paraconsistent logic · Annotated logic · Decision-making

1 Introduction

The application of Intelligent Systems reaches a very wide variety of fields. For instance, the speed and robustness of intelligent systems are increasingly demanded. The use of big data is more and more common, and experts are frequently needed to extract the maximum knowledge in helping the solutions to various problems. Such data often contains incomplete data, fuzzy and even conflicting information, and so on. The need to deal with such kind information is now one of the key points for a good response to the various inquiries required.

For the task, we need more generic and formal tools. In particular, classical logic is not suitable to handle the concepts quoted, at least directly. This paper shows in outline how to deal with paraconsistency and paracompleteness data without trivialization. For the task, it is employed a new class of non-classical logics, namely the paraconsistent annotated evidential logic Eτ [1, 10, 11].

© Springer International Publishing AG 2018
I. Czarnowski et al. (eds.), *Intelligent Decision Technologies 2017*,
Smart Innovation, Systems and Technologies 73, DOI 10.1007/978-3-319-59424-8_18

2 Paraconsistent and Paracomplete Logic Eτ

We focus in a particular paraconsistent and paracomplete logic, namely the paraconsistent annotated evidential logic Eτ - logic Eτ. The atomic formulas of the logic Eτ is of the type $p_{(\mu, \lambda)}$, where $(\mu, \lambda) \in [0, 1]^2$ and [0, 1] is the real unitary interval (p denotes a propositional variable). An order relation is defined on $[0, 1]^2$:

$(\mu_1, \lambda_1) \leq (\mu_2, \lambda_2) \Leftrightarrow \mu_1 \leq \mu_2$ and $\lambda_1 \leq \lambda_2$, constituting a lattice that will be symbolized by τ. A detailed account of annotated logics is to be found in [1, 11].

$p_{(\mu, \lambda)}$ can be intuitively read (among others): "It is assumed that p's favorable evidence degree (or belief, probability, etc.) is μ and contrary evidence degree (or disbelief, etc.) is λ." Thus, (1.0, 0.0) intuitively indicates total favorable evidence, (0.0, 1.0) indicates total unfavorable evidence, (1.0, 1.0) indicates total inconsistency, and (0.0, 0.0) indicates total paracompleteness (absence of information). The operator ~: $|\tau| \rightarrow |\tau|$ defined in the lattice ~ $[(\mu, \lambda)] = (\lambda, \mu)$ works as the "meaning" of the logical negation of Eτ.

The consideration of the values of the belief degree and disbelief degree is made, for example, by specialists who use heuristics knowledge, probability or statistics.

We can consider several important concepts (all considerations are taken with $0 \leq \mu, \lambda \leq 1$):

Segment DB - segment perfectly defined: $\mu + \lambda - 1 = 0$
Segment AC - segment perfectly undefined: $\mu - \lambda = 0$

Uncertainty Degree: $G_{un}(\mu, \lambda) = \mu + \lambda - 1$; Certainty Degree: $G_{ce}(\mu, \lambda) = \mu - \lambda$;

To fix ideas, with the uncertainty and certainty degrees we can get the following 12 regions of output: *extreme states* that are, False, True, Inconsistent and Paracomplete, and *non-extreme states*. All the states are represented in the lattice of the next figure: such lattice τ can be represented by the usual Cartesian system.

These states can be described with the values of the certainty degree and uncertainty degree using suitable equations. In this work, we have chosen the resolution 12 (number of the regions considered according to the Fig. 1), but the resolution is entirely

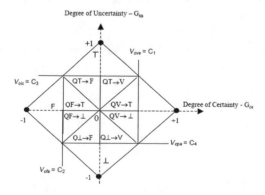

Fig. 1. Extreme and non-extreme states

dependent on the precision of the analysis required in the output, and it can be externally adapted according to the applications.

So, such limit values called Control Values are:

V_{cic} = maximum value of uncertainty control = C_3
V_{cve} = maximum value of certainty control = C_1
V_{cpa} = minimum value of uncertainty control = C_4
V_{cfa} = minimum value of certainty control = C_2

For the discussion in the present paper we have used: $C_1 = C_3 = 1/2$ and $C_2 = C_4 = -1/2$ (Fig. 1).

Table 1. Extreme and non-extreme states

Extreme states	Symbol	Non-extreme states	Symbol
True	V	Quasi-true tending to Inconsistent	QV→T
False	F	Quasi-true tending to Paracomplete	QV→⊥
Inconsistent	T	Quasi-false tending to Inconsistent	QF→T
Paracomplete	⊥	Quasi-false tending to Paracomplete	QF→⊥
		Quasi-inconsistent tending to True	QT→V
		Quasi-inconsistent tending to False	QT→F
		Quasi-paracomplete tending to True	Q⊥→V
		Quasi-paracomplete tending to False	Q⊥→F

In what follows, we present the algorithm para-analyser.

The primary concern in any analysis is to know how to measure or to determine the certainty degree regarding a proposition if it is False or True. Therefore, for this, we take into account only the certainty degree G_{ce}. The uncertainty degree G_{un} indicates the measure of the inconsistency or paracompleteness. If the certainty degree is low or the uncertainty degree is high, it generates an indefinite.

The resulting certainty degree G_{ce} is obtained as follows:

If: $V_{cfa} \leq G_{un} \leq V_{cve}$ or $V_{cic} \leq G_{un} \leq V_{cpa}$ $\Rightarrow$ G_{ce} = Indefinite
For: $V_{cpa} \leq G_{un} \leq V_{cic}$
If: $G_{un} \leq V_{cfa} \Rightarrow G_{ce}$ = False with degree G_{un}
$\quad V_{cic} \leq G_{un} \Rightarrow G_{ce}$ = True with degree G_{un}

The algorithm Para-analyzer is as follows:

```
*/ Definitions of the values */
Max_vcc = C₁ */ maximum value of certainty Control*/
Max_vctc = C₃ */ maximum value of uncertainty control*/
Min_vcc = C₂ */ minimum value of certainty Control */
Min_vctc = C₄ */ minimum value of uncertainty control*/
*/ Input Variables */
μ
λ
*/ Output Variables */
digital output = S1
Analogical output = S2a
Analogical output = S2b
* / Mathematical expressions * /
being:
0 ≤ μ ≤ 1 and 0 ≤ λ ≤ 1
G_un (μ; λ) = μ + λ - 1;
G_ce (μ; λ) = μ - λ
* / determination of the extreme states * /
if G_ce (μ; λ) ≥ C₁ then S₁ = V
if G_ce (μ; λ) ≥ C₂ then S₁ = T
if G_un (μ; λ) ≥ C₃ then S₁ = F
if G_un (μ; λ) ≤ C₄ then S₁ = ⊥
*/ determination of the non-extreme states * /
for 0 ≤ G_ce < C₁ and 0 ≤ G_un < C₂
if G_ce ≥ G_un then S₁ = QV→T
else S₁ = QT→V
for 0 ≤ G_ce < C₁ and C₄ < G_un ≤ 0
if G_ce ≥ | G_un | then S₁ = QV→⊥
else S₁ = Q⊥→V
for C₃ < G_ce ≤ 0 and C₄ < G_un ≤ 0
if |G_ce | ≥ | G_un | then S₁ = QF→⊥
else S₁ = Q⊥→F
for C₃ < G_ce ≤ 0 and 0 ≤ G_un < C₂
If |G_ce | ≥ G_un then S₁ = QF→T
else S₁ = QT→F
G_ct = S_2a
G_ce = S_2
*/ END */
```

3 Applications

The expert system considered has been applied in several problems: in what follows we sketch some of them (Fig. 2). The details of each issue the reader can be found in the references [2, 8, 11, 12].

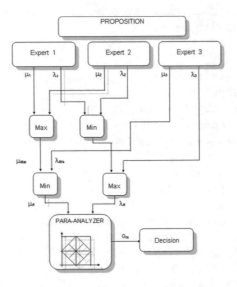

Fig. 2. Scheme of an expert system based on logic Eτ and para-analyzer algorithm

When we have a complex problem, we express it as a proposition p. All relevant parameters (factors, characteristics, etc.) are considered in the analysis. Also, we find a group of experts capable of giving their opinions about such parameters expressed as favourable evidence and also unfavourable evidence plus external data. Here lies a fundamental difference regarding the other logics. The majority of logical systems (including Fuzzy logic) consider the positive fact of p. In this way, the language of the logic Eτ is more faithful in describing the portion of reality in consideration.

Computer networks have two essential characteristics: the vast diversity of connecting devices and a significant variability of the physical distribution of equipment. Therefore, the performance analysis of a particular network based on absolute references or third parties may not be applicable in all circumstances, especially in highly complex and heterogeneous networks. Indeed, it carries a high degree of uncertainty, and the classical logic may not be appropriate to deal it. In [4] the paper aims to parameterize and evaluate the operating elements of heterogeneous networks, from the analysis of representative attributes, based on concepts of Logic Eτ.

The work [3] analyses IT management services quality focusing on its critical incidents as a strategic business factor and also to understand difficulties faced by professionals improving IT services. To assist managers in their decision-making process to improve IT services it was proposed a new method based on Logic Eτ.

Validation of the proposed method, it was compared to a case study of a Brazilian Foreign Trade company. The Logic Eτ was an important instrument in this study as it is a tool to analyse multiple takes of contradictory decisions, to converge on a single type of central decision-making and is suitable mainly for beginner's managers (first level) who need support for decision quickly and accurately.

After the results entered into the algorithm spreadsheet to the analyser and compared with surveys of a real case, reached the conclusion that the tool based on Logic Eτ presents a unified and efficient front of the comparison with the facts. The use of the expert's experience and the transformation of this qualitative information on quantitative data by the algorithm to-analyzer were essential for the validation of this study instrument.

Organisations rely on various IT (Information Technology) services for maintenance of their operations. Proper management of these services is of paramount importance because in many cases the services are subject to incidents, which may be defined as unplanned events that have the potential to lead to an accident.

Categorization and correct classification of incidents are of paramount importance, as they may cause stoppage of essential services and result in financial losses and even cause an impact on the organisation's image. Classification of incidents is the act of identifying the exact kind of incident and what components are involved, as well as determining the incident priority, which is, for example, classifying it as critical or not.

Depending on the size of the operation, the Service Desk can have many technicians of level 1 (first contact with the user), and differences in classification of incidents occur frequently, i.e. the same type of incident can be classified as critical by a technician and as not critical by the other. Therefore, classification may have inconsistent or contradictory results. The research problem on which this study is based is: there are differences between the classifications of incidents in the IT Service Desk teams.

For this situation, the classical Aristotelian logic does not treat those inconsistencies and contradictions correctly. Thus, the use of a non-classical logic, the Paraconsistent Logic, for example, would be the most appropriate in this situation.

In [5] an expert system based on logic Eτ was used to address inconsistencies in the classification of IT Incidents, helping managers to maintain the quality of services utilized by the organisation for the increase of the efficiency in the Service Desk area. ITIL (Information Technology Infrastructure Library) was considered as a management tool, which is a framework of better practices that deal with Incident Management in IT and the ISO/IEC 20000, which addresses this concept depicting Systems Certification of Services Management in IT.

In [6] the expert system was applied to present a method of prospective scenarios, proposing a new way of constructing technical and operational criteria, in such way that future studies can take the contradictions into consideration and can be not only reliable but also operationally efficient. Such method presents numerical output generated by the model so that they are easily understood by the decision makers (in work [6] we have considered 02 (two) economists 02 (two) executives and 02 (two) professors). It shows results of strategic topics between truth and falsehood, answering the following question: is it possible to develop future prospectives scenarios with contradictory and paracomplete data? The main advantages of using the Logic Eτ are due to the fact of the input parameters are set by the structure of the thinking of experts, consolidating a common logic translated into mathematical terms.

The paraconsistent decision-making employed was compared with other classical studies: with statistical methods [17] and Fuzzy decision method [16].

4 Aspects of Paraconsistent Neurocomputing

A suitable combination of Para-analyzer algorithm leads to built the paraconsistent artificial neural network [8]. In the paraconsistent analysis, the main aim is to know how to measure or to determine the certainty degree concerning a proposition if it is False or True to some degree. Therefore, for this, we take into account only the certainty degree G_{ce}. The uncertainty degree G_{un} indicates the measure of the inconsistency or paracompleteness. If the certainty degree is low or the uncertainty degree is high, it generates an indefinite.

The resulting certainty degree G_{ce} is obtained as follows:

If: $V_{cfa} \leq G_{un} \leq V_{cve}$ or $V_{cic} \leq G_{un} \leq V_{cpa}$ $\Rightarrow$ G_{ce} = Indefinite
For: $V_{cpa} \leq G_{un} \leq V_{cic}$
If: $G_{un} \leq V_{cfa} \Rightarrow$ G_{ce} = False with degree G_{un}
$V_{cic} \leq G_{un} \Rightarrow$ G_{cc} = True with degree G_{un}

The algorithm that expresses a basic Paraconsistent Artificial Neural Cell - PANC - is:

```
* /Definition of the adjustable values * /
V_cve = C₁ * maximum value of certainty control * /
V_cfa =C₂ * / minimum value of certainty control * /
V_cic =C₃ * maximum value of uncertainty control * /
V_cpa =C4 * minimum value of uncertainty control* /
* Input /Variables * /
μ, λ
* Output /Variables *
Digital output = S1
Analog output = S2a
Analog output = S2b
* /Mathematical expressions * /
begin:
0≤ μ ≤ 1 e 0≤ λ≤ 1
G_un = μ + λ - 1
G_ce = μ - λ
* / determination of the extreme states * /
if G_ce ≥ C₁ then S₁ = V
if G_ce ≥ C₂ then S₁ = F
if G_un ≥ C₃ then S₁ = T
if G_un ≤ C₄ then S₁ = ⊥
If not: S₁ = I - Indetermination
G_un = S2a
G_ce = S2b
```

A PANC is called *basic* PANC when given a pair (μ, λ) is used as input and resulting as output: G_{un} = resulting uncertainty degree, G_{ce} = resulting certainty degree, and X = constant of Indefinition, calculated by the equations $G_{un} = \mu + \lambda - 1$ and $G_{ce} = \mu - \lambda$ (Fig. 3).

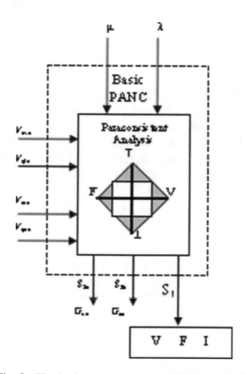

Fig. 3. The basic paraconsistent artificial neural cell

Such PANN was applied recently in the following topics:

In [9] algorithms based on the logic Eτ were interconnected into a network of paraconsistent analysis (PANnet). PANnet was applied to a dataset comprising 146 Raman spectra of skin tissue biopsy fragments of which 30 spectra were determined to represent healthy skin tissue (*N*), 96 were determined to represent tissue with basal cell carcinoma, and 19 were determined to be tissue with melanoma (MEL). In this database, the paraconsistent analysis was able to correctly discriminate 136 out of a total of 145 fragments, obtaining a 93.793% correct diagnostic accuracy. The application of logic Eτ in the analysis of Raman spectroscopy signals produces better discrimination of cells than conventional statistical processes and presents a good graphical overview through its associated lattice structure. The technique of logic Eτ-based data processing can be fundamental in the development of a computational tool dedicated to supporting the diagnosis of skin cancer using Raman spectroscopy.

In [7] presents the development of an MPC (Model Predictive Control) based on logic Eτ for a synchronous generator excitation control application, evaluating the performance of the Model Predictive Control for small signal stability analysis. Using the computational tool MATLAB®, from results obtained for the MPC, comparisons were made regarding results presented by a combination of AVR (Automatic Voltage Regulator) and a PSS (Power System Stabilizer), and results presented by a standard MPC.

PANN was also employed in the Alzheimer's disease [14, 15], and many other issues.

5   Conclusion

This paper shows at least in outline that annotated logics is a suitable formal tool to deal with vagueness, inconsistencies and paracompleteness. Annotated logics give a logical foundation for such mechanisms.

Issues, as described, are more and more common in intelligent systems, and annotated logics are capable of being the underlying logic for those logical formalisms giving robustness to theories based on them [13, 18]..

References

1. Abe, J.M.: Fundamentos da lógica anotada (Foundations of annotated logics). Ph.D. thesis, University of São Paulo, São Paulo (1992, in Portuguese)
2. Abe, J.M.: Some aspects of paraconsistent systems and applications. Logique et Analyse **15**, 83–96 (1997)
3. Pimenta Jr., A.P., Abe, J.M., Oliveira, C.C.: An analyzer of computer network logs based on paraconsistent logic. In: Umeda, S. et al. (eds.) APMS 2015, Part II, IFIP AICT, vol. 460, pp. 620–627. Springer, Cham (2015)
4. Pimenta Jr., A.P., Abe, J.M., Silva, G.C.: Determination of operating parameters and performance analysis of computer networks with paraconsistent annotated evidential logic Eτ. In: Nääs, I. et al. (eds) APMS 2016, IFIP AICT, vol. 488, pp. 3–11. Springer, Cham (2016)
5. Tavares, P.F., Silva, G.C., Sakamoto, L., Abe, J.M., Pimenta, A.P.: IT incident management and analysis using non-classical logics. In: Nääs, I. et al. (eds) APMS 2016, IFIP AICT, vol. 488, pp. 20–27. Springer, Cham (2016)
6. Reis, N.F., Oliveira, C.C., Sakamoto, L., Lira, A.G., Abe, J.M.: Paraconsistent Method of Prospective Scenarios (PMPS). In: Grabot, B., Vallespir, B., Gomes, S., Bouras, A., Kiritsis, D. (eds) APMS 2014, Part I, IFIP AICT, vol 438, pp. 76–84. Springer, Heidelberg (2014)
7. Oliveira, R.A.B., Abe, J.M., Rocco, A., Mario, M.C., da Silva Filho, J.I.: Modelo de Controle Adaptativo de Excitação de Gerador Síncrono Baseado em Lógica Paraconsistente, Science and Technology, vol. 5, pp. 182–192 (2016)
8. Da Silva Filho, J.I., Torres, G.L., Abe, J.M.: Uncertainty treatment using paraconsistent logic - introducing paraconsistent artificial neural networks, vol. 211. IOS Press (2011). doi:10.3233/978-1-60750-558-7-I
9. Da Silva Filho, J.I., Nunes, V.C., Garcia, D.V.N., Mario, M.C., Giordano, F., Abe, J.M. Pacheco, M.T.T., Silveira Jr., L.: Paraconsistent analysis network applied in the treatment of Raman spectroscopy data to support medical diagnosis of skin cancer. Med. Biol. Eng. Comput. **54**, 1–15 (2016). Springer. doi:10.1007/s11517-016-1471-3

10. Sylvan, R., Abe, J.M.: On general annotated logics, with an introduction to full accounting logics. Bull. Symbolic Logic **2**, 118–119 (1996)
11. Abe, J.M., Akama, S., Nakamatsu, K.: Introduction to annotated logics - foundations for paracomplete and paraconsistent reasoning. Intelligent Systems Reference Library, vol. 88. Springer, Berlin (2015). doi:10.1007/978-3-319-17912-4
12. Abe, J.M.: Paraconsistent intelligent based-systems: new trends in the applications of paraconsistency. In: Abe, J.M.(ed.) Intelligent Systems Reference Library, vol. 94. Springer, Cham (2015). doi:10.1007/978-3-319-19722-7
13. Akama, S.: Towards paraconsistent engineering. In: Akama, S. (ed.) Intelligent Systems Reference Library, vol. 110. Springer, Cham (2016). doi:10.1007/978-3-319-40418-9
14. Abe, J.M., Lopes, H.F.S., Anghinah, R.: Paraconsistent artificial neural network and alzheimer disease: a preliminary study. Dement. Neuropsychol. **1**(3), 241–247 (2007)
15. Lopes, H.F.S., Abe, J.M., Anghinah, R.: Application of paraconsistent artificial neural networks as a method of aid in the diagnosis of Alzheimer disease. J. Med. Syst. **34**(6), 1073–1081 (2009). doi:10.1007/s10916-009-9325-2
16. Carvalho, F.R., Abe, J.M.: A simplified version of the fuzzy decision method and its comparison with the paraconsistent decision method. In: AIP Conference Proceedings, vol. 1303, pp. 216–235 (2010). doi:10.1063/1.3527158
17. Carvalho, F.R., Brunstein, I., Abe, J.M.: Decision making method based on paraconsistent annotated logic and statistical method: a comparison, computing anticipatory systems. In: Dubois, D. (ed.) American Institute of Physics, Melville, vol. 1051, pp. 195–208 (2008). doi:10.1063/1.3020659
18. Nakamatsu, K., Kountchev, R.: New approaches in intelligent control, techniques, methodologies and applications. In: Nakamatsu, K., Kountchev, R. (eds.) Intelligent Systems Reference Library, vol. 107. Springer International Publishing, Cham (2016)

Towards REX Method for Capitalizing the Knowledge of a Corporate Memory

Hioual Ouided[1(✉)], Laskri Mouhamed Tayab[2], and Maifi Lyes[3]

[1] Computer Science Department, Abbès Laghrour University,
Khenchela, Algeria
hioual.ouided@univ-khenchela.dz
[2] LRI Laboratory, Computer Science Department,
BADJI Mokhtar University, Annaba, Algeria
[3] Physics Chemistry of Semi Conductor Laboratory, Physics Department,
Exact Sciences Faculty, Constantine University, Constantine, Algeria

Abstract. Knowledge management has become a strategic and vital issue for many companies in order to respond to an economic, technical and informational environment which evolves dynamically. The knowledge loss risk has led companies to look for a way to capitalize them for later reuse. The goal of knowledge capitalization, in an organization, is to favor the growth, transmission and conservation of knowledge in that organization. The construction of corporate memories, with a view to preservation and sharing, has become enough current practice. In this paper, we are interested in the capitalization of corporate knowledge using the REX (Return of EXperience) method. The objective of this method is to capitalize the knowledge that must be strategic and to favor the feedback of experience. The obtained results, based on a real medical case study, show that the method we have applied is promising in the field of knowledge capitalization.

Keywords: Corporate memory · Knowledge capitalization · REX method knowledge management

1 Introduction

Knowledge management refers to the management of all the knowledge and know-how in action mobilized by the company's stakeholders to enable them to achieve their objectives. Several stages have been identified in the knowledge management process (capitalization and knowledge sharing), it is an explicit tacit knowledge identified as crucial for the company, the sharing of the knowledge capital made explicit in the form of Memory and the appropriation and exploitation of part of this knowledge by the actors of the company [1]. The sharing and appropriation of corporate memories are still real bottlenecks within organizations. Knowledge management methods are not sufficient to enable effective ownership of knowledge by the company's stakeholders. However, the goal of knowledge capitalization is the sharing and re-use of experience in order to optimize the organizational learning process.

I. Czarnowski et al. (eds.), *Intelligent Decision Technologies 2017*,
Smart Innovation, Systems and Technologies 73, DOI 10.1007/978-3-319-59424-8_19

Today, knowledge modeling techniques are relatively well mastered and have been widely used. Among these, there are techniques which are based on logical repre-sentations, on the use of semantic networks, artificial neural networks, on the writing of rules, or on mixtures of these different approaches. There are several methods and tools to manage these different types of knowledge representation. In this study, we are interested in the capitalization of corporate knowledge using the REX (Return of EXperience) method [2]. The objective of this method is to capitalize the knowledge that must be strategic and to favor the feedback of experience. The remainder of this paper is structured as follows: In Sect. 2, we introduce the notion of corporate memory and the different categories of corporate knowledge. In Sect. 3, we present the cycle of knowledge capitalization. In Sect. 4, we explain the REX method and its different objectives. Next, in Sect. 5 and through an illustrative example, we present how REX is used in order to capitalize knowledge of a corporate and then we give some implementation results. Finally, the Sect. 6 presents a conclusion and future work.

2 Corporate Memory

Although the term "corporate memory" is new in the corporate landscape, its origins date back to the end of the 19th century [3] at the beginning of the 20th century, this term was forgotten to reappear in 1981. In [4] the first definition of an organizational memory in the sense that it is a memory that "establishes the cognitive structures of information processing within an enterprise". The corporate memory is defined in [1–4] as "an explicit and persistent representation of knowledge and information in an organization in order to facilitate their access and reuse by the appropriate members of the organization for their task". The construction of a corporate memory is based on the desire to "preserve reasoning's, behaviors and knowledge in order to reuse them later or as soon as possible, even in their contradictions and in all their variety" [5].

According to [6], the corporate memory management consists of detecting the needs of corporate memory, building, distributing, using, maintaining and evolve it.

There are several types of knowledge in an organization [7]: explicit knowledge and tacit knowledge [8]. Therefore, for any knowledge capitalization operation, it is important to identify the strategic knowledge to be capitalized [9]. These two types of knowledge are defined in the following section.

2.1 Tacit Knowledge

They are based on the origin idea Japanese "we know no more than we can say", they reside in the heads of people, and are little formalizable and difficult to communicate. They are, generally, individual character and need to be formalized in order to be shared. Tacit knowledge includes:

A Cognitive Aspect. It is translated by the mental models that humans form on the world [10]. It includes patterns, paradigms, beliefs and points of view.

A Technical Aspect. Concrete know-how is transmitted by word of mouth [11]. It is expressed in terms of informal technical expertise [8] or of capacity for action, adaptation and evolution [12].

2.2   Explicit Knowledge

Unlike the previous ones, this knowledge is transcribed on any supports. It is the knowledge of the corporate stored in the form of procedures, plans, manuals, maintenance manuals, troubleshooting manuals, technical-commercial documents, technical notes, databases, Expert systems, audio or video recordings, photos and films [11].

3   Knowledge Capitalization Cycle

Knowledge capitalization is defined in [13] as «the formalization of experience gained in a specific field». The principal purpose is to «locate and make visible the enterprise knowledge, be able to keep it, access it and actualize it, know how to diffuse it and better use it, put it in synergy and valorize it» [14].

The capitalization process consists of several stages focusing on the notion of strategic knowledge (see Fig. 1):

- The locating process: location of critical knowledge, i.e. explicit knowledge and tacit knowledge which are necessary for the decision processes and the progress of essential processes that form the core of the corporate activities [14].
- The Actualizing process: actualization of know-how and proficiency: it is necessary to appraise them, to update them, to standardize them and to enrich them according to the returns of experiments, the creation of new knowledge, and the contribution of external knowledge [14].
- The Enhancing process: added-value of know-how and proficiency: it is necessary to make them accessible according to certain rules of confidentiality and safety, to distribute them, to share them, to exploit them, to combine them and to create new Knowledge [14].
- The Preserving process: preservation of know-how and proficiency, when knowledge is explicable, it is necessary to acquire it, to model it, to formalize it and to preserve it; when knowledge is not explicable, the transfer of master-apprentice knowledge and communication networks between people, for example, should be encouraged [1].
- The Managing process: the interactions between the various problems mentioned above. It is there that the management of activities and processes is positioned to amplify the use and creation of knowledge in organizations.

Several methodologies are designed for the construction of "corporate memory" defined as "an explicit, disembodied and persistent representation of knowledge and

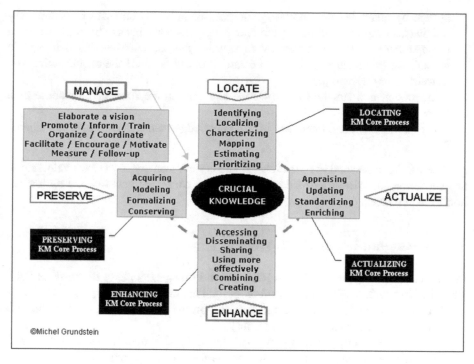

Fig. 1. Knowledge capitalization cycle [5]

information in an organization". In the next section, we present the method chosen for the construction of the corporate memory.

4 Return of EXperience (REX) Method

REX method is due to the capitalization of experiences archived during the enterprise activity realization. The finality is to build a knowledge base for future teams. The goal, of this method, consists in constituting "experience elements" extracted from an activity and to return these elements to a user to value them. Three types of experience elements exist: documented knowledge elements, experience elements issued from interview with experts and know-how elements issued from particular activity [2].

REX is divided into three stages which are:

1. Needs analysis and the identification of knowledge sources (specification and dimensioning of the future knowledge management system): It is necessary to take into account the needs of the organization; its objectives are to construct a representation of the existing situation by modeling the main activities of the concerned field. This modeling makes it possible to identify the flows of knowledge between these activities. It is, then, necessary to size the future knowledge management

system by analyzing the knowledge sources of organization (list of documentary holdings…) and then estimating the knowledge elements that can be produced from these sources. It is also necessary to identify the specialists in the field and to identify the valorization actions to be carried out to facilitate the capitalization and transfer of knowledge [2].

2. Construction and storage of knowledge elements in the form of a database or a corporate memory.
3. The establishment and operation of the knowledge management system is created [2].

The REX can respond to four different and complementary objectives: valorization of experience, knowledge extraction, creation of a corporate memory, and mutualization of experiences.

5 Implementation of REX Method

In this section, we apply the chosen REX method on an example of the medical field. This field has wide application in different organizations such as hospitals.

The basic principle of this method is based on the construction of the experience elements (EE) and the experiments memory.

We can divide our work into two parts: the first one is used for the collection and indexing of experiments and the second one is for access and exploitation of knowledge. We used an ontology of diseases that already exists because our aim is the application of the method REX.

5.1 Collection and Indexing of Experiences

This part deals with the reports recording of the completed experiments in the form of erased access by indexing files. There are several forms of knowledge representation such as: logical formalism, conceptual graph, frame… etc. in our work we used RDF form to createfiles, this means the creation of files in RDF form which will contain keywords that have semantic links to the content of the recorded experiments.

We mean by an RDF (Resource Description Framework) [15] file a graph template designed to describe Web resources and their metadata. These later are described in a formal way in order to allow automatic processing of such descriptions. RDF is the basic language of the semantic Web, and it is associated with a syntax, whose purpose is to allow a community of users to share the same meta data for the shared resources by annotating unstructured documents and serving them as an interface for structured applications and documents (e.g. databases, GED, etc.). RDF allows for a certain interoperability between applications exchanging informal and unstructured information on the Web.

Index Example. See (Fig. 2).

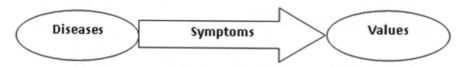

Fig. 2. Example of index

Example: Source Code

– Creating the query: This piece of code is for the creation of a request

```
String key = search_field.getText();
String Query = " SELECT ?maladie WHERE { ?y
<http://www.w3.org/2001/maladie-rdf/3.0#symptome> \"" +
key +"\" . \n ?y <http://www.w3.org/2001/maladie-
rdf/3.0#name> ?maladie . }";TextBox1.setText(Query);
```

– Opening an RDF file: *This* code fragment shows us how to open an RDF file

```
InputStream in = new FileInputStream(new File("D:/vc-db-
1.rdf"));
Create the model of enthology
Model mod-
el=ModelFactory.createMemModelMaker().createDefaultModel(
) ;
model.read(in,null);
System.out.println(model);
in.close();
```

– Execute the query: This code fragment shows us how to execute a query that we create

```
String queryString = TextBox1.getText()
com.hp.hpl.jena.query.Query
q = QueryFactory.create(queryString);
QueryExecution qe = QueryExecutionFactory.create(q, mod-
el);
ResultSet results = qe.execSelect();
```

- Get the result in xml: This code fragment shows us how to import the result and convert it to an XML file

```
TextBox2.setText("");
if (results.hasNext())
{

String s = ResultSetFormatter.asXMLString(results);
TextBox2.setText(s);
String xmlRecords = TextBox2.getText();}
```

- Extract the keywords from the XML file: This code fragment shows us how to extract the keyword from the XML result file

```
DocumentBuilder db = null;
try {
db = DocumentBuilderFacto-
ry.newInstance().newDocumentBuilder();
InputSource is = new InputSource();
is.setCharacterStream(new StringReader(xmlRecords));
Document doc = db.parse(is);
NodeList nodes = doc.getElementsByTagName("result");
String values = "";
for (int i = 0; i < nodes.getLength(); i++)
{
Element element = (Element) nodes.item(i);
values += element.getTextContent().trim() + "\n";
}
suggs.setText(values);
}
```

5.2   Experiments Access and Exploitation

This sub-section deals with the access to the accounts of the stored experiments and uses a semantic search. To simplify the realization, all by keeping the paradigm of the Rex method, the reports will contain observations by researchers on diseases:

- The report is saved in a text file with the subject name of the study topic.
- The index file (RDF) contains keywords that have direct links or semantic resemblance in the description of the disease (symptoms).

- Search: To do this step we have chosen SPARQL [16] which is a query language and protocol that allows us to search, add, modify, or delete RDF data available over the Internet. Its name is an acronym of "SPARQL Protocol and RDF Query Language".
- Open the file that corresponds more to the keywords added to the search.

  Below are some printed screens of our example implementation (Figs. 3 and 4).

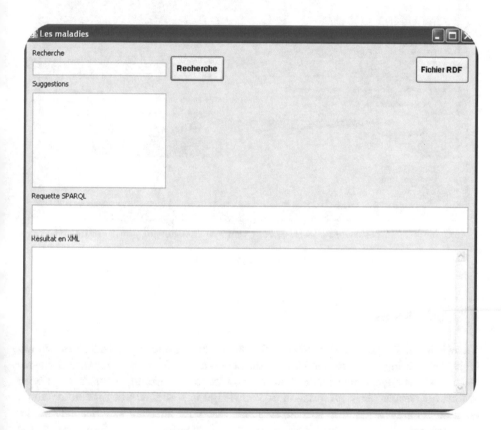

Fig. 3. First windows of the interface

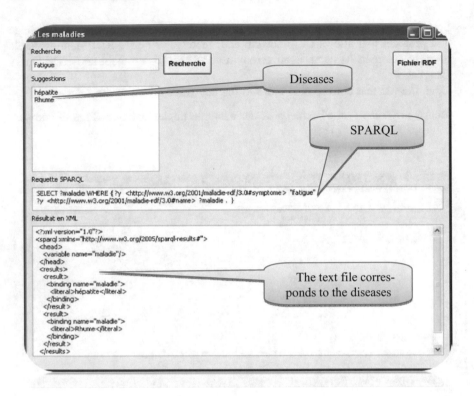

Fig. 4. Result windows

6 Conclusions

Knowledge management is a process that encompasses the capitalization of knowledge, the sharing and appropriation of that knowledge. Several techniques for capitalizing knowledge which inherent most knowledge engineering methods have been defined. However, the appropriation of knowledge is still a subject for further study. because its phase requires special attention since its success will depend on the effectiveness of the organizational learning and therefore the performance of the company in part. The work presented in this paper concerns the capitalization of the knowledge of the corporate memory by using the REX method. We have apply this on an example of the medical field. As future work, we wish to find a solution to the problem of the appropriation of knowledge.

References

1. Van Heijst, G., Van der Spek, R., Kruizinga, E.: Organizing corporate memories. In: Proceedings of KAW 1996, Banff, Canada, pp. 42.1–42.17 (1996)

2. Malvache, P., Prieur, P.: Mastering corporate experience with the REX method. In: Proceedings of ISMICK 1993, Compiègne, octobre 1993
3. Lehner, F., Maier, R., Klosa, O.: Organisational memory systems application of advances database and network technologies in organizations, Departments of Business Informatics III University of Regensburg, Paper No. 19, February 1998
4. Hedberg, B.: How organizations learn and unlearn. In: Nystrom, P., Starbuck, W.H. (eds.) Handbook of Organizational Design, vol. 1. Cambridge University Press, London (1981)
5. Pomian, J.: Mémoire d'entreprise, techniques et outils de la gestion du savoir. Ed. Sapientia, 1996. Prognosis and Diagnosis in the PROTEUS e-maintenance platform, IEEE Mechatronics & Robotics, E-maintenance Special Session, Aachen, Allemagne (2004)
6. Dieng, R., Corby, O., Giboin, A., Ribière, M.: Methods and tools for corporate knowledge management. Int. J. Hum.-Comput. Stud. **51**, 567–598 (1999). Academic Press
7. Baumard, P.: Tacit knowledge in organisations, organisations déconcertées: La gestion stratégique de la connaissances. Sage Publication, London (1996)
8. Nonaka, I., Takeuchi, H.: The knowledge-creating company, trad. fr., La connaissance créatrice. La dynamique de l'entreprise apprenante, Bruxelles, Paris, De Boeck université (1997)
9. Grundstein, M., Barthès, J.P.: An industrial view of the process of capitalizing. In: Actes du Congrès ISMICK 1996, Rotterdam, Pays Bas, pp. 258–264 (1996)
10. Johnson, L.: Mental Models Towards a Cognitive Science of Language Inference, and Consciousness. Cambridge University Press, Cambridge (1983)
11. Barthès, J.P.: Capitalisation et transfert des connaissances au sein des PMI. Journée Mémoire d'Entreprise. Europôle Méditerranéen de l'Arbois. Les Milles, 27 mars 1997
12. Grundstein, M.: Management des connaissances de l'entreprise: problématique, axe de progrès, orientations (2000)
13. Matta, N., Ermine, J.L., Aubertin, G., Trivin, J.Y.: Knowledge capitalization with a knowledge engineering approach: the MASK method. In: Proceedings of IJCAI 2001 Workshop on Knowledge Management and Organizational Memory (2001)
14. Grundstein, M.: Knowledge engineering within the company: an approach to constructing and capitalizing the knowledge assets of the company. In: Proceedings of Third Annual Symposium of the International Association of Knowledge Engineers, IAKE 1992, Washington, D.C. (1992)
15. RDF primer 2004. http://www.w3.org/TR/rdf-primer/
16. DuCharme, B.: Learning SPARQL. O'Reilly Media, Sebastopol (2011)

Solar Power Monitoring System "SunMieru"

Alireza Ahrary[1(✉)], Masayoshi Inada[2], and Yoshitaka Yamashita[3]

[1] Department of Computer and Information Sciences, SOJO University, Kumamoto, Japan
ahrary@cis.sojo-u.ac.jp
[2] FusionTech Inc., Kumamoto, Japan
inada@fusiontech.jp
[3] Sanwa Hi-Tech Co., Ltd., Kumamoto, Japan
yoshitaka_y@sanwa-hitech.com

Abstract. This paper proposes a monitoring approach based on single string real-time measurements. An electronic board was developed able to sense string voltage and string current along with open circuit voltage and short circuit current of a selected solar panel belonging to the string. Acquired information allowed real time monitoring of the power actually produced by each string. At the same time, information gained from the monitored solar panel allowed the definition of a production target which is dynamically compared with the actual value. Experiments performed on a medium size solar field allowed the evaluation of energy losses attributable to underperforming strings. The conversion of energy losses into money losses can be adopted to quantify the revenues of fault fixing.

Keywords: Photovoltaic · Solar monitoring · Real-time monitoring

1 Introduction

Photovoltaic (PV) is one of the technologies with the greatest future projection within the framework of renewable energies. Its numerous advantages, such as simple installation, high reliability, zero fuel costs, very low maintenance costs, and the lack of noise due to the absence of moving parts [1], have resulted in a high growth rate.

Nowadays PV represents the third-largest source of renewable energy after hydro and wind [2]. Since 2010, the world has added more solar photovoltaic capacity than in the previous four decades. Market researchers [3, 4] forecast another year of solar growth for 2016, when new installations are expected to reach 69 GW.

Optimizing the production process and the reliability of photovoltaic systems requires plants to increase production, as well as the lifetime and availability of all of their elements, while reducing operating and maintenance costs [5–8]. Since solar installed capacity is growing continuously, a small percentage improvement would imply significant net progress and, in general terms, a reduction in these facilities costs. Such systems cannot work efficiently if operations are not automated [9, 10].

Some of the advantages of solar monitoring system are; letting businesses and home owners to get a real-time readout of their solar panels with the associated economic benefit of making the best trade-off between switching between electrical and solar

© Springer International Publishing AG 2018
I. Czarnowski et al. (eds.), *Intelligent Decision Technologies 2017*,
Smart Innovation, Systems and Technologies 73, DOI 10.1007/978-3-319-59424-8_20

supply. Second is rapid problem identification and preemptive resilience to failure allowing qualified service technicians to quickly fix the problem. Third is self-repairing of the solar system through automated software when possible.

For this purpose, a new solar power monitoring system, named SunMieru, developed and installed in a PV Power Plant, is presented in this paper. The system developed here provides a detailed comprehensive real-time supervision of the performance of solar power components and detecting the output power of the system and letting owners to monitoring system on-line.

2 IoA Cloud-Based System Construction

We designed a system to collect the various information of the PV solar panel in the real-time environment through Machine to Machine (M2M) interface automatically. M2M interface is a microcomputer with various I/O to transmit the measured data to cloud system through the 3G network. Some related technologies in order to provide the better solutions in this field are described in follow.

2.1 Sensor Networks

The use of different specific-application designed sensor in a network array connected to a high-speed network, allow information exchange among different business partners. This sensor network is the core of multiple application layers that will be built over the information provided.

2.2 Cloud Computing

Many current applications, from storage to Software as a Service (ASA) are using this common affordable platform in order to reduce their hardware expenses. The information retrieved from the sensor networks could be processed or stored in the cloud for future purposes.

2.3 Control Area Network (CAN)

The data generated by the different agriculture machines, e.g. yield of pesticides, production, etc., can be stored in the cloud as well.

The structure of the M2M interface for proposed PV solar monitoring system is illustrated as shown in Fig. 1. The information from measuring sensors are collected in main board and transfer through the specially designed 3G shield to the cloud database.

The specification of specially designed M2M interface is shown in Table 1. The analog or digital signals can be transformed by M2M interface. And these signals will be transformed into the cloud database by 3G network.

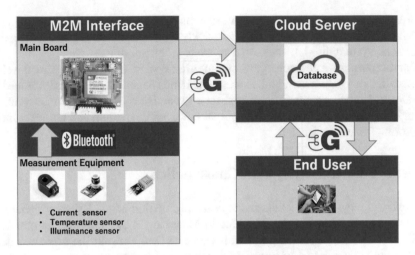

Fig. 1. System construction of "SunMieru"

Table 1. Specification of specially designed M2M interface

Item	Specification
Microcontroller	ATmega2560
Operating voltage	5 V
Input voltage	7–12 V
Digital I/O	54
Analog I/O	16
Flash memory	256 KB
Network	3G

3   User Interface

A system for visualizing the various information managements was developed. This system visualizes the data requested to the server, which is dedicated to obtain the information related to the sensor characteristics of the PV solar monitoring in real time. The information is requested and visualize in real time. Moreover, the system is design in order to correctly visualize the requested data in a legible design. The main HTML page is the client one, it is possible to access the M2M interface from integrated HTML pages. It is important to mention that coding pages contain the information to perform the sensor measurement requests, the sensor information is presented through the cloud website, and the measurements are accumulated using cloud services.

The end user access to the main page after logged from top page. The main page consists of sign up for new monitoring system, registration, GPS information of solar panel, history, graph and online display parts.

Example of the user interface is shown in Fig. 2, and the diagram of specially designed website is illustrated in Fig. 3.

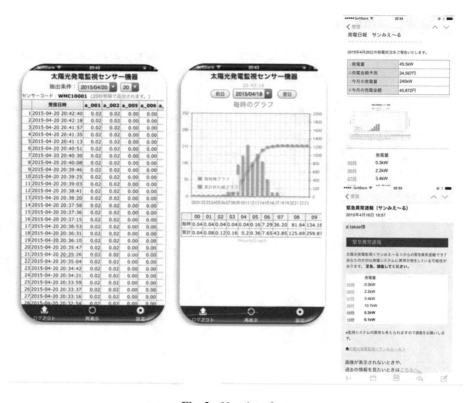

Fig. 2. User interface

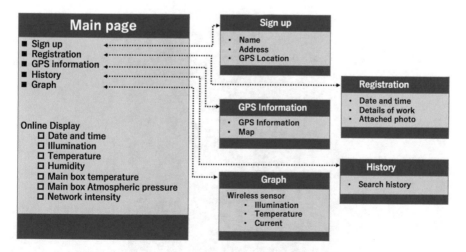

Fig. 3. Diagram of specially designed website

4   Experimental Results

The overview of proposed system is illustrated in Fig. 4. Solar power monitoring system, "SunMieru", which specially designed in this project is shown in Fig. 5.

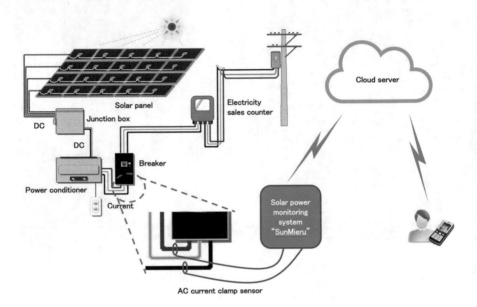

Fig. 4. Overview of proposed system

Fig. 5. Solar power monitoring system, "SunMieru"

Figure 6 shows the daily output pattern of PV systems by using proposed system. the output power of PV system is analyzed as the highest value from 12 PM to 3 PM, by using the saving data function of proposed system.

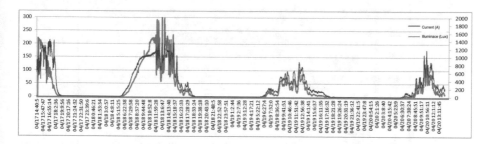

Fig. 6. Daily output pattern of PV systems by using proposed system

Figure 7 shows the output patterns of PV systems in different weather condition, such as clear, cloudy and rainy. Figures 8, 9 and 10 illustrate the relation between current vs. illuminance, current vs. temperature and current vs. atmospheric pressure for PV system by using proposed system.

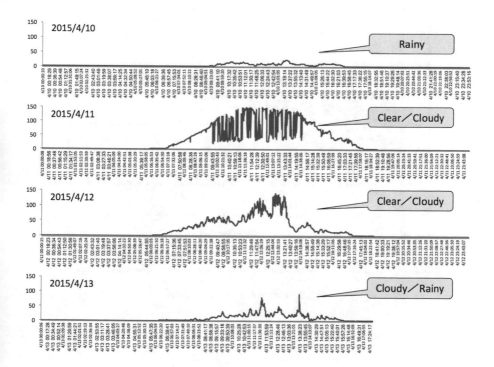

Fig. 7. Output patterns of PV systems in different weather condition

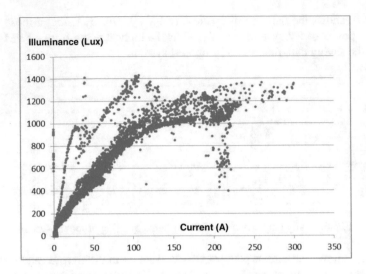

Fig. 8. Current vs. Illuminance

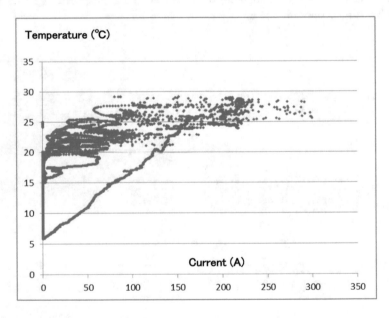

Fig. 9. Current vs. Temperature

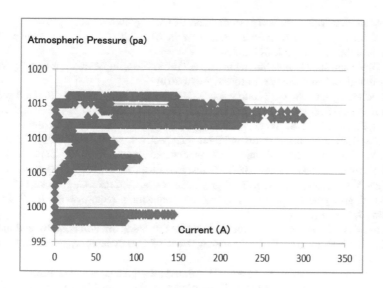

Fig. 10. Current vs. Atmospheric Pressure

Finally, it is confirmed that, predicting the output pattern of PV system through the proposed system can be effectively performed.

5   Conclusions

A new system has been implemented and set up in order to achieve the detailed and comprehensive supervision of all components of a Utility-Scale PV Power Plant connected to the grid. The system, with the incorporation of additional sensors (wired and wireless) and programmable controllers, monitors the production at specific points distributed around a PV plant's area. The system can be configured with different sampling times, which renders it highly versatile. The processing of all of the information provides real-time knowledge of the performance of all PV installation components and the presence of any failure in their operation. Due to the higher capacity of the cRIO programmable controllers selected, the PQ of the generated electrical signal can also be analyzed. The monitoring and supervising applications, which were also developed specifically to be integrated into the PV-on time system, enables complete real-time monitoring of the plant.

Acknowledgement.   This work was supported by the social and system-related industry commercialization support funded by Kumamoto prefecture.

References

1. Chine, W., Mellit, A., Pavan, A.M., Kalogirou, S.: Fault detection method for grid-connected photovoltaic plants. Renewable Energy **66**, 99–110 (2014)

2. European Photovoltaic Industry Association-EPIA. http://www.epia.org/home
3. Global Solar Installations to Reach 64.7 GW in 2016. http://solarindustrymag.com/global-solar-installations-to-reach-647-gw-in-2016
4. IHS Forecasts Global Solar Market to Top 69GW in 2016. http://www.pv-tech.org/news/ihs-forecasts-global-solar-market-to-top-69gw-in-2016
5. Petrone, G., Spagnuolo, G., Teodorescu, R., Veerachary, M., Vitelli, M.: Reliability issues in photovoltaic power processing systems. IEEE Trans. Ind. Electron. **55**, 2569–2580 (2008)
6. Zhang, P., Li, W., Li, S., Wang, Y., Xiao, W.: Reliability assessment of photovoltaic power systems: Review of current status and future perspectives. Appl. Energy **104**, 822–833 (2013)
7. Jain, N., Singh, S.N., Srivastava, S.C.: A generalized approach for DG planning and viability analysis under market scenario. IEEE Trans. Ind. Electron. **60**, 5075–5085 (2013)
8. Gungor, V.C., Sahin, D., Kocak, T., Ergut, S., Buccella, C., Cecati, C., Hancke, G.P.: A survey on smart grid potential applications and communication requirements. IEEE Trans. Ind. Inform. **9**, 28–42 (2013)
9. Meliones, A., Apostolacos, S., Nouvaki, A.: A web-based three-tier control and monitoring application for integrated facility management of photovoltaic systems. Appl. Comput. Inform. **10**(1–2), 14–37 (2014)
10. Koohi-Kamali, S., Rahim, N., Mokhlis, H.: Smart power management algorithm in microgrid consisting of photovoltaic, diesel, and battery storage plants considering variations in sunlight, temperature, and load. Energy Convers. Manag. **84**, 562–582 (2014)

Eye Movement Data Processing
and Analysis

Visual Analysis of Eye Movement Data with Fixation Distance Plots

Michael Burch[(✉)]

VISUS, University of Stuttgart, 70569 Stuttgart, Germany
`michael.burch@visus.uni-stuttgart.de`

Abstract. Eye tracking has become an increasingly important technology in many fields of research like marketing, psychology, human-computer interaction, and also in visualization. Understanding the eye movements of people while solving a given task can be of great support to improve a visual stimulus. The challenging problem with this kind of spatio-temporal data is the difficulty to provide a useful visualization that can provide an overview about the fixations with their durations and sequential order, the saccades with their orientations and lengths, but also the distances of several fixations in space. Traditional visualizations like gaze plots - showing the stimulus in its original form overplotted with the scan paths - typically produce vast amounts of visual clutter and make a visual exploration of the eye movement data a difficult task. In this paper we introduce the fixation distance plots that place the fixation sequences to a horizontal line of color coded and differently thick circles while showing additional saccadic information. Moreover, the user can apply distance thresholds that indicate if fixations are within a certain distance allowing to get an impression about the spatial stimulus information. We illustrate the usefulness of the approach by applying it to eye movement data from a formerly conducted eye tracking experiment investigating route finding tasks in public transport maps.

1 Introduction

Eye tracking devices steadily improve due to the progress in hardware technology [11,15,27]. This means that the recorded spatio-temporal data gets bigger and bigger demanding for novel algorithmically, visually, and perceptually scalable visualizations [2] and visual analytics techniques [1,3].

For example, traditional gaze plots [10,12,13,18,20] or scan path visualizations [21] produce vast amounts of visual clutter if the number of fixations in an eye movement trajectory gets really large. Consequently, it becomes difficult to judge if there are certain visual scanning patterns for at least one person. Visual attention maps [4,23,26], on the other hand, give an overview about the visual attention paid to a stimulus, but, negatively, the eye movement data is aggregated over time, space, and study participants or the average times of the temporal behavior is incorporated in the visual attention map design [6].

If more than one scan path has to be inspected visually, i.e., for example compared to others, we need a clutter-free and scalable visualization variant to get

© Springer International Publishing AG 2018
I. Czarnowski et al. (eds.), *Intelligent Decision Technologies 2017*,
Smart Innovation, Systems and Technologies 73, DOI 10.1007/978-3-319-59424-8_21

an overview about visual patterns that can be mapped to data patterns which finally, lead to insight detection, conclusion drawing, or hypothesis building [16]. Moreover, the dynamics of the data plays a crucial role, in particular, if comparison tasks have to be answered in different time intervals and on different levels of temporal or spatial granularity. Solving such tasks reliably is challenging, even impossible, with traditional eye tracking data visualizations like visual attention maps or gaze plots.

Moreover, if additional attributes come into play, that should also be visually reflected in a diagram, we need a visually scalable visualization technique that is free of visual clutter [22]. We identify the restriction to the spatial dimension of the stimulus as the major problem in a corresponding visual design for eye movement data since the freedom and flexibility to layout data elements and to visually encode them is typically lost if the stimulus content has to be incorporated as a spatially proportional one-to-one visual encoding.

In this paper we describe the fixation distance plots that map scan paths to horizontal lines with circles of different sizes and colors to indicate fixation durations. Moreover, the saccades are displayed as proportionally long lines preserving their original orientation combined with the distance circles. To visually explore the spatial (Euclidian) distances between fixations we use color coded parallel distance lines which are aligned with the respective start and end points of the fixations in a pairwise manner. Interaction techniques are applicable to filter the distance plots by selecting a distance threshold value or an interval. Moreover, data can be aggregated in the time dimension and the user can algorithmically search for study participants with characteristic distance features.

We illustrate the usefulness of our eye movement data visualization technique by means of an application example investigating data from a formerly conducted eye tracking study asking route finding tasks in public transport maps [5,19].

2 Related Work

There are various visualization techniques for eye movement data as surveyed by Blascheck et al. [2]. Also visual analytics approaches have been investigated to explore such spatio-temporal data [1,17], but mostly, they only show one or two aspects of the data without getting an overview about the proximity of fixations in a sequence. This is an important information since it provides insights about the scanning behavior based on certain well-defined small regions or scanning strategies that are far apart in a visual stimulus. This information is oftentimes lost if displayed together with the temporal information and the scan paths of several study participants.

For example, in visual attention maps [4,6,23,26] this information can be derived from inspecting the visual stimulus, but due to the aggregated and overplotted eye movement data, an observer cannot explore the individual scan paths together with the fixation proximities to understand applied visual task solution strategies among the participants [7]. Gaze plots [10,12,13,18,20] show the temporal behavior directly placed on top of the stimulus but visual clutter [22] will be the result due to overplotting and line crossings.

More recent techniques like parallel scan paths [21] are a way to get rid off the clutter by displaying each eye movement trajectory separately, but also in this approach the information about distances is lost. Color bands [8] are useful to visually encode scan paths to color coded bands of varying shapes supporting the identification of time-varying visual attention patterns. The strengths of the human's visual system [14,24,25] are exploited for comparison tasks which is difficult to do with standard fixation point-based approaches. But as a negative consequence of color bands there is no way to provide the distance information about the fixations, for example, with the goal to detect fixation clusters. The saccade plots [9] reflect fixation clusters as curved arcs placed vertically or horizontally outside a corresponding visual attention map, but the arcs only show the movement in either x- or y- direction and the sequence of fixations is typically lost without interaction techniques.

In this paper we describe a technique with which the eye movement trajectory is visually encoded as a sequence of differently large circles and different distances depending on the occurrence over time and the fixation durations. Moreover, the saccadic information is displayed on top of each of the circles to reflect the varying eye movement directions and saccade lengths. Finally, as a major contribution, we integrate fixation distance plots as a visual means to explore the eye movements based on fixation proximities which is provided as an overview-based representation below the scan path visualization aligned with the fixations. Color coded parallel lines describe the fixation distance patterns which support the comparison of visual scanning strategies between participants but also between stimuli.

3 Fixation Distance Plots

In this section we describe the fixation distance plots and illustrate how eye movement data is visually encoded. Then we describe how spatial distances are computed and how they are visually represented in a visually scalable way.

3.1 Eye Movement Data and Distances

We model eye movement data, i.e., $m \in \mathbb{N}$ scan paths consisting of fixations and saccades attached with additional data attributes as

$$S_i := \{p_{i,1}, \ldots, p_{i,n_i}\},\ 1 \leq i \leq m \in \mathbb{N}$$

where

$$p_{i,j} \in X \times Y \times \mathbb{N}.$$

The display space is expressed by the variables X and Y, i.e., the horizontal and vertical dimensions. $\mathbb{N}$ models the fixation duration to a point $p_{i,j}$, i.e., the time a study participant is fixating point $p_{i,j}$ until he moves to another point. The variable m expresses the number of study participants, i.e., people from which eye movement scan paths were recorded by asking a certain (same) task in a certain (same) stimulus. The spatial distance in our approach is obtained by computing pairwise Euclidian distances between all fixation points.

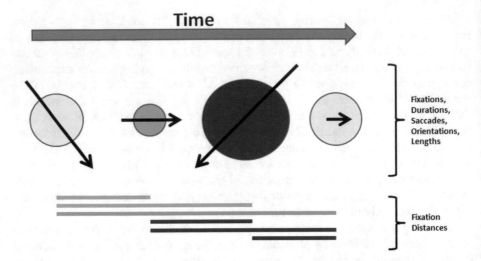

Fig. 1. A fixation distance plot showing one scan path in which the fixation duration is mapped to circle size and the saccades to lines with lengths and orientations. The pairwise distances are depicted as colored parallel lines below the fixation sequence.

3.2 Visual Encoding of Scan Paths

Figure 1 depicts a scenario in which an individual scan path is represented consisting of only 4 fixations for illustrative purposes. As we can see, the scan path is plotted to a one-dimensional horizontal line with the time axis pointing from left to right. Each fixation is placed at the corresponding point in time it occurs in the sequence. The duration is encoded in the circle size as well as in the color coding. It may be noted that two different metrics can be displayed here, in the size and in the color, as well as some more by using different shapes or even texture. The saccades between two fixations are mapped to black colored straight lines of varying lengths and orientations, proportionally reflecting the saccadic information in the eye movement data. Also here, the shapes of the lines and the color coding might serve as useful visual features for further data attributes, i.e., the circles are based on a glyph-based design combining several data variables into several visual variables in a single representation.

3.3 Distance Plots

The lower part of the plot in Fig. 1 visualizes the pairwise distances of the fixations aligning them with the circle centers indicating those fixations. The color coding reflects the distance values, i.e., the more red the color the closer are two fixations in the displayed stimulus. This approach allows to display the distances in a compressed and aggregated way as a visually scalable technique. Moreover, filtering and clustering can be applied directly on the plot, supporting an ordering of the color coded distance lines and a hierarchical structuring of the fixation sequence into hierarchical subsequences.

3.4   Interaction Techniques

We provide several interaction techniques with which the data can be visually transformed, i.e., different perspectives on the data can be generated.

- **Filtering:** Since the number of pairwise fixation distances can become large we support the filtering for only certain parallel distance lines. Either a threshold value can be selected or value intervals leading to a distance-based filtered plot. For the scan paths we support a filtering for fixation durations, time intervals, saccade orientations and lengths. Those can also be applied in combination while the filtered out data elements can be removed completely or only grayed out for preserving the context information.
- **Clustering:** The parallel distance lines are naively ordered and vertically stacked by the occurrences of start and end fixations. A clustering based on the distance values can provide a better grouping of identical or similar distance lines. The distances can also be used to group the eye movement sequence into subsequences that reflect neighbored fixation regions.
- **Distance features:** The generated pairwise distances can also be compared pairwisely by computing Jaccard coefficients of the distance values. This feature can give hints about similarities among the distance values of a certain number of participants' scan paths.
- **Aggregation:** Visual scalability can become an issue if too many data elements have to be visually encoded. For this reason we support distance value aggregation as well as scan path aggregation to provide a more scalable visualization for the unaggregated data elements.

3.5   Visual Patterns

The data-to-visualization mapping reflects several visual patterns that, if visually identified, can be remapped to the data, serving as data exploration tool.

- **Fixation patterns:** The thicknesses and color codings of the circles indicate the fixation durations while the distances between the circle centers reflect the occurrence time and the density expresses the occurrence frequencies in certain time intervals.
- **Saccade patterns:** The differently oriented black lines indicate the saccade directions and the lengths over time. These patterns can be inspected to get an impression about the movements of the eye in certain time intervals.
- **Distance patterns:** Color coded lines describe the distances of pairwisely compared fixations. Distance clusters can be observed by looking at specific color coded blocks which indicate fixation clusters. The color coding visually encodes the distances and the variance depicts changes in distance values.
- **Participant patterns:** Taking into account several of the fixation distance plots can give insights about different or similar scanning strategies among several participants. An analyst can inspect the fixation, saccade, but also the distance patterns to build hypotheses with this overview-based visualization.

- **Stimuli/Task patterns:** Apart from comparing participants and their scanning behavior, we can also visually explore the scanning behavior for several stimuli, for example, if the same task is asked investigating the performances and visual task solution strategies of several participants. Another scenario is imaginable if several different tasks are asked in the same stimulus like in the Yarbus [27] experiment on 'The unexpected visitor'.

4 Application Scenario

We apply the fixation distance plots to eye movement data from an eye tracking study investigating route finding tasks in metro maps [19]. Color coded and gray scale maps were shown as stimuli in the study while 40 participants took part. Each participant had to find a route from a highlighted start to a highlighted destination station while the interchange points had to be mentioned which is some kind of partly think-aloud study combined with eye tracking.

Figure 2 shows a fixation distance plot for one participant for the metro map of Antwerp in Belgium and one route from a start to a destination station. The scan path consists of approximately 50 fixations which are not equally distributed but show several differently long temporal gaps (1, 2, 3, 4). The fixation durations vary over time as well as the saccades which seem to be longer in the beginning and the end (5, 6), but shorter in-between (7) for a certain time period. Also the orientations and directions of the saccades are mostly from top to bottom (8) and back again but less from left to right which indicates a more top-down or bottom-up scanning behavior. The fixation distances indicate lots of short distances between nearly all fixations, but also a few outliers (yellow color) (9) that reflects that the eye moves over longer distances. This effect can be a result of the cross checking behavior of the task solution, i.e., the eyes move back over longer distances to confirm or reject the found route in the metro map study.

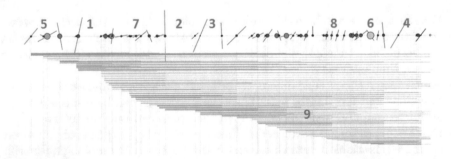

Fig. 2. A fixation distance plot for a route finding task in a public transport map study conducted by Netzel et al. [19]. The grayscale color coding indicates the fixation distance values of the pairwise comparisons. Blue color indicates short distances while yellow color reflects longer distances.

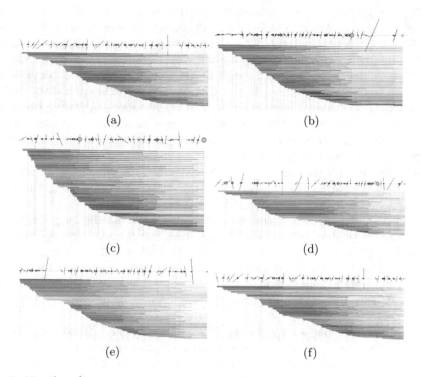

Fig. 3. Fixation distance plots for six eye tracking study participants. We can see different scanning strategies in each of the plots (a)–(f), but also some similarities. Black color expresses shorter distances while the distance increases with gray to white becoming distance lines.

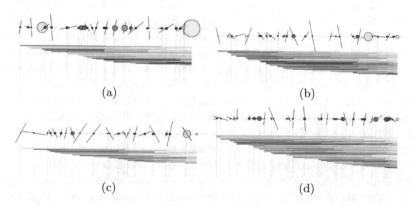

Fig. 4. Fixation distance plots for four study participants. The scan paths are much shorter (less fixations) than those in Fig. 3. Black color expresses shorter distances while the distance increases with gray to white becoming distance lines.

Figure 3 illustrates scan paths of six participants for the same route finding task in the map of Antwerp in Belgium. There are differences and similarities in the displayed fixation distance plots compared to those in Fig. 4. Hence, the ten participants can be classified into 2 groups by their visual scanning behavior. The 6 participants in Fig. 3 took much longer to answer the given route finding task than the 4 participants in Fig. 4. The black colored fixation distance lines indicate that people sometimes went back to points close to already visited ones reflecting a cross checking behavior. White areas between neighbored fixations in a sequence indicate that those are far away and hence, the eye jumps over a longer distance. For this specific scenario this means that people do not follow a metro line from start to end but instead, try to find the answer of the route finding task by subdividing it into subtasks. Those subtasks are given by the line parts between interchange points.

Moreover, the fixation durations are more or less equally long for most of the sequences (Fig. 3), but for the shorter ones, we can see some outliers, e.g., the larger yellow colored circles (Fig. 4(a), (b), and (c)) or the big gray one in Fig. 4(a). This indicates that people mostly do a rapid scan over the stimulus, while at some places they stop for a while. Looking at the corresponding stimulus, we can identify those stops as the interchange points or one of the highlighted stations.

The differently long scan paths can be explained by two different task solution routes for the displayed metro map stimulus. The longer ones can be mapped to routes following a metro line running to the north while the shorter ones take a solution route running to the south.

5   Limitations and Scalability

There are several limitations of our fixation distance plots. Those can be classified into algorithmic, visual, and perceptual issues which can be discussed as follows:

For computing all pairwise distances we have to compare all fixation point pairs and the Euclidian distances have to be stored. The longer the scan path the more comparisons have to be done, i.e., the algorithmic runtime grows in $O(n^2)$ if n is the number of fixations in a scan path. The visualization can grow in two dimensions which are the number of fixations or the time axis and as a consequence, the number of pairwise Euclidian distances. If too many fixation points are contained in a scan path, aggregation and filtering techniques should be applied to reduce the amount of displayed data and to obtain a visually scalable visualization. Perceptually, a data analyst has to explore the different color codings, circle sizes, and line orientations. Also the start and end points of the parallel running distance lines have to be perceptually aligned with the corresponding fixations. If the number of elements grows, these issues can become challenging perceptual problems in this visualization technique.

6   Conclusion and Future Work

In this paper we described the fixation distance plots as a way to provide an overview about fixation durations, fixation occurrences over time, and saccadic information like eye movement directions and lengths. As an additional enhancement we incorporate the pairwise fixation distances as color coded parallel lines aligned with the fixations. The additional view on the data visually reflects eye movement patterns, in particular, if fixation clusters are existing in the data, i.e., small stimulus regions in which the eye remains for a while but where the scan path consists of many fixations. Those visual patterns are useful as a tool to compare visual scanning strategies among several eye tracking study participants but also between several stimuli, either based on the same task or for comparing the scanning strategies for different tasks like in the work of Yarbus about 'the unexpected visitor' [27]. For future work, we plan to add more interactions as well as additional visual encodings of the eye movement data. Moreover, a user study might be of interest to evaluate the fixation distance plots.

Acknowledgements. The research in this paper was supported by DFG under grant WE 2836/6-1.

References

1. Andrienko, G.L., Andrienko, N.V., Burch, M., Weiskopf, D.: Visual analytics methodology for eye movement studies. IEEE Trans. Vis. Comput. Graph. **18**(12), 2889–2898 (2012)
2. Blascheck, T., Kurzhals, K., Raschke, M., Burch, M., Weiskopf, D., Ertl, T.: State-of-the-art of visualization for eye tracking data. In: EuroVis - STARs, pp. 63–82 (2014)
3. Blascheck, T., Burch, M., Raschke, M., Weiskopf, D.: Challenges and perspectives in big eye-movement data visual analytics. In: Proceedings of the 1st International Symposium on Big Data Visual Analytics, pp. 17–24 (2015)
4. Bojko, A.: Informative or misleading? Heatmaps deconstructed. In: Proceedings of Human-Computer Interaction, pp. 30–39. Springer, Heidelberg (2009)
5. Burch, M., Kurzhals, K., Weiskopf, D.: Visual task solution strategies in public transport maps. In: Proceedings of ET4S@GISCIENCE, pp. 32–36 (2014)
6. Burch, M.: Time-preserving visual attention maps. In: Proceedings of Conference on Intelligent Decision Technologies.. pp. 273–283 (2016)
7. Burch, M., Bott, F., Beck, F., Diehl, S.: Cartesian vs. radial - a comparative evaluation of two visualization tools. In: Proceedings of International Symposium on Advances in Visual Computing ISVC, pp. 151–160 (2008)
8. Burch, M., Kumar, A., Weiskopf, D., Mueller, K.: Color bands: visualizing dynamic eye movement data. In: Proceedings of 2nd Workshop on Eye Tracking and Visualization (2016)
9. Burch, M., Schmauder, H., Raschke, M., Weiskopf, D.: Saccade plots. In: Proceedings of the Symposium on Eye Tracking Research and Applications, ETRA, pp. 307–310 (2014)

10. Çöltekin, A., Fabrikant, S.I., Lacayo, M.: Exploring the efficiency of users' visual analytics strategies based on sequence analysis of eye movement recordings. Int. J. Geogr. Inf. Sci. **24**(10), 1559–1575 (2010)
11. Duchowski, A.T.: Eye Tracking Methodology - Theory and Practice. Springer, Heidelberg (2003)
12. Egusa, Y., Takaku, M., Terai, H., Saito, H., Kando, N., Miwa, M.: Visualization of user eye movements for search result pages. In: Proceedings of the 2nd International Workshop on Evaluating Information Access, EVIA (2008)
13. Goldberg, J.H., Helfman, J.: Visual scanpath representation. In: Proceedings of the Symposium on Eye-Tracking Research and Applications, ETRA, pp. 203–210 (2010)
14. Healey, C.G., Enns, J.T.: Attention and visual memory in visualization and computer graphics. IEEE Trans. Vis. Comput. Graph. **18**(7), 1170–1188 (2012)
15. Holmqvist, K., Nyström, M., Dewhurst, R., Jarodzka, H., van de Weijer, R.: Eye Tracking: A Comprehensive Guide to Methods and Measures. Oxford University Press, Oxford (2011)
16. Keim, D.A.: Solving problems with visual analytics: challenges and applications. In: Proceedings of Machine Learning and Knowledge Discovery in Databases - European Conference, pp. 5–6 (2012)
17. Kurzhals, K., Fisher, B.D., Burch, M., Weiskopf, D.: Evaluating visual analytics with eye tracking. In: Proceedings of the Fifth Workshop on Beyond Time and Errors: Novel Evaluation Methods for Visualization, BELIV, pp. 61–69 (2014)
18. Lankford, C.: Gazetracker: software designed to facilitate eye movement analysis. In: Proceedings of the Eye Tracking Research and Application Symposium, ETRA, pp. 51–55 (2000)
19. Netzel, R., Ohlhausen, B., Kurzhals, K., Woods, R., Burch, M., Weiskopf, D.: User performance and reading strategies for metro maps: an eye tracking study. Spat. Cogn. Comput. **17**(1–2), 39–64 (2017)
20. Räihä, K., Aula, A., Majaranta, P., Rantala, H., Koivunen, K.: Static visualization of temporal eye-tracking data. In: Proceedings of International Conference on Human-Computer Interaction - INTERACT, pp. 946–949 (2005)
21. Raschke, M., Herr, D., Blascheck, T., Ertl, T., Burch, M., Willmann, S., Schrauf, M.: A visual approach for scan path comparison. In: Proceedings of Eye Tracking Research and Applications, ETRA, pp. 135–142 (2014)
22. Rosenholtz, R., Li, Y., Mansfield, J., Jin, Z.: Feature congestion: a measure of display clutter. In: Proceedings of Conference on Human Factors in Computing Systems, pp. 761–770 (2005)
23. Spakov, O., Miniotas, D.: Visualization of eye gaze data using heat maps. Electron. Electr. Eng. **2**(74), 55–58 (2007)
24. Ware, C.: Information Visualization: Perception for Design. Morgan Kaufmann, San Francisco (2004)
25. Ware, C.: Visual Thinking: for Design. Morgan Kaufmann Series in Interactive Technologies, Paperback (2008)
26. Wooding, D.S.: Fixation maps: quantifying eye-movement traces. In: Proceedings of the Eye Tracking Research and Application Symposium, ETRA, pp. 31–36 (2002)
27. Yarbus, A.L.: Eye Movements and Vision. Plenum Press, New York (1967)

Which Symbols, Features, and Regions Are Visually Attended in Metro Maps?

Michael Burch[✉]

VISUS, University of Stuttgart, 70569 Stuttgart, Germany
`michael.burch@visus.uni-stuttgart.de`

Abstract. We conducted an eye tracking study with 40 participants to understand which visual objects like metro lines, stations, interchange points, specific symbols, or extra information like labels or legends are visually attended during a free examination question scenario. In this study we did not ask a specific question like a route finding task as in a previous eye tracking study, but we let the study participants freely inspect a displayed metro map system for 20 s each. We used 24 different metro maps with the same characteristics, but varied between color coded maps and gray scale ones. Understanding the visual scanning behavior of people while inspecting metro maps is an important, but also challenging task. But positively, the analysis of such eye movement data can support a map designer to produce better maps, in particular, to find out which regions are visually attended first or most frequently, maybe to guide the viewer. The visually attended regions and objects can be a key aspect in a metro map to make them easier and faster comprehensible and finally, useful for travellers in foreign and unknown cities all over the world. The major result from our eye tracking experiment is that the study participants significantly inspect symbols that pop out from the map like the airport signs or the map legends which belong to the key features in maps. Moreover, dense regions are more frequently attended than sparse ones. The visual attention maps of colored and gray scale maps look very similar.

1 Introduction

Metro maps are typically designed in a way to support travellers to find routes from a starting point to a destination in a foreign city [7]. In many cases a starting point for a journey is the airport, a larger train station, or any other prominent point in the system like a famous sight. Consequently, it is important to visually augment such characteristic points or regions by perceptually distinguishing graphical objects or signs which give a traveller the chance to read the metro map more rapidly.

Several map designs have been developed over the years, all focussing on making the finding of routes easier and more comprehensible, in particular, if the traveller is a layman, not using the maps very frequently as an every day task. Apart from the metro lines and the interchange points, typical map designs

© Springer International Publishing AG 2018
I. Czarnowski et al. (eds.), *Intelligent Decision Technologies 2017*,
Smart Innovation, Systems and Technologies 73, DOI 10.1007/978-3-319-59424-8_22

augment those line-based diagrams with additional symbols indicating tourist attractions like sights. It is questionable which graphical elements people visually attend if the only task is to have a look at a displayed map in order to freely explore it, i.e., if they have no specific task in mind. To understand the visual scanning strategies while freely examining maps we recorded people's eye movements [24] and analyzed the recorded data later on as also described by other eye tracking researchers [9,12].

In this paper we evaluate the recorded eye movement data of 40 participants being confronted with such a free examination question scenario. To reach our goal we show 24 different metro maps from public transport systems from cities all over the world. We additionally vary between color coded and gray scale maps (which might be cheaper to print) in an in-between study design and let people visually attend the maps for exactly 20 s.

As a major outcome of our eye tracking free examination question experiment we found that most of the people always visually attend the provided legends in metro maps. This is an important information to understand the map symbols and finally, to read the map and solve tasks with it. But people also had a look at the airport signs, but also other graphical elements like rivers or legends were inspected. Moreover, dense regions in the maps as well as long labels were visually attended. For the color coded metro lines people looked more on red color coded lines. As another major outcome we found out that the visual attention of color coded maps and gray scale maps is pretty similar which is a fascinating result since a between-subjects study design was used which separates people into two groups, those looking at color coded maps and those looking at gray scale maps avoiding learning effects.

2   Related Work

The design of metro maps is challenging since an uncluttered [22], aesthetically pleasing [4], readable [10], and intuitive [13] line-based diagram has to be computed or manually be designed [7]. Not only the metro lines with the stations, the color codings and textual labels, or the placement and trajectories of the stations and lines, i.e., the layout, are critical visual features, but also additional symbols that have the goal to enhance the map for the viewer.

The question arises how many of these symbols should be placed in a public transport map to not lead to visual clutter and a mass of additional information on the one hand, but to a not well-informed traveller on the other hand, if not enough symbols visually augment the map. In an eye tracking experiment, Netzel et al. [19] extended the work of Burch et al. [3,6] by recording the eye movements of study participants while they answered route finding tasks in public transport maps. They found that the geodesic path tendency [14] plays a crucial role in these tasks and that people tend to crosscheck their answers which is a typical visual scanning strategy for line-based diagrams in which the task can be subdivided into subtasks [5]. Moreover, people can be separated into two groups, i.e., either starting from the beginning of a route or from the end.

Although Netzel et al. found several insights when annotating [20] and analyzing the eye movement data, the visual attention of symbols or specific regions
has not been investigated in their experiment nor the pop out effects of several symbols, features, and regions [11,23]. There is an indication that people
sometimes have a look at the map legends and also at the metro lines' final destinations, but those results are rather task-driven and hence, possibly be required
to reliably solve the route finding task.

Tracking the eyes of people while inspecting metro maps is in particular
useful to evaluate the user attention [16], to understand the influence on the
readability of the public transport maps [2], or to distinguish between ambient
and focal attention [17]. Kiefer et al. [15] used eye tracking as a support system
while Ooms et al. [21] facilitated the exploration of spatio-temporal data. Also
visual interaction strategies can be analyzed with eye tracking techniques [8].
But although various applications of eye tracking exist, only a few take into
account the design and evaluation of metro maps. In this paper we investigate
the eye movements of people while doing a free examination task, in particular,
which symbols, features, and regions are visually attended. This might serve
as a pilot study for future follow-up eye tracking experiments investigating the
readability of metro maps.

3   Eye Tracking Study

We conducted an eye tracking study investigating how special symbols, features,
and regions integrated in the maps are visually attended. The study was designed
as an in-between subjects study design with the color coded and gray scale maps
dividing the participants into two equally large groups. In this paper we present
some results of this free examination scenario which were not published before.

3.1   Free Examination Question

The eye tracking study consisted of two parts, the first one in which route finding
tasks were asked and a second one, **the free examination scenario**. 24 metro
map stimuli were displayed in both, color coding and gray scale, one after the
other. The first part of the eye tracking study was already published by Netzel
et al. [19] while the free examination scenario data was not analyzed for visual
scanning strategies although the analysis of the data showed many insights,
probably supporting a map designer to enhance the map design. All the details
about the study design and the experimental setup are explained in the work of
Netzel et al. [19].

In the free examination scenario people were shown one static metro map in
either color coding or gray scale for exactly 20 s. The task was to just explore
the map without any further guidance. Consequently, people were not given a
detailed task like finding a route from a start station to a destination which lets
them the freedom to concentrate on specific regions, symbols, and features in
the map that pop out to some degree.

We used a Tobii T60 XL eye tracking system with a TFT screen resolution of 1920 × 1200 pixels. We used a regular 9-point calibration while the participants sat in front of the display at a distance of about 60 cm, given by the calibration feature of the eye tracking system. After the recording of the eye movements we checked the validation of the data by means of the integrated eye tracking function and based on that, we decided to use all of the recorded eye movements from all of the fourty participants.

Figure 1 shows a colored metro map of Barcelona in Spain. As we can see, the map contains lots of stations and lines, but is also enhanced by additional visual features and symbols like the airport sign or a map reading legend. Moreover, water is indicated as a blue region while sparse and dense regions can be found depending on the number of lines and stations in certain areas.

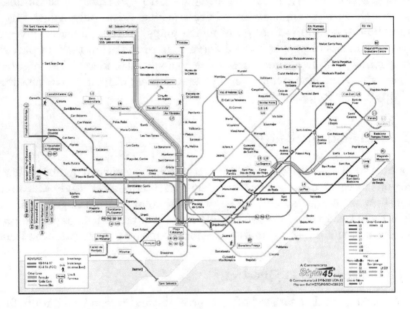

Fig. 1. The public transport map of Barcelona in Spain: There are specific regions, points, and symbols in the map which may have an impact on the understandability and applied reading strategies. But the question arises about which elements are attended at first, at last, and also most frequently.

3.2   Important Symbols and Regions

There is a list of specific symbols, features, and regions in a metro map worth investigating in an eye tracking study (see Fig. 1). Those will be described in the following:

– **Airport signs:** The airport is important since many passengers may arrive there planning a journey to a hotel close to the city center, and hence, often-times builds the starting point of a journey.

- **Legends:** If an unfamiliar map design is used, a view on the legends is necessary to understand the symbols and to read and interpret the map.
- **Rivers (Water):** Rivers, lakes, or the ocean can provide additional visual cues to the abstract and distorted map.
- **Metro lines:** The metro lines and their color coding can already drive the first impression of the viewer.
- **Metro stations:** Metro stations are a crucial information to support a viewer about important points to start or end a journey.
- **Interchange points:** If more complex journeys are done, one has to identify interchange stations, i.e., typically several metro lines have to be used.
- **Metro line ends:** Understanding the direction in which a metro line is operated can be obtained by looking at the final destination which is typically indicated at the ends of the corresponding metro lines.
- **Sight symbols:** If tourists are travelling they are mostly interested in sights, i.e., additional information in form of symbols should augment the map.
- **Dense regions:** Larger cities like Paris, London, or Tokyo typically contain many dense regions, mainly in the city centers.
- **Sparse regions:** In the outside regions, i.e., in the corner of the displays, in parks, or in the ocean, there is only a minimum amount of information to be displayed.
- **Labels:** To correctly interpret the map and to match it with other information (from a travel book) one needs to read textual information like station or metro line labels.
- **Outliers/strange features:** Unexpected features in a designed map can lead to misinterpretations and should be avoided by the map designer.

3.3   Results

We analyze the recorded eye movement data for typical phenomena like which symbols, features, or regions are visually attended most frequently, which ones in the beginning, which ones towards the end, for how long, and finally, which ones are attended several times in the 20 s. The provided percentage values are averaged over the study participants and the displayed stimuli. Consequently, by the repetition of the experiment we can get a feeling about which of the analyzed aspects in a metro map are most likely to be looked at and also how often, how long, and at what time.

To evaluate the data we had a look at the provided videos from the Tobii T60 XL eye tracking system overlaid with the fixations. From this information we build a table containing the percentage values.

For the fixation frequency we record if a symbol, feature, or region has been attended at least once. This is then divided by the number of displayed maps containing that specific feature. For example, an airport symbol has been visually attended in 76.3% of the cases, i.e., if 100 maps with an airport sign were shown, in 76.3 maps the airport sign has been looked at. For the visual attention at the beginning and at the end, this works similar with the restriction that the first or last look plays a deciding role here. Again all displayed maps are taken into

account and the percentage value is computed. For the length we compute the
average fixation duration in seconds for each symbol, feature, or region and
average that over all displayed maps. If someone is fixating something several
times (with fixation gaps in between) we count the maximum number of those
fixations and average them over all displayed maps. For example, a legend might
be inspected for 3 s in the beginning and again for 5 s. Finally, another 4 s are
spent to look at the legend. This means it has been looked at for 12 s in total
and 3 times. All of those results are displayed in Table 1.

Table 1. Visual attention to specific symbols, features, or regions like attention frequency, fixation at begin or at end, length of fixation, or times of fixation. All values
are computed over all displayed metro maps and all study participants.

Symbols	Freq. (%)	Begin (%)	End (%)	Length (sec.)	Times (n)
Airports	76.3	11.7	6.5	1.3	5.2
Legends	93.9	23.1	8.9	4.3	7.6
Rivers (Water)	76.5	4.3	1.0	1.9	0.8
Metro lines	73.5	6.9	4.8	2.7	16.5
Metro stations	70.1	4.5	3.4	1.3	9.8
Interchange points	65.3	3.3	1.6	0.8	2.2
Metro line ends	56.2	0.2	0.4	0.5	1.1
Sight symbols	45.3	0.4	0.2	0.8	3.4
Dense regions	67.3	5.4	2.4	2.3	6.5
Sparse regions	64.9	2.4	2.3	0.9	3.4
Labels	66.8	1.1	2.3	1.6	7.4
Strange features	45.2	0.3	0.0	0.5	1.3

As a major outcome of Table 1 we can observe that the provided legends
have been inspected at least once in 93.9% of all displayed maps. They were
attended at the very beginning of the free examination task in 23.1% and in
the end in 8.9%. The average fixation length is 4.3 s, i.e., much of the 20 s is
used to have a closer look at the legends while they are visually attended several
times (7.6 times in average). This means people probably look at a legend, try
to understand it, apply it to the displayed map, and go back to it several times.
These results show that map legends seem to be important visual features for a
metro map, but also some other symbols are of importance.

For example, metro lines are inspected here although no specific route finding
task was asked. They were looked at 16.5 times in average meaning people jump
to and fro frequently between the lines. Also dense regions are visually attended
for 2.3 s in average which is a relatively large amount of time compared to the
20 s of total time for the free examination scenario. We can also see that rivers
or water in general is frequently looked at which is an important information

like the river Thames in London giving some extra context information. We can also see that strange features (like design errors) have not been looked at 100% of the times, i.e., the viewers only found less than 50% of those strange objects (like typos in the labels, incomplete legends, or wrong arrows).

Also text labels indicating the stations' or metro lines' names are frequently attended (66.8). They are more looked on at the end (2.3) of the free examination task than in the beginning (1.1). Labels are also looked at several times (7.4), but it may be noted that we counted here all labels that appear in a map (which are many). This is different to an airport sign, for example, since this only occurs once or twice in a map and hence, has a lower probability to be looked at and also not that often. This fact also indicates that airport signs seem to be very prominent in the maps compared to traditional text labels.

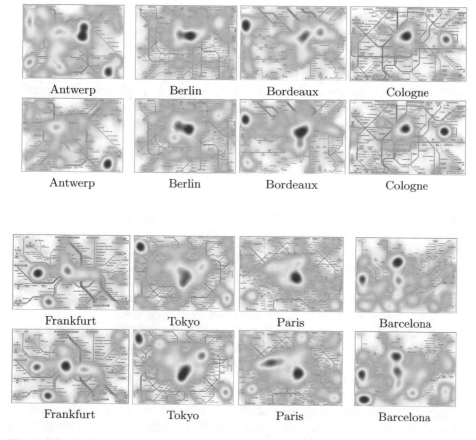

Fig. 2. The public transport maps of several cities in the world are differently visually attended. But, several graphical elements or regions in the stimuli all got more visual attention than others. We can find a similar attention behavior for both color coded (top row) as well as gray scale public transport maps (bottom row). We used an in-between subjects study design.

Also the center of the maps seems to be visually attended much more frequently. One reason for that is that dense regions (67.3) are typically occurring close to a map or city center, although the map design focusses on distorting the map in a way to reduce the densities but to still preserve the topological information, i.e., various stations and metro lines are moved away. This visual attention effect can be observed in nearly all displayed maps in this study.

We also compared colored and gray scale maps and found an interesting phenomenon. If we have a look at the visual attention maps for both scenarios of the free examination setting we can see that there are various similar hot spot regions (see Fig. 2). Although we used an in-between subjects study design, people frequently looked at similar symbols, features, and regions. Consequently, we might argue that color coding is not deciding for this free examination scenario as it is for a typical route finding task as found out by Netzel et al. [19].

4 Discussion

As described in the work of Yarbus [24] the task has a large impact on the visual scanning strategies, a phenomenon that we also found out in the route finding study part. Although we did not ask a specific task in the free examination scenario we are aware of the fact that a study participant might try to answer one or several tasks that are more based on the semantics of the displayed map and the fact that someone tries to interpret the map. Consequently, there is always one task for trying to understand the displayed map which might have an influence on the recorded eye movement data. But trying to understand the map is exactly the task that a map designer wishes to be performing well.

Another aspect here is the prior knowledge that someone might have if a map is shown. Probably, the study participant has already travelled to this city or is familiar with it. To explore this we also collected an information about which cities people already visited in their life and found out that there is not much overlap between already visited cities and the cities for which metro maps are displayed in the eye tracking study.

Also color deficiencies and text reading might be a confounding variable in this study which we tested by an Ishihara test and a Snellen chart. The gender of the participants might have an influence due to differently perceived aesthetics criteria (color codings, shapes, symbols and the like). We did not further analyze this effect, but it may be worth investigating in future scenarios.

Although we know that an improvement of the maps is questionable based on such a trivial experiment, we already obtained some feedback and interesting insights about which symbols are visually attended and which might be a hint for reading and understanding problems in a map. Consequently, these results may serve as pilot study for further follow-up experiments going in these directions. Also the complexity of the maps in terms of number of metro lines and stations might be considered as another independent variable in the study.

5   Conclusion and Future Work

In this paper we investigated a free examination question scenario in which public transport maps had to be inspected. Our study participants had 20 s of time to explore a displayed metro map stimulus. The intention of this study is to record and analyze the eye movements of people and to understand which symbols, features, or regions are more or less frequently attended and also which ones first and last. We also varied color coded and gray scale maps in an in-between subjects study design to find out if color has an influence on the visual attention. As a major result we found out that legends seem to be very important features in a map since those are attended very frequently, very often, and also very long. An explanation to that might be that a viewer needs such information to further be able to read the map, but he also has to jump back to the legend several times since several annotations might have to be looked up again. Also the metro lines, dense regions, or special symbols like the airport signs are frequently looked at. We also found out that the visual attention is similar for color coded as well as gray scale maps. For future work, we plan a follow-up experiment that takes into account the complexity of the maps but also the number of extra symbols which might augment the map but also produce more visual clutter if too many of them are displayed. Also the order of visual attention to specific symbols has not been analyzed in this paper, but is worth researching for future work. Moreover, the eye tracking study took place in a laboratory, but typically, metro maps are shown in a metro station, placed as a poster on a wall, becoming more to a real world scenario which might be investigated in the future by using eye tracking glasses making an analysis of the recorded data even more challenging [1,18].

Acknowledgements. The research in this paper was supported by DFG under grant WE 2836/6-1.

References

1. Blascheck, T., Burch, M., Raschke, M., Weiskopf, D.: Challenges and perspectives in big eye-movement data visual analytics. In: Proceedings of the 1st International Symposium on Big Data Visual Analytics (2015)
2. Brychtova, A., Çöltecin, A.: An empirical user study for measuring the influence of colour distance and font size in map reading using eye tracking. Cartographic J. **51**(4), 1–11 (2014)
3. Burch, M., Kurzhals, K., Weiskopf, D.: Visual task solution strategies in public transport maps. In: Proceedings of ET4S@GISCIENCE, pp. 32–36 (2014)
4. Burch, M.: The aesthetics of diagrams. In: Proceedings of the 6th International Conference on Information Visualization Theory and Applications, pp. 262–267 (2015)
5. Burch, M., Andrienko, G.L., Andrienko, N.V., Höferlin, M., Raschke, M., Weiskopf, D.: Visual task solution strategies in tree diagrams. In: Proceedings of IEEE Pacific Visualization Symposium, pp. 169–176 (2013)

6. Burch, M., Raschke, M., Blascheck, T., Kurzhals, K., Weiskopf, D.: How do people read metro maps? An eye tracking study. In: Proceedings of the 1st International Workshop on Schematic Mapping (Schematics) (2014)
7. Burch, M., Woods, R., Netzel, R., Weiskopf, D.: The challenges of designing metro maps. In: Proceedings of International Conference on Information Visualization Theory and Applications (2016)
8. Cöltecin, A., Fabrikant, S., Lacayo, M.: Exploring the efficiency of users' visual analytics strategies based on sequence analysis of eye movement recordings. Int. J. Geograph. Inf. Sci. **24**(10), 1559–1575 (2010)
9. Duchowski, A.T.: Eye Tracking Methodology - Theory and Practice. Springer, London (2003)
10. Garland, K., Beck, H.: Mr. Becks's Underground Map. Capital Transport, London (1994)
11. Healey, C.G., Enns, J.T.: Attention and visual memory in visualization and computer graphics. IEEE Trans. Vis. Comput. Graph. **18**(7), 1170–1188 (2012)
12. Holmqvist, K., Nyström, M., Dewhurst, R., Jarodzka, H., van de Weijer, R.: Eye Tracking: A Comprehensive Guide to Methods and Measures. Oxford University Press, Oxford (2011)
13. Horne, M.A.C.: Information design aspects of the London underground map (2012). http://www.megadyne.co.uk/UndMap.html
14. Huang, W., Eades, P., Hong, S.H.: A graph reading behavior: geodesic-path tendency. In: Proceedings of IEEE Pacific Visualization Symposium, pp. 137–144 (2009)
15. Kiefer, P., Giannopoulos, I., Kremer, D., Schlieder, C., Raubal, M.: Starting to get bored: an outdoor eye tracking study of tourists exploring a city panorama. In: Proceedings of the Symposium on Eye Tracking Research and Applications, pp. 315–318 (2014)
16. Kiefer, P., Giannopoulos, I., Raubal, M.: Using eye movements to recognize activities on cartographic maps. In: Proceedings of the 21st ACM SIGSPATIAL International Conference on Advances in Geographic Information Systems, pp. 488–491 (2014)
17. Krejtz, K., Duchowski, A.T., Cöltecin, A.: High-level gaze metrics from map viewing - charting ambient/focal visual attention. In: Proceedings of the 2nd International Workshop on Eye Tracking for Spatial Research, pp. 37–41 (2014)
18. Kurzhals, K., Fisher, B.D., Burch, M., Weiskopf, D.: Evaluating visual analytics with eye tracking. In: Proceedings of the Fifth Workshop on Beyond Time and Errors: Novel Evaluation Methods for Visualization, BELIV, pp. 61–69 (2014)
19. Netzel, R., Burch, M., Ohlhausen, B., Woods, R., Weiskopf, D.: User performance and reading strategies for metro maps: an eye tracking study. Special Issue on Eye Tracking for Spatial Research Spatial Cogn. Comput. Interdisc. J. **17**(1–2) (2016)
20. Netzel, R., Burch, M., Weiskopf, D.: Interactive scanpath-oriented annotation of fixations. In: Proceedings of the Ninth Biennial ACM Symposium on Eye Tracking Research and Applications, ETRA, pp. 183–187 (2016)
21. Ooms, K., Cöltecin, A., Maeyer, P.D., Dupont, L., Fabrikant, S.I., Incoul, A., der Haegen, L.V.: Combining user logging with eye tracking for interactive and dynamic applications. Behav. Res. Methods **47**(4), 977–993 (2015)
22. Rosenholtz, R., Li, Y., Mansfield, J., Jin, Z.: Feature congestion: a measure of display clutter. In: Proceedings of Conference on Human Factors in Computing Systems, pp. 761–770 (2005)
23. Ware, C.: Information Visualization: Perception for Design. Morgan Kaufmann, San Francisco (2004)
24. Yarbus, A.L.: Eye Movements and Vision. Plenum Press, New York (1967)

Gaussian Function Improves Gaze-Controlled Gaming

Cezary Biele[1]([✉]), Dominik Chrząstowski-Wachtel[1], Marek Młodożeniec[1],
Anna Niedzielska[1], Jarosław Kowalski[1], Paweł Kobyliński[1], Krzysztof Krejtz[1],
and Andrew T. Duchowski[2]

[1] National Information Processing Institute, Warsaw, Poland
cbiele@opi.org.pl
[2] School of Computing, Clemson University, Clemson, SC, USA

Abstract. Gaze as a gaming input modality poses interaction challenges, not the least of which is the well-known *Midas Touch* problem, when neutral visual scanning leads to unintentional action. This is one of the most difficult problems to overcome. We propose and test a novel method of addressing the Midas Touch problem by using Gaussian-based velocity attenuation based on the distance between gaze and the player position. Gameplay controlled by the new control method was rated highest and the most engaging among those tested, and was rated similarly to the mouse or keyboard in terms of joy of use and ease of navigation. We also showed empirically that this method facilitated visual scanning, without harming game performance compared to other gaze control methods and traditional input modalities. The novel method of game control constitutes a promising solution to problems associated with gaze-controlled human computer interaction.

Keywords: Eye tracking · HCI · Gaze controlled gaming · Midas Touch problem

1 Introduction

The video game industry continues to introduce novel devices dedicated to gaze-controlled gaming [19]. Despite the growing accessibility of eye-tracking devices in recent years, gaze still has not been popularized as an input modality in computer games. This lack of adoption is mostly due to limitations of gaze-controlled interaction with computer applications.

In most computer games, control is often indirect, where movement of a cursor or a character is controlled by moving a controller such as a mouse, and action is triggered by, e.g., the press of a mouse button. Another crucial component of a human computer interaction during game play is visual scanning of the screen. One of the main problems of gaze-controlled interaction is that such neutral eye scanning may be mistakenly interpreted as a meaningful gesture and lead to unintentional actions. This is known as the "Midas Touch" problem [8].

© Springer International Publishing AG 2018
I. Czarnowski et al. (eds.), *Intelligent Decision Technologies 2017*,
Smart Innovation, Systems and Technologies 73, DOI 10.1007/978-3-319-59424-8_23

We propose and test a novel method of addressing the Midas Touch problem. The novelty of the method stems from its use of the distance between the controlled character and the user's gaze as the input to a Gaussian-based function used to calculate the speed of character movement. This mechanism attenuates the character's movement when the user's gaze is far away from the character, i.e, when visually scanning the scene. Attenuation thus avoids causing unintentional character movement when the users' gaze is directed away from the character.

2   Background

Gaze-based human-computer interaction has been the subject of empirical research since the late 1980s. Numerous gaze-controlled computer applications have been tested, including various kinds of video games e.g., simple arcade games [3], platform adventure games [14], simulators [15], first person shooters [10], and massively multiplayer online role-playing games as well [7].

Consensus concerning interaction appears to be that gaze cannot simply replace the mouse as a pointing device. Introducing gaze control into a game often requires its redesign. The reason is that the mechanics of gaze and mouse are quite different. Mouse control functions in at least two modes: neutral mode (for pointing) and active mode, when the button is pressed (e.g., for dragging, selecting etc.). Producing such signals with gaze alone, without the help of an external device, is non-trivial and potentially error-prone. Various selection methods have been developed and tested: dwell-time, which is most common [1,5,21], blinking and winking [16,22], voluntary pupil size manipulation [4], and external triggers [23]. However, none is as quick and simple as pressing a button [9]. The second problem is precision: a cursor stands perfectly still indefinitely when its control device remains untouched, while the human eye is in constant motion. Moreover, due to technical limitations and the nature of gaze itself, eye-tracking measurement is burdened with error greater than that of any other input modality. This results in less precision and potential for targetting error.

Since there is no easy way to select objects with the eyes, gaze as input works well in games that require little or no "clicking" at all [3,15]. Krejtz et al. [11] have tested gaze control in such a game consisting of navigating a simple maze. Poorer performance and inferior gameplay enjoyment were likely caused, at least in part, by the Midas Touch effect: visual scanning with no intention to move was often interpreted as a signal to initiate involuntary movements of the game character. Inspired by visuo-spatial cueing, the proposed solution to this problem was to modify the character's locomotion mechanics with a cutoff function: movement would be triggered only by gaze located within a given radius of the characters position. Gaze fixations outside this radius would be interpreted as visual scanning and would not affect the game character.

Attentional cuing (e.g., by cueing at a location some distance away from the current gaze location) is also used to improve gaze-based human computer interaction. Cueing may be used to direct attention to certain parts of the screen as well as to remind the users where the active screen areas are located. For directing attention, however, research suggests that visuo-spatial cueing may not be effective (when viewing animation) [13] and that arrows acting as external indicators may be imprecise [2,12].

Instead of the cutoff function mentioned above the speed of the game character could be controlled via a normal distribution function. The input for such function would be the distance from the current gaze position to the game character. Evaluation of such a function is the main goal of the present study. We hypothesized that such a function would simultaneously facilitate effective game-play along with effective scanning of the visual field, increasing the user's gaming performance and subjective ratings of the game. We hypothesized that the presence of visual cues would also influence game performance and the game's subjective evaluation.

3 Empirical User Study

3.1 Participants

Twenty-four subjects (11 male and 13 female, aged $M = 26.8, SD = 5.43$) took part in the experiment voluntarily after signing a consent form. All participants reported they were familiar with computer games, and had good eyesight. Eye-tracking data for one participant was removed due to equipment failure during data recording in one of the trials.

3.2 Experimental Design: Dependent and Independent Variables

The user study employed a 5×2 mixed-factorial experimental design, with independent within-subjects factors of game control (at 5 levels) and visual cues (at 2 levels) as a between-subjects factor.

The 5 levels of game control included: keyboard, mouse, and three gaze control versions (for detailed description see Sect. 3.3) The 2 levels of visual cues indicated presence vs. absence of the arrows placed around the game character.

Dependent variables were:

1. *gaming performance:* indicated by the number of completed game tasks, included as an objective measure of gaming experience;
2. *gaming experience:* indicated by a Game Experience Questionnaire, which consisted of six items: enjoyment, naturalness, difficulty, engagement, game rating, and interaction rating, each evaluated on a 9-point Likert type scale. The lowest value of the scale was *definitely no* and highest was *definitely yes*. The questionnaire was adopted from Bednarik et al. [1];
3. *proportion of peripheral visual scanning:* indicated by the percentage of the fixations landing outside the center (500 px radius) area of the screen, included to measure visual attention distribution during gaming.

3.3   Experimental Procedure

During the experiment, participants played a simple arcade game *Skydiver* written in Python [18] and Pygame [20]. The goal of the game was to guide the game character towards checkpoints that appeared at various locations on the screen (see Fig. 1). The movement was driven by the current gaze/mouse position: the player character always moved towards the current gaze position, or by keyboard input. When the game detected that the character arrived at the checkpoint area, next checkpoint was shown and the timer was reset. If, however, the player was not able to reach the checkpoint in given time (approx. 10 s), the next checkpoint would appear nonetheless. In order to score a checkpoint the player had to remain inside the checkpoint for half a second. The total game duration was 120 s.

Fig. 1. Screenshot from the game (zoom in on a player character)

The game was controlled via 5 different methods: by mouse, keyboard, and three gaze-based control methods using different functions for game character speed, based on its distance from the user's gaze:

1. **Mouse or keyboard.**
   When the mouse or keyboard was used, the character moved at the constant speed of 300 px/s. In the keyboard version, the game character was moved in one of the four cardinal directions, according to the arrow key pressed. In the mouse version, the game character was moved towards the mouse cursor, but only if the mouse was in motion.
2. **Gaze – constant function.**
   The game character was always in motion with constant speed towards the user's gaze position.
3. **Gaze – cutoff function.**
   The cutoff variation worked similarly to the constant version, but only when the user's gaze was within a 500 px radius the game character, otherwise the game character stopped moving. The radius size was chosen in pre-test and was directly influenced by the screen resolution. This control variant is similar to the "snap clutch" mechanism that allows both looking around and controlling a character [6].

4. Gaze – Gaussian function.

This gaze-based control method transformed the game character's speed $g(x)$,

$$g(x) = \frac{1}{\sigma\sqrt{2\pi}}e^{-\frac{1}{2}\left(\frac{x-\mu}{\sigma}\right)^2} \tag{1}$$

by the Gaussian function, with x the distance between the user's gaze and the game character position.

Parameters for the Gaussian-based speed transformation, $\sigma = \sqrt{1/160}$, $\mu = 2$, were chosen empirically in pre-tests relative to the screen size so that the maximum value of $g(x)$ would be about 5 (speed 500 px/s), with peak at about 400, see Fig. 2.

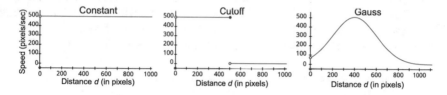

Fig. 2. Representation of three functions for the game character's speed, plotted as speed vs. distance.

The effect of the Gaussian-based speed transformation is such that speed is low at short distances, increases to its peak at about 400 px, and then decreases at larger distances. Speed transformation facilitated three types of actions:

1. precise game character positioning, i.e, when gaze landed near the character, the character remained almost still;
2. fast movement around the map, i.e, when gaze was moderately far away from the game character; and
3. slow movement, i.e, when looking far away the game character moved slowly allowing the user visual scanning of the screen, e.g., the game character was practically frozen when user was looking at the edge of the screen.

Each participant played five versions of the game (mouse, keyboard, constant, cutoff, gauss). The first version played was either the mouse or keyboard, followed by three different versions of the gaze-controlled game, presented in random order, following by the keyboard or mouse, depending on which one was presented at the outset. After playing each version of the game, a Gaming Experience Questionnaire was administered, focusing on evaluation of interaction with the game.

3.4   Apparatus

The experiment was conducted on a 22-inch LCD computer monitor with 1680×1050 resolution and 60 Hz refresh rate, connected to a laptop. For standard game control, a PC keyboard was used. Eye movements were recorded at 60 Hz with an SMI RED 250 eye tracking system using a chin-rest. SMI's BeGaze software was used for fixation and saccade detection with a dispersion-based event detection algorithm. The dispersion was 100 px, with minimum fixation duration set to 100 ms.

4   Results

Analyses of the data focused on three aspects: visual attention distribution, in-game performance indicators, and subjective game experience. The analyses followed the experimental design and were computed with R statistical software [17] using analysis of variance (ANOVA) with type III sum of squares correction, followed by a pairwise comparisons with Bonferroni correction when needed.

Performance. Performance was defined as the number of goals reached during the game (120 s). ANOVA revealed a main effect of game control $F(2.44, 53.57) = 82.01; p < 0.001, \eta^2 = 0.724$. Performance was highest for the gauss gaze-controlled game (M $= 31.54$, SE $= 1.11$), followed by mouse and keyboard (M $= 31.04$, SE $= 0.75$ and M $= 30.25$, SE $= 0.57$ respectively) (see Fig. 3). Performance was the worst during gaze-controlled constant and cutoff versions of the game (M $= 19.08$, SE $= 0.61$ and M $= 19.37$, SE $= 0.66$ respectively).

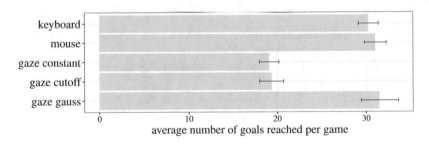

Fig. 3. Effect of game control and visual cues on the game performance. Whiskers represent boundaries of confidence interval for $p < 0.05$.

Post-hoc analyses revealed that performance in cutoff and constant gaze-controlled conditions was significantly lower than performance for any of the remaining game control types (see Table 1).

Visual cues had no impact on performance, with main effect of visual cues not significant $F(1, 22) = 3.55; p > 0.05$. The interaction effect between game control and visual cues was also not significant $F(2.44, 53.57) = 0.48; p > 0.05$.

Table 1. Comparison of game performance for different control type (post-hoc analyses t(4,92)).

	Keyboard	Mouse	Gaze-constant	Gaze-cutoff
Mouse	0.80			
Gaze-constant	11.24*	12.04*		
Gaze-cutoff	10.95*	11.74*	0.29	
Gaze-gauss	1.30	0.50	12.54*	12.25*

*denote statistically significant difference $p < 0.05$

In order to examine the role visual cues had on the gaze-controlled versions of the game, we conducted another ANOVA using only the gaze-controlled games and visual cues (3×2 mixed scheme). Analysis revealed a main effect of gaze control $F(1.31, 28.71) = 79.35; p < 0.001, \eta^2 = 0.702$, as well as of visual cues $F(1, 22) = 5.75; p < 0.05, \eta^2 = 0.083$. Contrary to our hypotheses, performance was slightly better for games played without visual cues ($M = 24.47, SE = 2.15$ and $M = 22.19, SE = 1.89$ respectively).

Game Experience Evaluation. In order to examine the influence of game control and visual cues on subjective game experience (measured using Game Experience Questionnaire), we conducted a 5 (game control) $\times$ 2 (visual cues) analysis of variance (ANOVA). ANOVA showed that a main effect of game control on all six dimensions of game experience: enjoyment ($F(2.72, 59.87) = 7.897, p < 0.001, \eta^2 = 0.163$); naturalness ($F(2.83, 62.29) = 12.662, p < 0.001, \eta^2 = 0.264$); difficulty ($F(3.52, 77.35) = 23.32, p < 0.001, \eta^2 = 0.399$); engagement ($F(2.72, 59.79) = 10.385, p < 0.001, \eta^2 = 0.134$); game rating ($F(3.15, 69.30) = 8.232, p < 0.001, \eta^2 = 0.145$) and interaction rating ($F(3.08, 67.67) = 7.047, p < 0.001, \eta^2 = 0.153$) (see Table 2). Effects of visual cues or interaction did not reach significance for any of the six dimensions.

Post-hoc analyses (see Table 2) revealed that the Gaussian speed transformation function was assessed by participants: (1) as natural in usage as the keyboard and mouse and more natural than other gaze-controlled games; (2) at a similar level of enjoyment as the keyboard and mouse and more enjoyable than other gaze-controlled games; (3) at a similar level of difficulty as the mouse and keyboard and much easier than other gaze-control methods; (4) as most engaging among all five variants tested. Game steering with the Gaussian function was the most preferred in general and also the most preferred method of interaction.

All participants were asked which of the presented game control methods they prefer. The majority (75%) preferred the gaze-controlled game with Gaussian transformation. The second choice was the keyboard (8.32%), and all other were evaluated *ex aequo* (4.17%).

Proportion of Peripheral Visual Scanning. In order to examine the differences in visual distribution, we conducted an ANOVA using the percentage of fixations landing outside of screen center as the dependent variable (see Sect. 3.2).

Table 2. Means and standard errors M(SE) of the ratings in game experience questionnaire

	Enjoyment	Naturalness	Difficulty	Engagement	Game rating	Interaction rating
a. Mouse	5.75	6.17^{c}	3.42^{cd}	5.13	5.38	5.21
	0.46	0.42	0.47	0.50	0.47	0.51
b. Keyboard	6.46^{cd}	7.25^{cd}	1.79^{cd}	4.88	5.58	5.38
	0.44	0.39	0.41	0.47	0.39	0.40
c. Gaze constant	4.75^{be}	4.13^{abe}	6.67^{abe}	5.17	4.63	5.04
	0.58	0.52	0.44^{abe}	0.62	0.56	0.61
d. Gaze cutoff	4.83^{be}	4.54^{be}	5.63^{abe}	5.67	5.50	5.42
	0.46	0.51	0.50	0.52	0.51	0.56
e. Gaze gauss	7.29^{cd}	7.00^{cd}	3.33^{cd}	7.42^{abcd}	7.33^{abcd}	7.71^{abcd}
	0.32	0.36	0.47	0.36	0.31	0.34

*Superscript letters denote statistically significant differences $p < 0.05$ (post-hoc tests with Bonferroni adjustments)

The main goal of the analysis was to determine how the different control types of the game influence visual scanning of the peripheries of the screen. The method of analysis was similar to previously reported ANOVAs. This analysis revealed a main effect of game control type $F(2.99, 65.81) = 13.14, p < 0.001, \eta^2 = 0.254$. The percentage of fixations landing outside the center of the screen was the highest for the gauss gaze-controlled game, followed by gaze cutoff, gaze constant, mouse and keyboard versions of the game (see Table 3).

Post-hoc analyses revealed significant differences between the gaze-controlled game with Gaussian transformation and all other variants on the percentage of fixations landing in the screen periphery.

Visual cues had no impact on visual distribution, with main effect of visual cues not significant $F(1, 21) = 1.11, p > 0.05$. Interaction between game control and visual cues was not significant $F(3.07, 64.54) = 0.89, p > 0.05$.

Table 3. Percentage of fixations landing outside of the center of the screen in different game control variants

a. Keyboard	b. Mouse	c. Gaze constant	d. Gaze cutoff	e. Gaze gauss
3.98 (0.61)^{de}	4.82(0.62)^{e}	5.55(0.46)^{e}	6.46(1.00)^{ae}	9.57 (0.69)^{abcd}

*Superscript letters denote statistically significant differences $p < 0.05$ (post-hoc tests with Bonferroni adjustments)

5   Discussion

In the presented study we tested a novel implementation of game control using gaze input with a mechanism for prevention of unintentional user actions. We hypothesized that this would lead to better gameplay both subjectively and objectively and that it would facilitate visual scanning.

Gaze control. Our experiment confirmed our hypotheses concerning gaze control with Gaussian speed transformation. Player performance was higher for this type of movement than for any other gaze controlled games. Also the game score for gaze-gauss, mouse and keyboard games exceeded 30 which was the maximum score achievable in the time allotted. Both other gaze-controlled games without the Gaussian function were the more difficult, likely due to unintended game character movement.

It cannot be ruled out that the performance score in the game may have been influenced not only by steering method itself, but also by the speed of the game character achievable for different input modalities. The maximum velocity differed between traditionally controlled and gaze-controlled versions of the game, since it was adjusted individually for each method to assure best playability. Although speed increase in the traditionally controlled versions of the game could possibly increase performance, it is also possible that higher speed would impair target aiming and hence decrease performance. For future studies, more attention should be paid to comparability of input modalities in terms of difficulty. It is also particularily interesting how the parameters of gaze-gauss function (i.e. kurtosis and skewness) would influence speed and in effect performance.

Subjective game ratings are in line with performance, also supporting our hypotheses. Gaze control with Gaussian transformation was rated as the highest in terms of interaction and gameplay among those tested. The Gaussian-based game was rated as enjoyable, natural, easy, and engaging as the keyboard version of the game.

Gaze distribution analysis clearly showed that the Gaussian function facilitated visual scanning. The amount of looking at the peripheral sections of the screen was the highest in the gaze gauss condition. We interpret this as an indicator of the Gaussian function allowing visual scanning while not impairing steering efficiency. Lower percentage of peripheral visual scanning for the traditional controls may have been caused by the fact that this type of game was not as engaging as gaze gauss variant. The size of the area marking center of the screen was set to the size of the active area in gaze cutoff game, which may have been too big to distinguish between visual scanning and focusing on the center on the screen for the traditional controls, which do not require as much gaze movements as gaze controlled games.

Visual cues. Contrary to our hypotheses regarding visual cues, their presence did not influence visual attention distribution and negatively impacted game performance, compared to other gaze-controlled variants of the game. While it is not possible to identify the exact cause of the negative influence of visual cues on performance, one of the reasons may be that they were too obstructive.

References

1. Bednarik, R., Gowases, T., Tukiainen, M.: Gaze interaction enhances problem solving: effects of dwell-time based, gaze-augmented, and mouse interaction on problem-solving strategies and user experience. J. Eye Mov. Res. **3**(1), 3, 1–10 (2009)
2. De Koning, B.B., Tabbers, H.K., Rikers, R.M.J.P., Paas, F.: Attention cueing as a mean to enhance learning from an animation. Appl. Cogn. Psychol. **21**(6), 731–746 (2007)
3. Dorr, M., Pomarjanschi, L., Barth, E.: Gaze beats mouse: a case study on a gaze-controlled breakout. Psychol. J. **7**(2), 197–211 (2009)
4. Ekman, I.M., Poikola, A.W., Mäkäräinen, M.K.: Invisible eni – using gaze and pupil size to control a game. In: CHI 2008 Extended Abstracts on Human Factors in Computing Systems, CHI EA 2008, pp. 3135–3140. ACM (2008)
5. Isokoski, P., Joos, M., Spakov, O., Martin, B.: Gaze controlled games. Univ. Access Inf. Soc. **8**(4), 323–337 (2009)
6. Istance, H., Bates, R., Hyrskykari, A., Vickers, S.: Snap clutch, a moded approach to solving the midas touch problem. In: Proceedings of the 2008 Symposium on Eye Tracking Research and Applications, ETRA 2008, pp. 221–228. ACM, New York (2008)
7. Istance, H., Hyrskykari, A., Vickers, S., Chaves, T.: For your eyes only: controlling 3d online games by eye-gaze. In: Gross, T., Gulliksen, J., Kotzé, P., Oestreicher, L., Palanque, P., Prates, R.O., Winckler, M. (eds.) IFIP Conference on Human-Computer Interaction, pp. 314–327. Springer, Heidelberg (2009)
8. Jacob, R.J.: What you look at is what you get: eye movement-based interaction techniques. In: CHI 1990 Conference Proceedings of Human Factors in Computing Systems, pp. 11–18. ACM Press (1990)
9. Kasprowski, P., Harezlak, K., Niezabitowski, M.: Eye movement tracking as a new promising modality for human computer interaction. In: 2016 17th International Carpathian Control Conference (ICCC), pp. 314–318, May 2016
10. Kenny, A., Koesling, H., Delaney, D., McLoone, S., Ward, T.: A preliminary investigation into eye gaze data in a first person shooter game. In: 19th European Conference on Modelling and Simulation, ECMS (2005)
11. Krejtz, K., Biele, C., Chrzastowski, D., Kopacz, A., Niedzielska, A., Toczyski, P., Duchowski, A.: Gaze-controlled gaming: immersive and difficult but not cognitively overloading. In: Proceedings of the 2014 ACM International Joint Conference on Pervasive and Ubiquitous Computing: Adjunct Publication, UbiComp 2014 Adjunct, pp. 1123–1129. ACM, New York (2014)
12. Kriz, S., Hegarty, M.: Top-down and bottom-up influences on learning from animations. Int. J. Hum.-Comput. Stud. **65**(11), 911–930 (2007)
13. Lowe, R.K., Boucheix, J.M.: Cueing complex animation: does direction of attention foster learning processes? Learn. Instr. **21**(5), 650–663 (2011)
14. Muñoz, J., Yannakakis, G.N., Mulvey, F., Hansen, D.W., Gutierrez, G., Sanchis, A.: Towards gaze-controlled platform games. In: 2011 IEEE Conference on Computational Intelligence and Games, CIG 2011, pp. 47–54. IEEE (2011)
15. Nielsen, A.M., Petersen, A.L., Hansen, P.J.: Gaming with gaze and losing with a smile. In: Proceedings of the Symposium on Eye Tracking Research and Applications, ETRA 2012, pp. 365–368 (2012)
16. Ohno, T., Mukawa, N., Kawato, S.: Just blink your eyes: a head-free gaze tracking system. In: CHI 2003 Extended Abstracts on Human Factors in Computing Systems, CHI EA 2003, pp. 950–957. ACM, New York (2003)

17. R Development Core Team: R: A Language and Environment for Statistical Computing. R Foundation for Statistical Computing, Vienna, Austria (2011). ISBN 3-900051-07-0
18. van Rossum, G.: Python tutorial, Technical report CS-R9526. Centrum voor Wiskunde en Informatica (CWI), Amsterdam (1995)
19. San Agustin, J., Mateo, J.C., Hansen, J.P., Villanueva, A.: Evaluation of the potential of gaze input for game interaction. Psychol. J. **7**(2), 213–236 (2009)
20. Shinners, P.: Pygame (2014). http://pygame.org/
21. Sibert, L.E., Jacob, R.J.K.: Evaluation of eye gaze interaction. In: Proceedings of the SIGCHI Conference on Human Factors in Computing Systems, CHI 2000, pp. 281–288. ACM, New York (2000)
22. Špakov, O.: Eyechess: the tutoring game with visual attentive interface. In: Alternative Access: Feelings and Games, pp. 81–86 (2005)
23. Stellmach, S., Dachselt, R.: Look and touch: gaze-supported target acquisition. In: Proceedings of the SIGCHI Conference on Human Factors in Computing Systems, CHI 2012, pp. 2981–2990. ACM, New York (2012)

Measurements of Contrast Detection Thresholds for Peripheral Vision Using Non-flashing Stimuli

Michał Chwesiuk[(✉)] and Radosław Mantiuk

West Pomeranian University of Technology, Szczecin,
al. Piastow 17, 70-310 Szczecin, Poland
{mchwesiuk,rmantiuk}@wi.zut.edu.pl

Abstract. In this work the change of the contrast detection threshold with eccentricity were measured for a range of eccentricities from 0 to 27°. A common approach used for such measurements is to display a flashing stimulus presented by a fraction of a second in the observer's peripheral viewing area. This condition prevents the registration of the results after unintended moving the eyes towards the stimulus and lowering the recorded thresholds. In contrast to this methodology, our stimuli are not modulated over time. We display stimuli continuously and use eye tracker to control the observers' gaze direction. We prove that results of the psychophysical experiments based on this approach are consistent with the previous work.

Keywords: Contrast detection thresholds · Gaze-dependent contrast thresholds · Gaze-contingent display · Gaze-dependent rendering · Eye tracking · Psychophysical experiments

1 Introduction

Our sensitivity to contrasts is reduced with the eccentricity, i.e. with the angular distance from the gaze direction [12]. Models of this feature of human vision are developed based on the data from psychophysical experiments, in which stimuli is presented to observers in her/his peripheral vision for a short time on the order of milliseconds [2,7–11]. Flashing the stimuli ensures that observers do not turn their eyes toward the stimuli and, in this way, unintentionally increase the sensitivity. However, such condition is unlikely to be found in typical viewing scenarios, because natural scenes do not flash. Even moving object are presented to viewers for longer time in a continuous manner. Another evidence of the drawback of this methodology is that in the different studies, the absolute values of contrast detection thresholds varied substantially among these studies [8]. The most likely reason of this effect appears to be different flash duration used in various experiments [8].

In this work, we conduct a similar psychophysical experiment but using non-flashing stimuli. The contrast detection threshold is measured for a number of the eccentricities ranging from 0 to 27°. We created sin-grating stimulus of a

© Springer International Publishing AG 2018
I. Czarnowski et al. (eds.), *Intelligent Decision Technologies 2017*,
Smart Innovation, Systems and Technologies 73, DOI 10.1007/978-3-319-59424-8_24

2 cpd (cycles-per-degree), which was continuously displayed on the screen in the horizontal or vertical orientation. Observers were asked to look straight to the marker and judge the orientation of stimulus seen in their peripheral vision. This condition is unnatural for humans, because we instinctively look away in the direction of the observed object. Therefore, we used eye tracker to test whether the observers changed their viewing direction. If such condition is detected, the stimulus was cleared and then redrawn with a random orientation.

We test the effectiveness of this methodology by performing a case study, in which the orientation of the stimulus is not changed after changing the viewing direction. Under such conditions sensitivity to contrasts should be higher, because observes could see the stimulus in the foveal vision for a short time before it was cleared. The obtained results revealed this relationship between these two experimental methodologies.

In Sect. 2 we review previous work related to the gaze-dependent contrast detection threshold measurements. In Sect. 3 the details of the conducted experiments are presented. We discuss the achieved results in Sect. 4.

2   Previous Work

The peripheral contrast detection thresholds have been measured in a number of studies. Robson and Graham [10] used 4 cycles patches of horizontal grating. This stimulus was displayed for 100 ms. Cannon [2] used vertical sin-grating patches presented to the right of fixation for 2 s, including 350 ms rise and fall times. Thomas [11] used a patch presented for 1 s, either with an abrupt onset and offset or ramped on and off over the whole second. Pointer and Hess [9] presented horizontally oriented sinusoidal grating patches in Gaussian envelopes. This stimuli were displayed for 250 ms using the Gaussian window with the temporal spread. Mullen [7] measured the detection threshold for chromatic stimuli. The sin-grated patch was displayed continuously. The results show that at each spatial frequency color contrast sensitivity declines with eccentricity approximately twice as steeply as luminance contrast sensitivity.

Our approach is inspired in particular by Peli et al. [8]. They measured the threshold contrast required for discrimination between horizontal and vertical sinusoidal grating patches (Gabor functions). Measurements were taken at the fovea and at temporal eccentricities of 2.5, 5.1, 10.3, and 22.8°. Thresholds at each eccentricity were measured for five spatial frequencies, 1, 2, 4, 8, and 16 cpd. Stimuli contained about 4 cycles, but only approximately two cycles were visible because of the rapid decline of the Gaussian envelope. The background luminance was equal to $37.5 \, cd/m^2$. The stimulus was presented for 0.5 s with an abrupt onset and offset.

3   Experiment Design

3.1   Stimuli

In our experiment the stimuli consisted of vertical or horizontal sine-gratings attenuated by a Gaussian envelope (see Fig. 1). To render stimuli, we used the

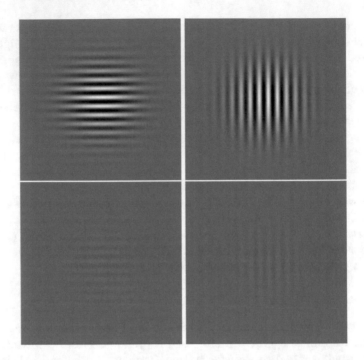

Fig. 1. Examples of the horizontal (left) and vertical (right) sin-gratings used in our experiments. The contrast of the stimuli was reduced from $logC_{10} = 0.3$ (top row) to $log_{10}C = -1$ (bottom row). In our experimental setup, frequency of each stimulus is equal to 2 cpd.

CreateProceduralGabor() function from the Psychtoolbox package[1], which is a Matlab toolbox for creating psychophysical experiments [1]. This function allows to specify Michelson contrast: $c = (I_{max} - I_{min})/(I_{max} + I_{min})$, where I_{max} and I_{min} correspond to maximum and minimum luminance of the Gabor patch, respectively. We report all data in terms of threshold contrast C, which is a relative modulation of the sine-grating: $C = c * L_b/L_{max}$, where L_b is the luminance of the background and L_{max} is the maximum luminance of the display.

We set the background luminance L_b to 60 cd/m^2, when L_{max} was equal to 120 cd/m^2. The luminance levels of the Gabor pattern were selected to avoid luminance levels lower than 1 cd/m^2 and higher than 120 cd/m^2, at which the display calibration was unreliable.

3.2 Display

The experiment were run using Sony PVM-A250 TRIMASTER EL, 1920 × 1080 pixel resolution, high quality reference OLED display. It offers good luminance reproduction with a 10-bit OLED panel. This bit-depth resolution is necessary

---

[1] http://docs.psychtoolbox.org/CreateProceduralGabor.

for near-threshold detection experiments, in which perceivable spatial resolution is measured. We used the native display calibration to sRGB color profile. Correctness of the calibration was confirmed using the Minolta CS-100A luminance meter.

3.3 Procedure

During experiment the stimuli were observed from a fixed distance of 90 cm, which gave an angular resolution of 57 pixels per visual degree.

The experimental procedure is presented in Fig. 2. Observer was sitting in the front of the green cross marker presented in Fig. 2a. She/he was asked to look at this marker plotted on the grey background. The Gaussian noise followed by the sin-grating has been drawn on the left side of the screen in an arbitrary angular direction. Observer task was to recognize the horizontal or vertical direction of the sin-grating by pressing the *up* or *right* keys on the custom-built control panel. If observer looks away from the marker, it's color turned to red and the sin-grating was cleared from the screen (Fig. 2b). The stimulus was redrawn in randomly chosen orientation, when observer began to look at the marker again (Fig. 2c). We captured the gaze direction using 60 Hz eye tracker (remote Eye-Tribe device with average accuracy of 0.5°) [14]. We set the acceptable deviation from the desired viewing direction to 2°, i.e. for lager deviation the stimulus was hidden.

The procedure was repeated for eccentricities of 5, 10, 15, 20, and 27°. We also measured sensitivity at fovea (i.e. eccentricity equal to 0°).

Fig. 2. Screenshots from our experiment. Green (or red) cross points out the desired viewing direction. The blue square depicts the gaze location captured by eye tracker.

To find the threshold magnitude of the sin-grating, we used the QUEST adaptive procedure [13]. The QUEST procedure is based on the assumptions about the distribution of responses near the threshold and an actual shape of the psychometric function [6]. Many trials are repeated while varying the magnitude of the stimulus. The magnitude of a current trial is determined on the basis of the observer's responses in previous trials. In our experiment, the stimuli magnitude was the degree of the contrast in log10 units. QUEST adaptively determines the degree of contrast for the next trial based on the observer's correct or incorrect response for the current trial. We used the QUEST implementation from Psychtoolbox (version 3).

The main assumption of the above procedure is that observer cannot turn the eyes toward stimulus and, in this way, increase her/his sensitivity to contrast by replacing the peripheral vision with the foveal vision. To test this assumption we performed the second experiment, in which orientation of the stimulus was not changed after the eye movement detected by eye tracker. The stimulus was still cleared but after moving the eyes again to the marker, the stimulus was redrawn in the same orientation. This modification allowed for looking at the stimulus for smaller eccentricity, which, of course, was inconsistent with the objectives of the experiment.

Our eye tracker needed about 17 ms to capture the gaze direction. Another 17–33 ms was consumed by the 60 Hz display to clear the image to the background gray and display the stimulus. This latency was enough to turn the eyes and see the stimulus even for the largest eccentricity of 27° [3]. This effect was significant in the case study without the orientation modification. However, it did not affect the results in the actual experiment with the random orientation modification.

3.4 Participants

We asked 6 volunteer observers to conduct the experiment (age between 21 and 47 years, average age 27.67, 2 females, 4 males). While there were no time limitations to our study, the average observer finished the experiment in approximately 10 min. Observers declared normal or corrected to normal vision and correct colour vision. Before the experiment we briefly described to each participant the motivation behind the detection threshold measurement but not the details of our strategy. In particular, they did not know if the stimulus orientation was changed after unintended gaze relocation, i.e. they did not know whether it was the case study or the actual experiment.

4 Results

The goal of our experiments was to measure the contrast detection threshold for the peripheral vision using the non-flashing stimuli. We also compare the achieved results with the contrast discrimination model reported in the literature.

4.1 Detection Threshold

The blue plot in Fig. 3 presents results of our experiment (the threshold values averaged over all observers are additionally depicted in Table 1). The contrast detection threshold for the foveal vision (zero eccentricity) is equal to $C = 0.013$, which can be expressed as the sensitivity $S = 1/C = 76.92$ or $log_{10}S = 1.89$. This sensitivity is comparable with the sensitivity for 2 deg stimuli and 60 cd/m^2 background luminance reported in the literature (in the recent studies Kim et al. [4]

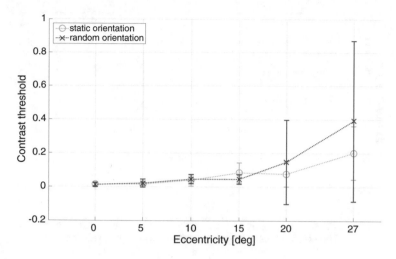

Fig. 3. Contrast detection thresholds for 2 cpd stimuli averaged over all observers. Error bars depict the standard deviation of the measurements.

Table 1. Statistics from the contrast detection experiments.

Eccentricity [deg]	0	5	10	15	20	27
Random orientation [log10 units]						
Detection threshold	0.013	0.022	0.046	0.046	0.148	0.393
Standard deviation	0.013	0.023	0.027	0.028	0.249	0.477
Minimum value	0.002	0.005	0.012	0.028	0.041	0.049
Maximum value	0.038	0.071	0.082	0.107	0.712	1.162
Static orientation [log10 units]						
Detection threshold	0.016	0.014	0.036	0.079	0.074	0.181
Standard deviation	0.009	0.009	0.035	0.059	0.075	0.159
Minimum value	0.005	0.004	0.009	0.02	0.022	0.044
Maximum value	0.028	0.034	0.115	0.159	0.235	0.455
Threshold difference [log10 units]:	−0.003	0.008	0.010	−0.033	0.074	0.212

report a value of $log_{10}S = 1.5$). The thresholds slightly increase for small eccentricity from 5 to 15° but they are clearly higher for 20 and 27°. As it was expected the thresholds are growing exponentially.

The green plot in Fig. 3 shows the results of the experiment in which orientation of the stimulus was not changed after detection of the observer's eyes movement. As can be seen the threshold values for eccentricities of 20° and 27° are clearly lower than thresholds measured in the previous experiment. It indicates that eye tracker plays crucial role in the experimental methodology.

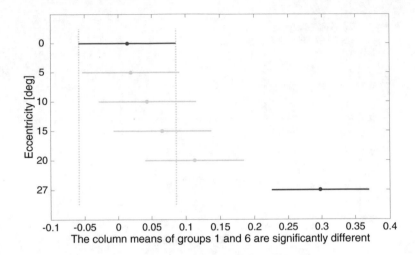

Fig. 4. The pairwise comparison results from ANOVA revealing a statistically significant difference between the results of the experiment with static and randomly modified orientation of the stimulus. The mean contrast threshold for each eccentricity is represented by a dot, and the confidence interval is represented by a line extending out from this dot. Two means for different eccentricities are significantly different if their intervals are disjoint.

A 2-way analysis of variance (ANOVA) was used to gauge the statistical difference between both experiments. The dependent variable was the measured contrast thresholds. The independent variables were experiment type (static or random orientation) and eccentricity. ANOVA reveals a statistically significant main effect of eccentricity ($p = 0, F = 9.77$), experiment type ($p = 0.0012, F = 4.57$), and interaction between eccentricity and experiment type ($p = 0.0207, F = 1.98$). In Fig. 4 the confidence intervals for measured eccentricities are presented. The plot depicts that results achieved in the experiments are significantly different for eccentricity of $27°$.

4.2   Model

As it has been justified in Peli et al. [8], it is difficult to directly compare the results of the peripheral contrast threshold measurement experiments because of a bias introduced by the experimental condition (e.g. different flash time). For this reason, following methodology presented in Peli et al., we fit the model of the contrast constancy based on our experimental results and then compare this model to model presented in the literature. As a reference model we chose results from Peli et al. because in this work there is a comparison of the different studies on the peripheral contrast detection thresholds indicating that the model is valid against other studies.

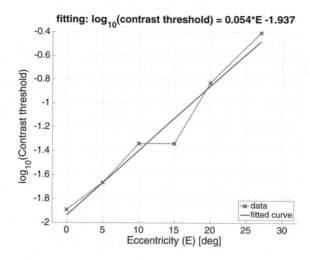

Fig. 5. Averaged contrast threshold plotted on the logarithmic scale. The magenta line shows fitting of the measured data based on the linear polynomial.

The *contrast constancy* holds for a wide range of frequencies, suggesting that sensitivity is constant across the spectrum at superthreshold [5]. However, lower-contrast features could disappear when their contrast is below the threshold level, when the object moves e.g. to the observer. Some mechanism must exist in the Human Visual System (HVS) to compensate for this effect, because these changes from visible to invisible, would affect the perception of images more than the variations in contrast at suprathreshold levels. When an object moves closer to the observer, the spatial frequency of various features in the object decreases. At the same time, the overall size of the object's retinal image increases. Therefore many of these features now can fall on retinal areas farther from the fovea, where contrast sensitivity is lower. Thus, the requirements for invariance could be satisfied if contrast thresholds were to vary as the product of the spatial frequency and the retinal eccentricity.

On a logarithmic scale the threshold varies linearly with the eccentricity (see Fig. 5). If we want features stay invariant with distance changes, the thresholds should be related to the eccentricity in a specific way:

$$log_{10}C = m * E + b, \qquad (1)$$

defined by the values of m and b parameters. Especially, the slope of the line (m) is important because b (contrast threshold at eccentricity of 0°) can vary depending on the stimuli and experimental procedure [8].

In Fig. 5 we fitted our experimental data to this model (magenta line). In Fig. 6 our fitting is compared with the data from Peli's orientation identification experiment [8]. We shifted the lines along the Y axis to $b = 0$ to obtain the visually consistent plots (and compensate the experimental bias). As can be seen in the plot our model (magenta line) matches results from Peli et al.

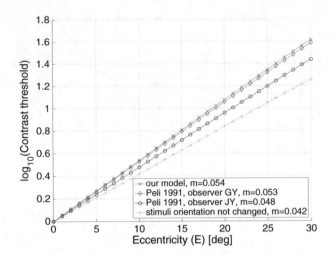

Fig. 6. Eccentricity-dependent contrast constancy model for 2 cpd stimulus obtained from Peli et al. 1991 [8] (blue and black lines) and our experiment (magenta line). The green line shows model for test experiment, in which stimuli orientation was not modified after eye movement.

(blue line) for observer GY. The difference between the m coefficients is equal to 0.001. Taking into account the average value for both GY and JY observers, this difference is equal to 0.0035.

5   Conclusions and Future Work

To capture the effect of eccentricity on contrast detection, new contrast threshold detection measurements were conducted using the non-flashing stimuli and eye tracker. The goal was to verify whether the results of such methodology are consistent with the previous works, in which the stimulus was presented in the periphery of vision for a short time. Both methodologies (with flashing stimuli and with eye tracker) prevent the registration of the results after unintended moving the eyes, however, our solution is more natural for typical viewing conditions. The results achieved in the preliminary studies for 2 cpd stimuli are consistent with the results reported in the previous work.

In future work we plan to measure the thresholds for a wider range of the stimuli frequencies, further periphery, and for chromatic stimuli. We also plan to replace the Gabor pattern with the complex stimuli in the form of the renderings of the three dimensional objects. These measurements should be more reliable than the experiments with flashing stimuli that introduce the measurement bias. We expect that better models of the peripheral contrast threshold detection are possible to achieve, especially for the mentioned complex stimuli, for which sufficient observation time is crucial for near-threshold contrast detection [3].

Acknowledgments. The project was funded by the Polish National Science Centre (decision number DEC-2013/09/B/ST6/02270).

References

1. Brainard, D.H.: The psychophysics toolbox. Spat. Vis. **10**, 433–436 (1997)
2. Cannon, M.W.: Perceived contrast in the fovea and periphery. JOSA A **2**(10), 1760–1768 (1985)
3. Chwesiuk, M., Mantiuk, R.: Acceptable system latency for gaze-dependent level of detail rendering. In: Magalhaes, L.G., Mantiuk, R. (eds.) EG 2016 - Posters. The Eurographics Association (2016)
4. Kim, K.J., Mantiuk, R., Lee, K.H.: Measurements of achromatic and chromatic contrast sensitivity functions for an extended range of adaptation luminance. In: IS&T/SPIE Electronic Imaging, pp. 86511A–86511A. International Society for Optics and Photonics (2013)
5. Kulikowski, J.: Effective contrast constancy and linearity of contrast sensation. Vis. Res. **16**(12), 1419–1431 (1976)
6. Mantiuk, R., Tomaszewska, A.M., Mantiuk, R.: Comparison of four subjective methods for image quality assessment. Comput. Graph. Forum **31**(8), 2478–2491 (2012)
7. Mullen, K.: Colour vision as a post-receptoral specialization of the central visual field. Vis. Res. **31**(1), 119–130 (1991)
8. Peli, E., Yang, J., Goldstein, R.B.: Image invariance with changes in size: the role of peripheral contrast thresholds. JOSA A **8**(11), 1762–1774 (1991)
9. Pointer, J., Hess, R.: The contrast sensitivity gradient across the human visual field: with emphasis on the low spatial frequency range. Vis. Res. **29**(9), 1133–1151 (1989)
10. Robson, J., Graham, N.: Probability summation and regional variation in contrast sensitivity across the visual field. Vis. Res. **21**(3), 409–418 (1981)
11. Thomas, J.P.: Effect of eccentricity on the relationship between detection and identification. JOSA A **4**(8), 1599–1605 (1987)
12. Wandell, B.A.: Foundations of Vision. Sinauer Associates, Sunderland (1995)
13. Watson, A.B., Pelli, D.G.: Quest: a Bayesian adaptive psychometric method. Percept. Psychophys. **33**(2), 113–120 (1983)
14. Wolski, K., Mantiuk, R.: Cross spread pupil tracking technique. J. Electron. Imaging **25**(6), 063012–063012 (2016)

Visual World Paradigm Data: From Preprocessing to Nonlinear Time-Course Analysis

Vincent Porretta[1(✉)], Aki-Juhani Kyröläinen[2], Jacolien van Rij[3], and Juhani Järvikivi[1]

[1] University of Alberta, Edmonton, AB T6G 2E7, Canada
`porretta@ualberta.ca`
[2] University of Turku, 20014 Turku, Finland
[3] University of Groningen, 9712 EK Groningen, The Netherlands

Abstract. The Visual World Paradigm (VWP) is used to study online spoken language processing and produces time-series data. The data present challenges for analysis and they require significant preprocessing and are by nature nonlinear. Here, we discuss *VWPre*, a new tool for data preprocessing, and generalized additive mixed modeling (GAMM), a relatively new approach for nonlinear time-series analysis (using *mgcv* and *itsadug*), which are all available in R. An example application of GAMM using preprocessed data is provided to illustrate its advantages in addressing the issues inherent to other methods, allowing researchers to more fully understand and interpret VWP data.

Keywords: Visual World Paradigm · Generalized Additive Mixed Modeling

1 Introduction

The Visual World Paradigm [VWP, 1] is an eye-tracking method used to study real-time spoken language processing [see 2 for overview]. VWP data require significant preprocessing prior to analysis, and the nonlinear, time-series nature of the data presents challenges for data analysis. To avoid this complexity, the data are often simplified, resulting in the potential loss of information. Here, we introduce a new tool for preprocessing and demonstrate how the time-series data can be analyzed using nonlinear regression in R [3]. Using data from a study examining the effect of foreign accentedness on spoken word recognition [4], we present: discussion of the nature of VWP data and required preprocessing using the R package *VWPre* [5]; discussion of Generalized Additive Mixed Modeling (GAMM) for time-course analysis using *mgcv* [6]; example GAMM analyses demonstrating a factorial variable interaction with time as well as a continuous variable interaction with time; and visualization of estimated effects using *itsadug* [7].

2 The Visual World Paradigm

2.1 The Paradigm

The motivation underlying VWP is that listeners subconsciously direct their gaze to visual representations of language as they processes the acoustic signal [1, 8]. VWP has

I. Czarnowski et al. (eds.), *Intelligent Decision Technologies 2017*,
Smart Innovation, Systems and Technologies 73, DOI 10.1007/978-3-319-59424-8_25

been used to study various aspects of spoken language processing, such as: acoustic, lexical, sentence, and discourse processing [9–12]; visual context in language comprehension [8]; predictive language processing [13]; and language development [14]. This paradigm is useful for understanding the dynamics of spoken language processing, by examining the time-course of looking behavior as speech unfolds. The analysis of VWP data examines the likelihood over time of eye gaze falling on objects in the visual scene relative to the onset of the spoken stimulus. Researchers must define interest areas (IAs) around the objects related to the auditory stimulus as well as signal its onset, in order to relate the gaze data to the critical portion of the speech. The resulting data indicate the IAs in which the gaze has fallen relative in time to the spoken stimulus; however, these data are generally not "analysis-ready". To examine the change in eye gaze over time, the proportion of looks to each of the defined IAs is typically calculated within a narrow span of time (i.e., bin).

2.2 Preprocessing and Statistical Considerations

To obtain the probability of looking at a particular IA over time relative to the auditory stimulus, a minimum expertise working with a programming language is required. This is not a trivial matter for most language researchers [see 15 for discussion]. Thus, we present *VWPre*, which facilitates the preparation of VWP data collected with SR Research eye-trackers for analysis and visualization. It performs all required formatting (e.g., message and time-series alignment) and calculations (e.g., proportions and empirical logits), and contains functions for both interactive data examination and the creation of customizable, publication-ready time-series plots. All functions are fully documented along with vignettes (available in R) illustrating their use. The basic functionality of the package is discussed and illustrated in further sections.

It is necessary to understand the nature of VWP data in order to apply the most appropriate statistical method. Two aspects, illustrated in Fig. 1, stand out in this regard. First, VWP experiments are concerned with online language processing and produce time-series data in which the sequential measurements tend to be correlated with each other. This is referred to as autocorrelation which violates the assumptions of many statistical tests [16]. Second, VWP data often present a nonlinear functional form, posing challenges for statistical methods that assume a linear relationship. These aspects are generally either disregarded or submitted to statistical "work-arounds", in order to avoid

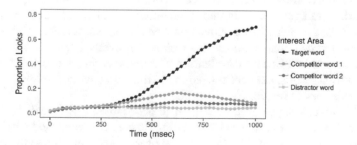

Fig. 1. Grand average (with SE bars) of proportion of looks by word type from stimulus onset

their problematic nature. In what follows, we advocate for a method that can account for these aspects and allows for a deeper understanding of the data.

3   Generalized Additive Mixed Modeling (GAMM)

3.1   Description of the Method

GAMM is a relatively new nonlinear regression method [17, 18] and has been applied to psycholinguistic data [4]. Methods like analysis of variance (ANOVA) or generalized linear mixed-effects regression (GLMER) assume a linear relationship between predictors and the response variable. This assumption is often unacknowledged, and assuming linearity in its absence can underestimate the strength of the relationship or fail to find a significant pattern at all. GAMM does not assume linearity, though can find a linear form if supported by the data. Also, GAMM allows for multidimensional nonlinear interactions between two (or more) continuous predictors (see Sect. 4.4), to produce a (possibly wiggly) surface.

GAMM, in *mgcv*, strikes a balance between model fit and the smoothness of the curve using either error-based or likelihood-based methods in order to avoid over- or under-fitting. Thus, the data guide the functional form [17]. A *p*-value is provided for each smooth term, indicating whether or not the curve is significantly different from zero. Random effects (i.e., intercepts and slopes) can be included to capture dependencies between repeated measurements, such as within/between subjects and items [see 19 for discussion]. Furthermore, GAMM allows for random structure that captures nonlinearity —so-called factor smooth interactions. These can capture adaptive patterns within and/or across trials. As nonlinear trends are difficult to capture with a single parameter, visualization of model estimates is more important in GAMM analyses than with linear regression methods. Generally speaking, the significance of a predictor is determined using a combination of the *p*-value of the smoothing parameter, model comparison procedures, and visual inspection of the functional form.

3.2   Advantages of GAMM

For VWP analysis, ANOVA and GLMER generally require the removal of time as a variable by averaging looks within a specified window in order to avoid the issue of repeated measures through time. This removes one of the most interesting aspects of the Visual World Paradigm—the continuous time-course of effects. Moreover, selecting a time window for averaging is not without its own issues, as it is rather arbitrary and the size and position can greatly impact the results. An overly large window may mask existing effects; and shifting a "non-significant" window earlier or later in time may then return significant results. Choices related to the time window thus introduce so-called "researcher degrees of freedom".

Growth Curve Analysis [GCA, 20], an implementation of GLMER, attempts to address the aforementioned issues by including time as an unaveraged predictor using polynomial functional forms. However, this approach does not address the problem of autocorrelation; inadequate fit of correlated observations causes structure to remain in the residuals of the model, which in turn results in an increased risk of overconfidence. GLMER cannot account for this autocorrelation and interpretation of the coefficients that specify the polynomial functions is generally difficult. While GCA does not require averaging, GAMM provides additional benefits over GCA. First, GAMM easily models the nonlinear curves typical of these time-series data. By default, thin plate regression splines are used instead of polynomials, as they are more flexible in fitting more complex nonlinear patterns. Second, modeling multidimensional continuous interactions is straightforward, something not convenient or advisable in GCA. Third, GAMM (in mgcv) provides the option to include an autoregressive model to account for autocor-related residuals (in Gaussian-family models). Specifically, setting the AR1 correlation parameter informs the model of correlation in the errors, so that the confidence of the estimates can be adjusted accordingly.

4   Applying GAMM to VWP Data

4.1   Demonstration and Data

In what follows, we demonstrate and discuss various aspects of a GAMM analysis of VWP data. Specifically, we present: the basic preprocessing work-flow; discussion around the choice of response variable (empirical logit vs. binomial); an analysis demonstrating a factorial interaction using empirical logit transformed data; and an analysis demonstrating a two-way continuous interaction using binomial data. The data used here is a subset from an experiment examining the effect of foreign accentedness on word recognition using native and Chinese-accented English [4]. In a "visual word" variant of VWP, participants heard an accented token and found its written form among four options. Accentedness was operationalized as a continuous variable based on a prior ratings study, and is utilized here for demonstrating a continuous interaction with time. To demonstrate a factorial interaction, this variable has been discretized by selecting only the native English talker (Low Accent) and the most accented Chinese talker (High Accent).

4.2   Preprocessing and Pre-analysis Considerations

The 1000 Hz data were output as a Sample Report from the SR Research Data Viewer software, using the predefined IAs and an interest period relative to the stimulus onset. Note that VWPre also contains functions for defining IAs and critical stimulus alignment (not discussed here). First, the columns required for preprocessing were verified (function `prep_data`). Second, samples falling outside of any IA were recoded as NA (`relabel_na`) and the encoding of the predefined IAs was checked (`check_ia`). Third, the time-series was created and aligned to the message (`create_time_series`). Fourth, the data from the eye(s) chosen for analysis was prepared (`select_recorded_eye`). Fifth, the proportion of looks to each IA was calculated using the desired bin size (`bin_prop`). Here, a 20 ms bin was used, downsampling the data to 50 Hz. Table 1 illustrates snippets of the input and output of this process.

Table 1. Fragment of unprocessed input data (Left) and (partial) resulting processed data output from `bin_prop` (Right)

SUBJECT	TRIAL	IA	MESSAGE	TIMESTAMP	Event	Time	IA_0_C	IA_1_C	IA_2_C	IA_0_P	IA_1_P	IA_2_P
14001	1	NA	NA	1047403	14001	-20	20	0	0	1	0	0
14001	1	NA	TargetOnset	1047404	14001	0	20	0	0	1	0	0
14001	1	NA	NA	1047405	14001	20	20	0	0	1	0	0
...	...	...	...	... →	...	...	...	...	...	...	...	...
14001	1	NA	NA	1048129	14001	700	20	0	0	1	0	0
14001	1	Target	NA	1048130	14001	720	6	14	0	0.3	0.7	0
14001	1	Target	NA	1048131	14001	740	0	20	0	0	1	0
14001	1	Target	NA	1048132	14001	760	0	20	0	0	1	0

The next step regards the preparation of the response variable. Of common interest is the probability of looking at a specific IA. Because proportions are inherently bound between 0 and 1, the scale of the response variable must be chosen prior to analysis. There are two approaches, each with advantages and limitations, that determine the distributional family required for modeling. First, the binomial distribution models proportions by means of the combination of successes and failures (i.e., number of samples falling within a given IA vs. those falling outside) within a given bin. While this closely mirrors the inherent scale of the data, binomial GAMM cannot account for autocorrelated errors. Further downsampling and more precise random structure can compensate for this, though likely not eliminating it. Here, the binomial response variable was created using 20 ms bins (`create_binomial`). Second, the Gaussian distribution can be used to control for autocorrelated errors, however, proportions must first be transformed to an unbound measure via the empirical logit transformation with weights for variance estimation [21]. However, using this transformation requires special consideration. The transformation is typically done using the number of observations per bin (inherently linked to the sampling rate and bin size) and a constant (preventing the return of $\pm\infty$). Both are user-specified and impact the results of the calculations. *VWPre* contains an interactive interface (`plot_transformation_app`) to visualize the effect of both values on the calculations. Additionally, this transformation changes the scale of the data and may impact the distribution. Here, the transformation was performed using 20 samples per bin and a constant of 0.5 (`transform_to_elogit`).

For advanced users, a function (`fasttrack`) is provided to quickly preprocess the data in a single function call. After preprocessing, the data can be visualized (see Fig. 1) using the customizable plotting function (`plot_avg`) aiding in the selection of the time range to be modeled. Because of time required to program a saccade based on auditory input (200 ms) [22] and the average duration of the stimuli (514 ms), a window of 200–800 ms post stimulus onset was chosen. Additionally, to investigate online word recognition, looks to the target IA was chosen for analysis. Alternatively, as suitable for some research questions, one could model the difference in looks between two IAs (i.e., looking preference), though this has advantages and limitations in addition to those above. Below, we present two examples illustrating the empirical logit and binomial approaches.

4.3   Gaussian Model with Transformation and Factor Interaction

In this demonstration, the response variable (IA_1_ELogit) is empirical logit looks to the target. For model evaluation and visualization, we use functions available in *itsadug* (see vignettes for illustration). The model included factor smooth interactions for Time by Subject, factor smooth interactions for Time by Item, as well as a smooth for Time by Accent Level. Additionally, we have included Accent Level as a parametric component, which is necessary to estimate the time curve for each level of accentedness. Lastly, we have included the variance estimates as weights in the model. After fitting this model we determined an appropriate value for the AR1 parameter (start_value_rho provides a sensible initial value), in this case $\rho = 0.867$, to account for autocorrelation in the residuals (i.e., error). Furthermore, it is necessary to include a logical vector which signals the starting point of each unique time series in the data.

Visualization of the residuals and model comparison indicated that the AR1 model improved model likelihood. Model comparison (compareML) tests the inclusion of predictors by comparing Maximum Likelihood scores. Here, all predictors led to a more likely model. Lastly, as the residuals are assumed to be normally distributed, we removed data points whose residual was greater than 2.5 standard deviations from the mean (0.3% data loss). This final model call is given in Listing 1.

```
ELogit3 <- bam(IA_1_ELogit ~ AccentLevel + s(Time, by =
        AccentLevel) + s(Time, Subject, bs = "fs", m =
        1) + s(Time, Item, bs = "fs", m = 1), data =
        ELdat, weights = 1/IA_1_wts, rho = 0.867,
        AR.start = ARstart, subset =
        abs(scale(resid(ELogit2))) < 2.5)
```
Listing 1. Example Gaussian R code

The model summary is provided in Table 2 and the estimated effects are visualized in Fig. 2A, back-transformed to probability scale (plot_smooth). Low Accent resulted in greater likelihood of looking to the target word than did High Accent. The difference in predicted values (plot_diff), Fig. 2B) becomes significant between approximately 640 and 800 ms.

Table 2. Generalized additive mixed model reporting parametric coefficients (Part A) and effective degrees of freedom (edf), reference degrees of freedom (Ref.df), F and p values for the smooth and random effects (Part B)

A. Parametric coefficients	Estimate	Std. error	t-value	p-value
Intercept	−0.9789	0.1433	−6.8297	<0.0001
Accent Level (High)	−0.1825	0.1723	−1.059	0.2896
B. Smooth terms	edf	Ref.df	F-value	p-value
Smooth for Time - Low Accent	1.0007	1.0008	69.6958	<0.0001
Smooth for Time - High Accent	4.615	5.2353	9.0157	<0.0001
Random effect for Subjects	226.3421	323	5.0057	<0.0001
Random effect for Items	475.2971	718	3.743	<0.0001

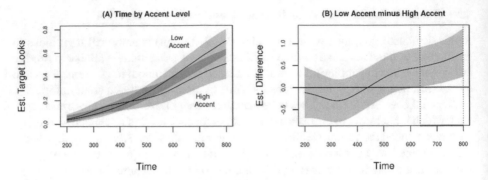

Fig. 2. A: Time by Accent level Interaction on probability scale with 95% CI. B: Difference in target looks between accent levels with 95% CI and vertical lines indicating significance

4.4 Binomial Model with Continuous Interaction

Next, we illustrate the use of a binomial response variable (IA_1_Looks), i.e., the number of samples falling within the target IA vs. those falling outside. While similar to the example above, the model call (Listing 2) contains a few notable differences. First, the interaction between Time and Accent Rating is modeled as a surface using a tensor product. Second, the binomial family is specified; note that these models may take considerably longer to compute. Lastly, a binomial model does not allow for the inclusion of the AR1 parameter to control for autocorrelation or trimming.

```
Binom1 <- bam(IA_1_Looks ~ te(Time, Rating) + s(Time,
       Subject, bs = "fs", m = 1) + s(Time, Item, bs =
       "fs", m = 1), data = Binomdat, family = "binomial")
```
Listing 2. Example Binomial R code

Model comparison was done as above, and indicated that our predictors led to a more likely model. The summary information of the model is provided in Table 3. The estimated effect of the interaction between Time and Accent Rating is visualized in Fig. 3 as a contour surface, back-transformed to probability scale (fvisgam). The probability of looking to the target generally increased through time. However, higher ratings lead

Table 3. Generalized additive mixed model reporting parametric coefficients (Part A) and effective degrees of freedom (edf), reference degrees of freedom (Ref.df), F and p values for the smooth and random effects (Part B)

A. Parametric coefficients	Estimate	Std. error	t-value	p-value
Intercept	−2.9372	0.31	−9.4756	<0.0001
B. Smooth terms	edf	Ref.df	F-value	p-value
Tensor product for Time and Rating	7.9892	8.1686	213.9012	<0.0001
Random effect for Subjects	295.2186	323	63275.341	<0.0001
Random effect for Items	1241.4035	1438	72689.937	<0.0001

to a decrease in probability of target looks, and this is particularly noticeable in the latter half of the time-course.

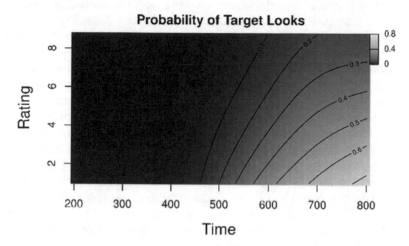

Fig. 3. Time by Accent Rating interaction on probability scale with model-predicted contour lines (back-transformed). Lighter shading indicates increased target looks

5   Conclusion

This paper illustrates the preprocessing and analysis of VWP time-series data, and how this process is facilitated using the packages *VWPre*, *mgcv*, and *itsadug* in R. These allow one to quickly prepare VWP data for time-series analysis, as well as analyze the data in a way that more closely matches their nature. A schematic work-flow for preprocessing and analyzing VWP time-series data (Fig. 4) is provided and meant to illustrate the necessary steps and decisions involved. *VWPre* makes doing this type of research more accessible for researchers by facilitating and streamlining data preparation and visualization. However, special care should be taken during preprocessing, as the

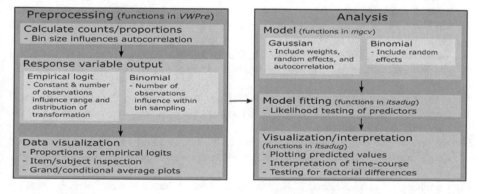

Fig. 4. Work-flow for preprocessing and analyzing VWP time-course data

researcher must understand the reasoning behind and consequences of the selection of the response variable. This selection influences the modeling procedure, as illustrated above. GAMM (as implemented in *mgcv*) provides a more precise method for examining when in time differences arise during online processing (e.g., as in Fig. 2B). This type of precision is difficult to reliably obtain through comparison of means within a user-defined window.

Additionally, GAMM affords researchers the opportunity to investigate how the time-course of looking behavior is influenced by linguistic variables that are inherently continuous. Because of this complexity, the researcher should undertake this type of modeling carefully, as—unlike more traditional methods—GAMMs must be evaluated using a combination of summary statistics, model comparison, and visualization of the estimated effects. *itsadug* makes performing this type of complex time series analysis more straightforward by providing researchers with tools that improve model comparison and visualization of model results. With these tools, the researcher may gain more fine-grained insight into possible nonlinear effects which could otherwise remain obscured by methods that compare group means and/or assume linearity. In sum, this approach provides the opportunity to ask new and interesting questions which can further our understanding of the time-course of language processing.

References

1. Cooper, R.M.: The control of eye fixation by the meaning of spoken language: a new methodology for the real-time investigation of speech perception, memory, and language processing. Cogn. Psychol. **6**, 84–107 (1974)
2. Huettig, F., Rommers, J., Meyer, A.S.: Using the visual world paradigm to study language processing: a review and critical evaluation. Acta Psychol. (Amst)**137**, 151–171 (2011)
3. R Development Core Team: R: A Language and Environment for Statistical Computing. R Foundation for Statistical Computing, Vienna (2016)
4. Porretta, V., Tucker, B.V., Järvikivi, J.: The influence of gradient foreign accentedness and listener experience on word recognition. J. Phonetics **58**, 1–21 (2016)
5. Porretta, V., Kyröläinen, A.-J., van Rij, J., Järvikivi, J.: VWPre: tools for preprocessing visual world data (2016)
6. Wood, S.N.: mgcv: mixed GAM computation vehicle with GCV/AIC/REML smoothness estimation (2016)
7. van Rij, J., Wieling, M., Baayen, R.H., van Rijn, H.: itsadug: interpreting time series and autocorrelated data using GAMMs (2015)
8. Tanenhaus, M.K., Spivey-Knowlton, M.J., Eberhard, K.M., Sedivy, J.E.: Integration of visual and linguistic information in spoken language comprehension. Science **268**, 1632–1634 (1995)
9. Nixon, J.S., van Rij, J., Mok, P., Baayen, R.H., Chen, Y.: The temporal dynamics of perceptual uncertainty: eye movement evidence from Cantonese segment and tone perception. J. Mem. Lang. **90**, 103–125 (2016)
10. Allopenna, P.D., Magnuson, J.S., Tanenhaus, M.K.: Tracking the time course of spoken word recognition using eye movements: evidence for continuous mapping models. J. Mem. Lang. **38**, 419–439 (1998)
11. Chambers, C.G., Tanenhaus, M.K., Magnuson, J.S.: Actions and affordances in syntactic ambiguity resolution. J. Exp. Psychol. Learn. Mem. Cogn. **30**, 687–696 (2004)

12. van Rij, J., Hollebrandse, B., Hendriks, P.: Children's eye gaze reveals their use of discourse context in object pronoun resolution. In: Holler, A., Goeb, C., Suckow, K. (eds.) Empirical Perspectives on Anaphora Resolution. De Gruyter, Berlin (2016)
13. Kamide, Y., Altmann, G.T.M., Haywood, S.L.: The time-course of prediction in incremental sentence processing: evidence from anticipatory eye movements. J. Mem. Lang. **49**, 133–156 (2003)
14. Järvikivi, J., Pyykkönen-Klauck, P., Schimke, S., Colonna, S., Hemforth, B.: Information structure cues for 4-year-olds and adults: tracking eye movements to visually presented anaphoric referents. Lang. Cogn. Neurosci. **29**, 877–892 (2014)
15. Dussias, P.E., Valdés Kroff, J., Gerfen, C.: Using the visual world to study spoken language processing. In: Jegerski, J., Van Patten, B. (eds.) Research Methods in Second Language Psycholinguistics, pp. 93–126. Routledge, New York (2014)
16. Baayen, R.H., van Rij, J., Cecile, D., Wood, S.N.: Autocorrelated errors in experimental data in the language sciences: some solutions offered by generalized additive mixed models. In: Speelman, D., Heylan, K., Geeraerts, D. (eds.) Mixed Effects Regression Models in Linguistics. Springer, Berlin (2016)
17. Hastie, T.J., Tibshirani, R.J.: Generalized Additive Models. Chapman & Hall/CRC, London (1990)
18. Wood, S.N.: Generalized Additive Models: An Introduction with R. Chapman & Hall/CRC Press, Boca Raton (2006)
19. Baayen, R.H., Davidson, D.J., Bates, D.M.: Mixed-effects modeling with crossed random effects for subjects and items. J. Mem. Lang. **59**, 390–412 (2008)
20. Mirman, D., Dixon, J.A., Magnuson, J.S.: Statistical and computational models of the visual world paradigm: Growth curves and individual differences. J. Mem. Lang. **59**, 475–494 (2008)
21. Barr, D.J.: Analyzing "visual world" eyetracking data using multilevel logistic regression. J. Mem. Lang. **59**, 457–474 (2008)
22. Fischer, B.: Saccadic reaction time: Implications for reading, dyslexia, and visual cognition. In: Rayner, K. (ed.) Eye Movements and Visual Cognition, pp. 31–45. Springer, New York (1992)

Examining the Impact of Dental Imperfections on Scan-Path Patterns

Pawel Kasprowski[1]([✉]), Katarzyna Harezlak[1], Pawel Fudalej[2],
and Piotr Fudalej[3,4]

[1] Silesian University of Technology, ul. Akademicka 16, 44-100 Gliwice, Poland
pawel.kasprowski@polsl.pl
[2] Warsaw, Poland
[3] Department of Orthodontics and Dentofacial Orthopedics,
School of Dental Medicine, University of Bern, Bern, Switzerland
[4] Department of Orthodontics, Faculty of Medicine and Dentistry,
Institute of Dentistry and Oral Sciences, Palacky University Olomouc,
Olomouc, Czech Republic

Abstract. The aim of the study was to assess if dental imperfections influence scan-paths recorded during eye tracking sessions of naive observers. The participants observed 15 faces with 8 of them having various dental imperfections. All observations were recorded using an eye tracking device and spatio-temporal information about gaze positions was stored.

Our assumption was that there is dependency between eye movement patterns while observing a face and dental imperfections of that face. There were several eye movement pattern related features analyzed as independent variables and it was checked if there is an effect on level or type of imperfection.

Analyses revealed that participants paid significantly more attention to the mouth area when a face they observed characterized with not perfect teeth. It leads to the conclusion that scan-paths recorded for some number of casual observers may be used to judge the level of imperfection of a face.

Keywords: Eye movement · Face observation · Dental imperfections

1 Introduction

It is well known that eye movement analysis may reveal a lot of information both about a scene being observed and about an observer. As faces are one of the most important visual objects analyzed by humans, eye movements during face observation are a subject of a lot of research [22].

There are two groups of eye tracking based research taking into account face observations. The first one analyzes observers. An analysis of the way how a person observes other people's faces may reveal interesting facts about this person.

P. Fudalej—Private Practice.

© Springer International Publishing AG 2018
I. Czarnowski et al. (eds.), *Intelligent Decision Technologies 2017*,
Smart Innovation, Systems and Technologies 73, DOI 10.1007/978-3-319-59424-8_26

For instance, it may be used to diagnose various diseases [23]. For example, it is well known that autistic people do not like to look at eyes [5]. Similarly, people with Alzheimer disease tend to forget faces and have problems with recognition of familiar faces [18]. The behavioral pattern of faces observations may also be used to identify observers [4,13] or to check if they know the person they are looking at [12]. There are also studies distinguishing eye movement patterns depending on observer's race or age [3,15]. The data may also be used to check what was the purpose of observation - eye movement patterns are different when people try to assess age, gender or race of faces [19] what is in line with the well-known Yarbus observation [25].

The second group of research analyses the image of a face itself, based on eye movement patterns recorded for some number of naive observers. Analysis of such a group of patterns (named scan-paths) may reveal various information. There are studies showing that patterns are different depending on facial expressions [7], stigmas [17], or deformities [10,21].

The research presented in this paper belongs to the latter group. It is a part of a bigger research project aiming at utilizing eye tracking data in dental evaluations of faces. The task of this particular research was to check if eye movement patterns of naive observers looking at a person's face may be used to asses a level of dental imperfection.

The main contribution of the paper is an easily reproducible experiment proving that eye movement patterns of naive observers may be used as a source of information about possible imperfections of the faces, which were looked at. Possible drawbacks of that method were also presented with examples taken from the dataset used. Such an evidence may be used in future to asses a threshold level of imperfection visible to laymen.

2   Background and Related Work

Gaze positions are recorded by an eye tracker in some intervals, e.g. 20 ms. Eye movements may be divided into sequences of fixations - moments when the eye is relatively still - and saccades - a rapid movement to another fixation location. Fixations are the most interesting elements of a scan-path, because brain acquires information about a scene only during the fixation [9]. Therefore, in a preprocessing stage, raw scan-paths are usually converted into a sequence of fixations. Each fixation characterizes with its location and duration.

It is known that eyes are the most important part of a face and attract most attention [11]. Therefore, usually a first or second fixation is placed near one of the eyes [8]. In general eyes, nose and mouth are typical areas of interest and a scan-path while observing a face resembles a triangle (Fig. 1).

With growing possibilities for cosmetic treatment, attractiveness becomes an important concern for many people. The need for perfection results in growing amounts of cosmetic procedures. Even little imperfections can become a serious problem - especially for women [16]. On the other hand, what really matters is the way how other people perceive us. If the imperfection is visible only to

Fig. 1. A scan-path of a typical face observation

specialists, it is useless to correct it with cumbersome and expensive treatments. That is why an objective judgment of the imperfection impact is desired.

Eye tracking may be used as a tool to collect objective information how others perceive our faces. It is known that people pay closer attention to objects which are surprising and not common [9]. It may be supposed that when looking at a face with a visible imperfection they will unintentionally gaze at it. And indeed, in [10] it has been shown that there are significant differences in fixation locations and times for faces with visible defects. Similarly, it occurred that managers conducting face-to-face interviews with applicants with visible cheek stigmas attended more to the cheek and as the result rated applicants lower [17]. There are also eye tracking experiments showing that cleft lip and nose deformity [21] or protruding ear [14] influences fixation patterns of observers. There were also objective differences in the way observers directed their attention to facial features when viewing normal and paralyzed faces [6]. In the recent study, significant deviations in the scan-paths of laypersons between viewing faces of subjects that need orthodontic treatment and normal subjects were noted [24].

3   Method

Every participant of the experiment looked at 15 female faces. Each face appeared for 8 s on a 26″ display. The participants did not know a purpose of the experiment, they were instructed just to look at subsequent faces. There were overall 10 participants involved. To mitigate a 'casual observer' problem (see Summary section for details) there were only male students of Computer Science, age 20–30, and only pictures of young women used in the current research.

Eye movements of the participants were registered during each session with The Eye Tribe eye tracker [2]. It was located below a display, every session started with prior calibration of the eye tracker. Eye positions were registered with a frequency 60 Hz, so each observation gave about 480 raw data points. These points were then transformed to fixations. A dispersion based IDT algorithm was used [20] with 1° dispersion threshold and 50 ms minimal fixation length.

The mouth and teeth were visible on each of the 15 faces. The faces were annotated by a dental expert with one of 5 *levels* of 'imperfection', with 0 denoting normal region of the mouth and 4 denoting a serious imperfection. Among all faces, seven were classified as normal (level 0), three with small crowding as level 1, two faces with diastema as level 2, three as level 3 (diastema, narrow upper dental arch or significant crowding) and one as level 4 (missing upper front teeth). Additionally, the faces were divided into three *types*. Faces with levels 0 or 1 were annotated as 'normal serious' (3 faces) or 'normal smiling' (6 faces) and faces with levels 2–4 were annotated as 'imperfect smiling' (6 faces). These two classifications referred later to as *level* and *type* became two dependent variables in subsequent statistical analyses.

The aim of the research was to check if there are any differences in eye movement patterns when people look at faces with correct and imperfect teeth. To analyze the time spent on gazing on imperfections there was one area of interest (AOI) defined that covered the region of mouth. There were seven independent variables analyzed for each session:

- fixNum - number of fixations,
- fixavgdur - average duration of fixation,
- sacavglen - averaged length of saccades,
- aoi-timeto - time to first fixation on mouth,
- aoi-time - the overall time spent gazing on mouth,
- aio-time-3s - the overall time spent gazing on mouth during the first 3 s of observation,
- nolook - equals 1 when a participant did not look at all towards the mouth region and 0 otherwise.

4   Results

All independent variables were measured for each of 150 sessions and averaged for all *levels* (Table 1) and *types* (Table 2).

The next step was the analysis if the differences for variables between faces with normal and imperfect mouth area are significant (Fig. 2). As none of the variables exhibited normal distribution according to the Shapiro-Wilk test, it was decided to use the non-parametric Kruskal-Wallis test for both dependent variables. Table 3 presents the results for both *levels* and *types*.

The results show that the mouth region attracts observers' attention significantly longer when there are visible dental imperfections (*aoi-time* variable).

Table 1. Mean values and standard deviations for different *levels* of imperfection

level	fixNum	fixavgdur [ms]	sacavglen [px]	aoi-timeto [ms]	aoi-time [ms]	nolook [%]	aoi-time-3s[ms]
0	11.27 (4.51)	779.29 (590.27)	6.62 (1.74)	5178.61 (3804.06)	330.84 (401.85)	34 (48)	201.34 (295.97)
1	12 (4.56)	701.98 (475.95)	7.37 (1.82)	4394.93 (3572.45)	513.53 (573.13)	23 (43)	216.2 (302.43)
2	10.4 (3.37)	706.52 (250.56)	6.22 (1.17)	3341.8 (3034.89)	576.5 (414)	10 (32)	442.8 (490.15)
3	10.4 (3.21)	750.61 (343.76)	6.46 (1.63)	3218.43 (3190.58)	724.97 (766.82)	13 (35)	392.8 (416.77)
4	9.7 (5.03)	1497.96 (2310.97)	5.88 (2.25)	3997.8 (3667.26)	793.6 (492.53)	20 (42)	587.9 (619.77)

Table 2. Mean values and standard deviations for different *types* of faces.

type	fixNum	fixavgdur [ms]	sacavglen [px]	aoi-timeto [ms]	aoi-time [ms]	nolook [%]	aoi-time-3s[ms]
Normal serious	12.5 (4.44)	676.52 (656.95)	6.98 (1.42)	4959.33 (3663.16)	347.67 (357.52)	30 (47)	248.9 (364.81)
Normal smiling	10.93 (4.33)	788.22 (496.38)	6.83 (1.93)	5400.02 (3726.2)	420.72 (539.94)	35 (48)	177.42 (273.77)
Imperfect	10.52 (4)	876.4 (1009.68)	6.33 (1.73)	3191.98 (3187.93)	639.03 (621)	13 (34)	409.32 (442.6)

Table 3. The results of the Kruskal-Wallis test for *levels* and *types*. Statistically significant differences denoted with a colored cell and (*).

variable	levels		types	
	KW test	p-value	KW test	p-value
fixNum	3.046	0.55	4.165	0.125
fixavgdur	3.592	0.464	5.636	0.06
sacavglen	7.489	0.112	3.977	0.137
aoi-timeto	7.961	0.093	16.443	0 (*)
aoi-time	15.783	0.003 (*)	9.214	0.01 (*)
nolook	6.662	0.155	7.825	0.02 (*)
aoi-time-3s	8.694	0.069	10.699	0.005 (*)

The effect was also significant for *types* when only the first 3 s of observation were taken into account (*aoi-time-3s* variable). The observers tend to look at imperfections faster (*aoi-timeto*), but the effect was significant only for *types*. Observation of faces with imperfections characterized also by fewer but longer fixations, and shorter saccades, however the effects were not significant (Fig. 2).

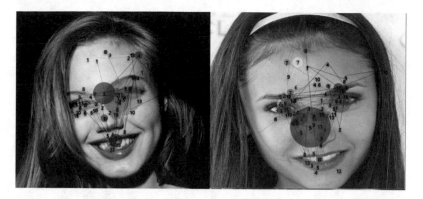

Fig. 2. Two faces from the dataset with fixations recorded for all observers. Eyes and nose are the most fixated regions, however, there is also a significant number of long fixations on mouth on the left picture (classified as 'imperfect smiling', level 3).

Even while for *levels* the significant effect was visible only for *aoi-time* variable, the other variables exhibited strong Pearson correlation with *level* value: 0.74 for the *aoi-time-3s*, −0.52 for the *aoi-timeto* and −0.55 for the *nolook*.

Table 4. The Mann-Whitney test for each pair of types. IMP-face with imperfection, SMIL-smiling face, SER-serious face. Significant differences after applying the Bonferrioni correction are denoted with a colored cell and (*).

variable	IMP-SMIL	IMP-SER	SMIL-SER
fixNum	0.645	0.039	0.122
fixavgdur	0.701	0.017	0.063
sacavglen	0.129	0.06	0.807
aoi-timeto	0 (*)	0.003 (*)	0.55
aoi-time	0.008 (*)	0.016 (*)	0.958
nolook	0.006 (*)	0.062	0.639
aoi-time-3s	0.001 (*)	0.073	0.438

The next step was a pairwise comparison between 'normal serious', 'normal smiling', and 'imperfect' types. The results presented in Table 4 show that there are significant differences between 'imperfect' faces and both 'correct smiling' and 'correct serious' faces. In the same time there are no significant differences between correct faces regardless of their facial expressions.

Figure 3 presents values of *aoi-time* for each face and *level* of imperfection. Pearson correlation between these two values is 0.82. It is visible that times for imperfect faces are generally higher, however a deviation among pictures is observable. The most notable outlier is the face number 5 which, despite of being classified as level 1, received high *aoi-time* value. It occurred, that it was a picture of Kate Middleton (The Duchess of Cambridge), who is known for her 'perfectly imperfect smile' specially prepared by dentistry experts [1].

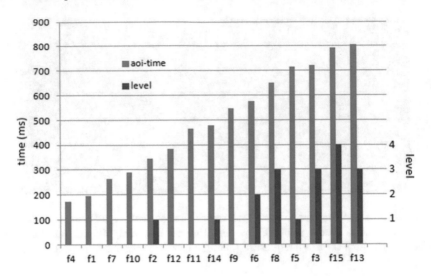

Fig. 3. Values of *aoi-time* and *level* for each face.

5   Summary

The results presented in this experiment show that it is possible to distinguish normal and imperfect faces based on eye tracking results registered for casual observers looking at these faces. It was shown that even for relatively small number of observers it is possible to obtain significant results.

Of course this research is only a preliminary screening, more experiments are required to obtain reliable results. The problem of 'dental imperfection' itself is not simple. For instance, one of the faces which was classified with level equal to 0 (no imperfections) was with braces. It occurred that *aoi-timeto* variable for this face had the lowest value among all faces, what indicates that the braces quickly attracted observers' attention. However, *aoi-time* variable was one of the lowest, what means that observers did not sustain their attention for long.

Another problem is a definition of a 'casual observer'. While it is obvious that for neutral judgments about dental imperfections the observers should not be connected with dentistry, there are other factors which could also be taken into account such as the observers past experience, gender and even sexual preferences. The only way to be independent of these factors is to prepare a considerably big and diverse group of participants. To overcome this problem there were only male students of computer science and only pictures of young women used in the current research.

In the future we plan to conduct a wider research taking into account more pictures with carefully estimated imperfections. It would also be interesting to compare scan-paths of professional dentists with laymen.

References

1. Smile like a princess: the secret to Kate Middleton's perfect pearly whites revealed. http://www.dailymail.co.uk/femail/article-2066489/Kate-Middleton-smile-Secret-Duchess-Cambridges-perfect-pearly-whites.html
2. The Eye Tribe eye tracker. https://theeyetribe.com/
3. Blais, C., Jack, R.E., Scheepers, C., Fiset, D., Caldara, R.: Culture shapes how we look at faces. PLoS One 3(8), e3022 (2008)
4. Cantoni, V., Galdi, C., Nappi, M., Porta, M., Riccio, D.: GANT: gaze analysis technique for human identification. Pattern Recogn. 48(4), 1027–1038 (2015)
5. Dalton, K.M., Nacewicz, B.M., Johnstone, T., Schaefer, H.S., Gernsbacher, M.A., Goldsmith, H., Alexander, A.L., Davidson, R.J.: Gaze fixation and the neural circuitry of face processing in autism. Nat. Neurosci. 8(4), 519–526 (2005)
6. Dey, J.K., Ishii, L.E., Byrne, P.J., Boahene, K., Ishii, M.: Seeing is believing: objectively evaluating the impact of facial reanimation surgery on social perception. Laryngoscope 124(11), 2489–2497 (2014)
7. Eisenbarth, H., Alpers, G.W.: Happy mouth and sad eyes: scanning emotional facial expressions. Emotion 11(4), 860 (2011)
8. Guo, K., Smith, C., Powell, K., Nicholls, K.: Consistent left gaze bias in processing different facial cues. Psychol. Res. 76(3), 263–269 (2012)
9. Holmqvist, K., Nyström, M., Andersson, R., Dewhurst, R., Jarodzka, H., Van de Weijer, J.: Eye Tracking: A Comprehensive Guide to Methods and Measures. OUP, Oxford (2011)
10. Ishii, L., Carey, J., Byrne, P., Zee, D.S., Ishii, M.: Measuring attentional bias to peripheral facial deformities. Laryngoscope 119(3), 459–465 (2009)
11. Janik, S.W., Wellens, A.R., Goldberg, M.L., Dell'Osso, L.F.: Eyes as the center of focus in the visual examination of human faces. Percept. Motor Skills 47(3 Pt 1), 857–858 (1978)
12. Kasprowski, P.: Mining of eye movement data to discover people intentions. In: International Conference: Beyond Databases, Architectures and Structures, pp. 355–363. Springer (2014)
13. Kasprowski, P., Harezlak, K.: The second eye movements verification and identification competition. In: 2014 IEEE International Joint Conference on Biometrics (IJCB), pp. 1–6. IEEE (2014)
14. Litschel, R., Majoor, J., Tasman, A.J.: Effect of protruding ears on visual fixation time and perception of personality. JAMA Fac. Plast. Surg. 17(3), 183–189 (2015)
15. Liu, S., Quinn, P.C., Wheeler, A., Xiao, N., Ge, L., Lee, K.: Similarity and difference in the processing of same-and other-race faces as revealed by eye tracking in 4-to 9-month-olds. J. Exp. Child Psychol. 108(1), 180–189 (2011)
16. Lopez, Y., Le Rouzic, J., Bertaud, V., Pérard, M., Le Clerc, J., Vulcain, J.M.: Influence of Teeth on the smile and physical attractiveness: a new internet based assessing method. Open J. Stomatol. 3, 52–57 (2013)
17. Madera, J.M., Hebl, M.R.: Discrimination against facially stigmatized applicants in interviews: an eye-tracking and face-to-face investigation. J. Appl. Psychol. 97(2), 317 (2012)
18. Mendez, M.F., Mendez, M., Martin, R., Smyth, K.A., Whitehouse, P.J.: Complex visual disturbances in Alzheimer's disease. Neurology 40(3 Part 1), 439–439 (1990)
19. Nguyen, H.T., Isaacowitz, D.M., Rubin, P.A.: Age-and fatigue-related markers of human faces: an eye-tracking study. Ophthalmology 116(2), 355–360 (2009)

20. Salvucci, D.D., Goldberg, J.H.: Identifying fixations and saccades in eye-tracking protocols. In: Proceedings of the 2000 Symposium on Eye Tracking Research and Applications, pp. 71–78. ACM (2000)
21. van Schijndel, O., Litschel, R., Maal, T.J., Bergé, S.J., Tasman, A.J.: Eye tracker based study: perception of faces with a cleft lip and nose deformity. J. Cranio-Maxillofacial Surg. **43**(8), 1620–1625 (2015)
22. Tanaka, J.W., Gordon, I.: Features, configuration, and holistic face processing. In: The Oxford Handbook of Face Perception, pp. 177–194 (2011)
23. Vidal, M., Turner, J., Bulling, A., Gellersen, H.: Wearable eye tracking for mental health monitoring. Comput. Commun. **35**(11), 1306–1311 (2012)
24. Wang, X., Cai, B., Cao, Y., Zhou, C., Yang, L., Liu, R., Long, X., Wang, W., Gao, D., Bao, B.: Objective method for evaluating orthodontic treatment from the lay perspective: an eye-tracking study. Am. J. Orthod. Dentofac. Orthop. **150**(4), 601–610 (2016)
25. Yarbus, A.L.: Eye movements during perception of complex objects. In: Yarbus, A.L. (ed.) Eye Movements and Vision. Springer, New York (1967)

Touch Input and Gaze Correlation on Tablets

Pierre Weill-Tessier$^{(\boxtimes)}$ and Hans Gellersen

School of Computing and Communications, Lancaster University, Lancaster, UK
{p.weill-tessier,h.gellersen}@lancaster.ac.uk

Abstract. Prior work has shown gaze behaviour correlates with mouse movement in conventional desktop interfaces. As many devices now employ direct touch instead of a mouse, we have conducted a first study analysing how gaze correlates with touch input. We collected touch and gaze data on a tablet from 24 participants, on search, shopping, and "link-following" tasks representative of typical online use of tablets. Our analysis shows a gaze fixation typically leads touch input by 337 ms within 197 pixels. We observed a stronger correlation in the context of link-following, in comparison with touch on other HTML objects and virtual keyboard.

Keywords: Touch input · Eye tracking · Gaze · Hand-eye correlation · Tablet interaction

1 Introduction

When interacting with computing devices, manual input is highly connected with how users visually inspect UI content. The correlation between manual input (using the mouse as a proxy for the hand) and gaze has been of particular interest in many research efforts [3,5,10], to better understand visual attention across the variety of computing devices we use on a daily basis, and to propose new concepts that enhance the interaction. However, besides the increasing popularity of tactile devices, correlation between touch input and gaze has, to our knowledge, not been studied yet.

In this work, we investigate how tapping correlates with gaze on a tablet device. We conducted a study with 24 participants and collected data related to touch input, gaze and tapped targets. Our study focused on Internet tasks, as browsing is a typical task widely used as study context for measuring mouse-eye correlation [8] and commonly performed by tablet users.

Analysis of the data indicates the following results: (1) gaze precedes touch with similar spatial and temporal features as observed with the mouse, and (2) the distance kept between the gaze and the touch varies across users, and is influenced by the learning and anticipation effects of the tasks.

2 Related Work

Before touch input modality became widespread, mouse served as a proxy for manual input on computers. Early work on gaze and mouse input correlation in

© Springer International Publishing AG 2018
I. Czarnowski et al. (eds.), *Intelligent Decision Technologies 2017*,
Smart Innovation, Systems and Technologies 73, DOI 10.1007/978-3-319-59424-8_27

Human-Computer Interaction can be found in [13], where Smith et al. studied the correlation in target selection tasks and found several patterns, and in [5], where Chen et al. found correlation patterns applied to web browsing tasks and an average correlation of 0.58. Liebling and Dumais [10] explored the correlation of eye and mouse in everyday computer work tasks. They confirmed the eyes lead the mouse, but nuanced this paradigm, indicating it occurs only two thirds of the time, as "[this] depends on the type of target and the familiarity with the application". We contribute to the understanding of the correlation between manual input and gaze in a natural environment by studying touch input instead of the mouse.

Browsing Internet is a common activity on tablets. Literature related to Internet usage often focuses on search tasks and Search Engine Results Page (SERP). Search queries can be categorised as informational, navigational or transactional - navigational being less common than transactional, and even less than informational [4,12]. Search tasks usually consist of answering informational questions [7], which can be combined with navigational questions [8]. Another classic activity is on-line shopping, with or without instruction regarding the items to buy [2,6]. Instead of focusing our work on a particular type of Internet based activity, we selected several in order to reflect the tablet's use in real life.

3   Study

3.1   Participants

We collected data from 24 participants (9 female, age $\mu = 31.4$, $\sigma = 11$). All of them were familiar with Internet browsing and experienced with tablets and touch devices (3.7 and 3.9 of average on a 5-point Likert scale). Some participants needed visual correction during the study (7 wore glasses, 3 wore contact lenses). All but one were right-handed.

3.2   Tasks

We prepared 3 tasks to cover different Internet related activities, while keeping similarities with other studies, and maintaining naturalness in the tablet's usage.

The **search** task comprises 10 questions which participants were asked to answer, by finding the relevant information on Internet with the means of their choice. This task echoes a common tablet activity, as often found in research literature. We chose 5 *informational* and 5 *navigational* questions, inspired by similar research.

The choice of a **shopping** task is driven by the interaction with richer content website style[1] and with forms (therefore typing). We asked participants to simulate the purchase of at least 10 different items. This process required them to fill forms.

---

[1] We chose the website of a leading supermarket in the UK, Sainsbury's.

The **game** task ("Wikipedia game") is thought of as a way to generate data from hyperlinks, in a non systematic manner, with deeper cognitive and reading demands. In Wikipedia, participants were asked to reach a specific article from a specific source article by only following internal hyperlinks. A basic description of the articles' topic was provided for help. They could play 2 rounds.

3.3 Apparatus

We favoured tablets for the data collection over other touch devices because of their reasonable size, prevalence, and compatibility with eye trackers. We used a Microsoft Surface Pro 3 (2160×1440 pixels resolution). We chose the Tobii X2-60 eye-tracker (60 Hz), designed for studies on smaller devices. It comes with a stand designed for interacting with the device without occluding the eye tracker. Figure 1 shows how the stand and eye-tracker were set for the data collection.

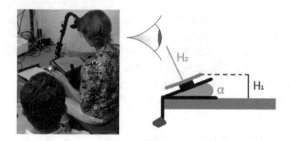

Fig. 1. Eye-tracker and stand configuration ($\alpha \approx 25°$, $H_1 \approx 12$ cm and $H_2 \approx 40$ cm).

3.4 Data and Implementation

The data collection consisted of retrieving the following information: touch input, on-screen gaze position, and tapped object characteristics. Touch data is collected through different steps. Initially, we parse the display's HID reports and send the touch time-stamp and coordinates to a Java application. The Java application interprets them as touch gestures, and records them onto log files. We wrote an application which retrieves gaze data samples from the eye-tracker, and writes in a log file their time-stamp, normalised on-screen position and validity code. The application also runs a 9-point calibration (22-pixel radius) before each task and a drift evaluation afterwards. We implemented a web browser (based upon Internet Explorer 11) in order to easily get feedback from it and offer a basic UI for all participants. The browser had a dimension of 1440×960 pixels[2], with a view port of 1440×914 pixels, topped by a navigation bar (Fig. 2). The tapped objects have 3 distinct natures: HTML, browser or keyboard element. HTML and browser targets are tracked via the browser which writes related

---

[2] This is the dimension of the full screen in the non high DPI mode on the tablet.

Fig. 2. Browser's navigation bar part (truncated).

information (time-stamp, nature, position and size) into a log file. We made a specific application to track keyboard elements and write related information (time-stamp and key code) into another log file. In a post-hoc step, we then merge these different log files with the taps log files into a single file, based on the time-stamps.

4   Results

4.1   General Results

We collected in total 574 675 touch data samples and 1 869 705 gaze data samples. We tracked 3 types of touch gesture actions: taps (72%), pans (28%) and zooms ($< 1\%$). The tablet was large enough for the participants to use comfortably (seldom zooms). The tapped object distribution is as follows: keyboard (68.9%), HTML (26.5%), browser (4.6%). Fixations are computed post-hoc with OGAMA[3] on a spatial detection threshold of 22 pixels ($\sim 0.56°$ of visual angle).

4.2   Similarity with Mouse Studies

We learnt from studies in psychology that, when pointing at objects, gaze leads the hand [1]. This is also verified in HCI studies with the mouse as manual input. We question how touch input compares with the mouse counterpart.

As we consider the **tapping** part of the whole target selection process, touch input can be assimilated with a punctual event in time and space. Similar to mouse/gaze studies, we check where gaze is located when a tap arises: the difference between tap and fixations positions around tap moment. We use the fixation's start moment for the temporal dimension.

Figure 3(a) shows gaze approaches the selected target *before* the tap is actually performed. This estimation is obtained by using a Generalized Additive Model (cubic spline). We estimate gaze precedes the tap by 0.337 s, within 197 pixels - similar to what is reported in gaze/mouse correlation studies (about 0.3 s and 160 pixels [3]). The spatial difference can be explained by the decrease of pointing precision with finger, and a target size generally greater than in [3]).

4.3   Influence of Participants, Tasks and Targets on the Correlation

In the previous section, we describe a coarse estimation for *all* participants, tasks and target types. But do they influence the correlation?

---

[3] http://www.ogama.net/.

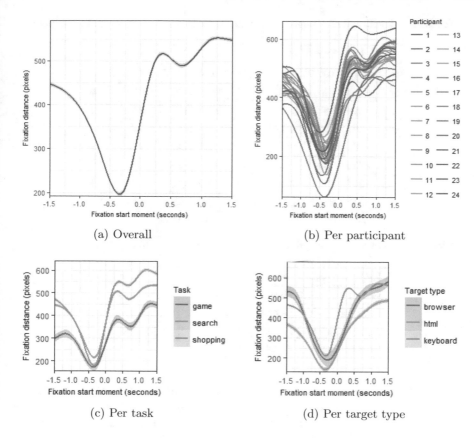

Fig. 3. Fixation start moment vs. fixation distance (relative to tap moment/position).

We estimate the relationship between touch input and gaze for each participant, task and tapped object type with the same model as in Sect. 4.2. In order to compare them, we report the descriptive statistics related to the **minimum point** given by the estimation model.

Estimations for each participant are plotted in Fig. 3(b), and the minima's value reported in Table 1. For all participants, gaze leads the tap. However, disparity is observed between each participant. This difference is more important in space ($\Delta = 218.8$ pixels, $\sigma = 50.2$ pixels) than in time ($\Delta = 249$ ms, $\sigma = 57$ ms). We deduce that the distance within which a user keeps gaze away from the target is a personal feature, and influences the correlation.

Estimations for each task are plotted in Fig. 3(c) and the minima's value reported in Table 2 (left part). Again, we observe that gaze precedes touch, and that tasks influence the spatial dimension ($\Delta = 43.4$ pixels, $\sigma = 23.3$ pixels) rather than the temporal dimension ($\Delta = 17$ ms, $\sigma = 9$ ms). For the search task, the minimum has a greater Y-axis value than for the other two tasks. We can interpret this as a consequence of SERPs being systematically queried by the participants for

Table 1. Fixation start moment vs. fixation distance (minima, per participant)

P[a]	S.M.[b]	Dist.[c]	P[a]	S.M.[b]	Dist.[c]	P[a]	S.M.[b]	Dist.[c]
#1	−377	106.9	#9	−402	246.4	#17	−325	200.1
#2	−432	135.1	#10	−333	209.9	#18	−24	183.6
#3	−324	185.8	#11	−411	221.5	#19	−339	173.9
#4	−291	184.3	#12	−304	249.2	#20	−368	62.2
#5	−365	181.3	#13	−323	137.1	#21	−347	176.9
#6	−314	234	#14	−363	192.6	#22	−405	191.2
#7	−274	206	#15	−408	265.5	#23	−417	216.1
#8	−313	226.8	#16	−305	243.4	#24	−489	281

[a]Participant [b]Start moment (ms) [c]Distance (pixels)

Table 2. Fixation start moment vs. fixation distance (minima, per task/target type)

Task	S.M.[a]	Dist.[b]	Target type	S.M.[a]	Dist.[b]
Search	−335	214.3	Keyboard	−332	210
Shopping	−339	177.9	HTML	−343	141.9
Game	−352	170.9	Browser	−286	189.5

[a]Start moment (ms) [b]Distance (pixels)

that task. Users mainly follow the first link after scanning a few [9]. Thus, taps are possibly already "prepared" to be performed while users still scan the page, and then tap without a need to acquire the target again. The task nature has therefore a clear influence on the correlation.

Estimations for each tapped object type are plotted in Fig. 3(d) and the minima's value reported in Table 2 (right part). Gaze still precedes touch, and the temporal difference varies less than in space (respectively $\Delta = 57$ ms, $\sigma = 3$ ms; $\Delta = 68.1$ pixels, $\sigma = 34.94$ pixels). Fixations for the keyboard and browser objects seem to happen earlier before the touch, with a farther distance to the target. We explain this by the potential learning effect in typing and using the browser. Participants take less time "searching" the target, and do not need to visually focus on it as they know where it is. Thus the tapped target has a role in the correlation, depending on its likelihood to be known in advance.

4.4   A Particular Fixation: $F_{Closest}$

We focus on the description of the specific fixation, $F_{Closest}$, that arises *before* the tap moment *at the closest* to the tap position. Describing $F_{Closest}$ confronts the estimations given earlier and may serve as a baseline for further studies.

For each tap, we retrieve $F_{Closest}$ in a window starting −0.6 s before the tap moment[4].

---

[4] Value based on the minimum point reported in Sect. 4.2 and the difference within participants in Sect. 4.3 ($-0.337 - 0.249 \approx -0.6$).

Figures 4(a) and (b) respectively show the histograms of $F_{Closest}$'s start moment (to the tap moment) and distance (to the tap position), and the associated quartiles values. The modes are -0.313 s for the start moment and 34 pixels for the distance. No correlation was found for the HTML targets between their size and $F_{Closest}$'s distance/time.

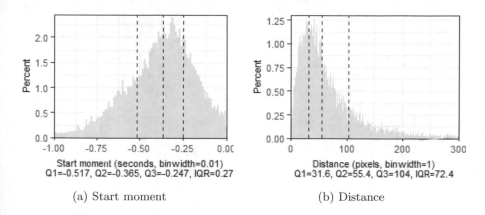

(a) Start moment (b) Distance

Fig. 4. $F_{Closest}$'s start moment/distance histograms and quartiles.

We study the spatial distribution of $F_{Closest}$ around the touch points, plotted in Fig. 5(a). The standard deviation on the X-axis is 205.4 pixels, and 116.4 pixels on the Y-axis. Both mean and median positions show an offset which can be explained: web users tend to look more at the top-left part of the page [11].

Figure 5(b) represents $F_{Closest}$'s mean and median positions of each participant. Although they are spread around the global mean and median positions, they remain generally offset towards the top-left direction, most notably for the mean positions (92% of participants, against 67% for the median positions).

Figure 5(c) illustrates $F_{Closest}$'s mean and median positions for each task. For both search and shopping tasks, we notice the mean and median positions do not vary more than 8 pixels around the overall values in each direction. For the game task, we observe that $F_{Closest}$'s mean and median positions are closer to the tap point. In this task, the targets' position (mostly links) cannot be "learnt" nor anticipated, contrary to the other tasks. For the search task, learning effect of selecting the first link(s) in the SERP, or a tap anticipation as discussed in the Sect. 4.3 can appear. For the shopping task, learning effect may come from the commercial website interaction. We suppose the learning effect and the tap anticipation brought by a task can influence the distance between gaze and tap: when there are none of these effects, gaze acquires targets with a closer distance.

$F_{Closest}$'s mean and median positions for each tapped object type are represented in Fig. 5(d). There is an expected difference for the browser elements. Being situated in the top part of the screen, $F_{Closest}$'s mean and median positions are not likely to show an offset on the top-left side of the screen. Browser

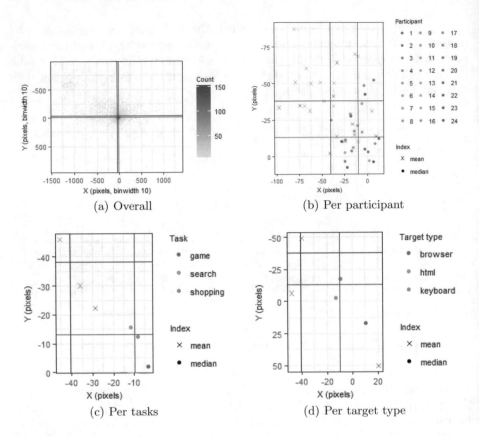

Fig. 5. $F_{Closest}$'s mean and median positions around tap position (red intersection: overall mean $(-40.5, -38.1)$ pixels; green intersection: overall median $(-9.8, -13.1)$ pixels).

elements allow navigation and trigger changes in the view-port situated below, hence an opposite offset direction. The case of the HTML elements shows a very small vertical offset (less than 7 pixels for both mean and median values) indicating that gaze is more often vertically aligned with the targets. We can interpret this result as an effect of reading: most HTML targets are links ($\approx 40\%$) and text input fields ($\approx 26\%$).

5   Discussion

Our results give a coarse description of the correlation between touch and gaze in a natural context. Understanding this correlation can lead the way to improved gaze estimation methods, based on touch, notably, for eye tracking calibration. Even finer results were limited by our study design. We did not consider touch within the whole web page content, and therefore cannot understand how elements other than the tapped one influence the correlation. Also, preserving the naturalness of the tasks in the study led to an heterogeneous data set: participants tapped on different elements since they were free to browse at will to complete tasks.

6 Conclusion

We have devised a data collection in order to study the correlation between gaze and touch input on tablets. To reflect the naturalness of the tablet usage, and align our study with other works, we designed 3 Internet tasks as context. Our work confirms similar results found in previous gaze/mouse correlation studies: gaze precedes the touch by 0.337 s, within 197 pixels. We described a specific fixation we coined $F_{Closest}$ (the fixation that is the closest from the target position, before the tap actually happens). We show that $F_{Closest}$'s median position can act as a gaze estimator, especially for a task which generates few learning and tap anticipation effects. Our study only focused on the tapping part of the gesture movement. Our future work will investigate a wider scope in order to cover the complete touch gestural mechanism.

References

1. Abrams, R.A., Meyer, D.E., Kornblum, S.: Eye-hand coordination: oculomotor control in rapid aimed limb movements. J. Exp. Psychol. Hum. Percept. Perform. **16**(2), 248–267 (1990)
2. Atterer, R., Schmidt, A.: Tracking the interaction of users with AJAX applications for usability testing. In: Proceedings of the SIGCHI Conference on Human Factors in Computing Systems, CHI 2007, pp. 1347–1350. ACM, New York (2007). http://doi.acm.org/10.1145/1240624.1240828
3. Bieg, H.J., Chuang, L.L., Fleming, R.W., Reiterer, H., Bülthoff, H.H.: Eye and pointer coordination in search and selection tasks. In: Proceedings of the 2010 Symposium on Eye-Tracking Research and Applications, ETRA 2010, pp. 89–92. ACM, New York (2010). http://doi.acm.org/10.1145/1743666.1743688
4. Broder, A.: A taxonomy of web search. SIGIR Forum **36**(2), 3–10 (2002). http://doi.acm.org/10.1145/792550.792552
5. Chen, M.C., Anderson, J.R., Sohn, M.H.: What can a mouse cursor tell us more?: correlation of eye/mouse movements on web browsing. In: CHI 2001 Extended Abstracts on Human Factors in Computing Systems, CHI EA 2001, pp. 281–282. ACM, New York (2001). http://doi.acm.org/10.1145/634067.634234
6. Chudá, D., Krátky, P.: Usage of computer mouse characteristics for identification in web browsing. In: Proceedings of the 15th International Conference on Computer Systems and Technologies, CompSysTech 2014, pp. 218–225. ACM, New York (2014). http://doi.acm.org/10.1145/2659532.2659645
7. Guo, Q., Jin, H., Lagun, D., Yuan, S., Agichtein, E.: Towards estimating web search result relevance from touch interactions on mobile devices. In: CHI 2013 Extended Abstracts on Human Factors in Computing Systems, CHI EA 2013, pp. 1821–1826. ACM, New York (2013). http://doi.acm.org/10.1145/2468356.2468683

8. Huang, J., White, R., Buscher, G.: User see, user point: gaze and cursor alignment in web search. In: Proceedings of the SIGCHI Conference on Human Factors in Computing Systems, CHI 2012, pp. 1341–1350. ACM, New York (2012). http://doi.acm.org/10.1145/2207676.2208591

9. Jansen, B.J., Spink, A.: Web Mining. IGI Global (2005). http://www.igi-global.com/chapter/analysis-document-viewing-patterns-web/31146

10. Liebling, D.J., Dumais, S.T.: Gaze and mouse coordination in everyday work. In: Proceedings of the 2014 ACM International Joint Conference on Pervasive and Ubiquitous Computing: Adjunct Publication, UbiComp 2014 Adjunct, pp. 1141–1150. ACM, New York (2014). http://doi.acm.org/10.1145/2638728.2641692

11. Nielsen, J.: Horizontal attention leans left (2010). https://www.nngroup.com/articles/horizontal-attention-leans-left/

12. Rose, D.E., Levinson, D.: Understanding user goals in web search. In: Proceedings of the 13th International Conference on World Wide Web, WWW 2004, pp. 13–19. ACM, New York (2004). http://doi.acm.org/10.1145/988672.988675

13. Smith, B.A., Ho, J., Ark, W., Zhai, S.: Hand eye coordination patterns in target selection. In: Proceedings of the 2000 Symposium on Eye Tracking Research and Applications, ETRA 2000, pp. 117–122. ACM, New York (2000). http://doi.acm.org/10.1145/355017.355041

Modeling Search in Web Environment: The Analysis of Eye Movement Measures and Patterns

Irina Blinnikova$^{(\boxtimes)}$ and Anna Izmalkova$^{(\boxtimes)}$

Lomonosov Moscow State University, Moscow, Russia
blinnikovamslu@hotmail.com, mayoran@mail.ru

Abstract. In the current study we analyzed several factors and their interaction on the speed and spatial organization of visual search in a modeled graphical interface environment. The participants had to find the target stimulus in a 9 × 9 matrix with 81 images, commonly used in web design. Search time and eye movement data were recorded. In addition to traditional measures we analyzed intersaccadic angles and saccade directions. The stimuli were either black and white, or colored (the Chromaticity factor). The Chromaticity factor did not exert direct influence on the search time, and there was little effect of chromaticity on eye movement characteristics, apart from fixation count in the area of target words. The target could be presented either as an image or as a word (the Target template factor). The Target template factor exerted significant influence on search time, fixation duration and on saccadic amplitude and velocity. Moreover, we identified sequential and non-sequential visual search patterns, based on the combination of intersaccadic angle and saccade direction measures, which proved to differ in subjects with high and low impulsivity.

Keywords: Eye movements · Visual search · Cognitive styles · Intersaccadic angle · Saccade direction

1 Introduction

The task of visual search implies detecting a specific target in a display filled with distractors. The problem of visual search has been one of the most significant issues in cognitive psychology and cognitive ergonomics since the 1980s. However, today researchers analyze the process of searching for realistic, semantically rich objects in complex visual contexts [see 5]. This is promoted by the development of the Internet technologies. For the developers of internet resources it is very important to understand the architecture and mechanisms of the processes providing the support information search in a complex environment. The determination of the conditions that increase the search efficiency is important as well. One of the very important tasks nowadays is predicting search efficiency on the basis of subject's eye movements. In our research we modeled visual search in Web environment and analyzed the influence of several factors on search efficiency, basic measures and complex eye movement patterns.

Visual search in Web environment and its determinants. Analysis of the literature on visual search in the Internet environment allows to single out three groups of factors

© Springer International Publishing AG 2018
I. Czarnowski et al. (eds.), *Intelligent Decision Technologies 2017*,
Smart Innovation, Systems and Technologies 73, DOI 10.1007/978-3-319-59424-8_28

determining the Web search strategy [2]: features of the subject (e.g., cognitive styles) [9], features of the target stimulus (e.g., target template format) [10], features of the search context (e.g., color of the stimuli) [13].

The current research was aimed at disclosing the mechanisms of detecting graphic elements among similar distractor-stimuli. Symbolic images (icons) have long been used in human-computer interaction. At first, the icons base was relatively small, easy to memorize and use. However, with the development of internet-based technologies, it has become virtually infinite, as anyone can create his or her own web page and fill it with various graphic content. Therefore, the interest in studying the mechanisms of visual search, perception, memorizing, and identifying pictograms has significantly increased [6].

Icons possess physical characteristics and semantic characteristics. In his work, J. Wolfe describes dozens of "physical" categories (e.g. movement, orientation, shape, size, color, etc.[1]) [15]. Our research was aimed at disclosing the role of grayscale/colored presentation of the stimuli. In the visual search for icons semantic characteristics are connected with the ability to determine the meaning of the searched element[2]. One of the possible ways to manipulate semantic parameters is with a search query, namely, the target template of the searched element [10]. The target template can be set either as an image, or as a word. If the target stimulus is set as a picture, a strictly defined task for visual search is modeled. If the target stimulus is set verbally, a less defined task for visual search is modeled. A search of this kind occurs when the subject does not know precisely what the target looks like, but can predict it by constructing mental images [7]. Mental activity of this kind is described by specialists in the field of information search as search query reformulation.

The physical characteristics of the website space and the format of the target template, apparently, are very important determinants of a search. However, it is essential to take into account the individual traits of the subjects. One of the important factors that influence Web searching is that of users' cognitive styles[3], which is currently under research in information science. Previous research has shown that users' cognitive styles play an important role in Web searching. However, only limited studies have showed the relationship between cognitive styles and Web search behavior. Most importantly, it is not clear which components of Web search behavior are influenced by cognitive styles. One of the most accurate methods of measuring cognitive styles was proposed by Kagan et al. in 1964 [8] - the Matching Familiar Figures Test (MFFT). Two variables are measured—latency (time taken to respond) and accuracy (number of errors). Kagan et al. classified people into two groups on the basis of their scores on these two variables relative to the median—reflective individuals (long latency, high accuracy) and

---

[1] Those characteristics are typically called low-level characteristics of cognitive processing.

[2] It is often described as the higher-level characteristics of cognitive processing.

[3] Cognitive style is said to be the relatively stable strategies, preferences and attitudes that determine an individual's 'typical modes of perceiving, remembering and problem solving' [8, 9]. They are thought of as the modes by which humans approach, acquire and process information, as well as including the consistent ways in which an individual memorizes and retrieves information.

impulsive individuals (short latency, low accuracy). This measure has been used since to study user behavior in Web environment [9].

Eye movements in visual search. Traditional studies on visual search normally attempted to minimize the number of eye movements in the process. However, the size and complexity of natural scenes have created the need for the analysis of eye movement activity and other behavioral components of the search process.

The research of Web pages visual search indicates that the search rate is connected with the way the virtual environment is organized and with the parameters of eye movements. In one study [12], eye movement measures of visual search were registered on 22 web pages. Eye movement characteristics were shown to depend on the web site's style, on information organization, and the subject's gender. In another study [3], it was illustrated that icon search was connected with the general design of a web page.

In the literature on eye movements in addition to basic measures there are attempts to allocate more complex integrative indices. One of such indices is based on the comparison of successive saccades. Any saccade can be represented as a vector, and the interaction of two vectors can be described by the angle between them. The magnitude of this angle determines the changes of the direction of eye movements. This index gives an integral description of the looking process or, in other words, the nature of space scanning.

Another index deals with the relation between the duration of fixation and the amplitude of saccades. Velichkovsky et al. [14] distinguished two types of information processing in visual search: "Overview scanning" (ambient processing) and "Focused inspection" (focal processing). Ambient processing is characterized by shorter fixation duration (<250 ms) and longer saccadic amplitude ($>4°$), whereas in focal state fixation durations are longer (>250 ms) and saccadic amplitude is shorter ($<4°$). Thus, the ratio of fixation durations and saccadic amplitudes enable determining the nature of information processing in visual search.

In this study, we analyzed the influence of the three groups of factors on eye movements and on the efficiency of the search strategy (efficiency was determined by the search rate). The strategies were identified on the basis of eye movement characteristics.

Hypotheses:

H1: Search time and eye movement characteristics depend on the chromaticity of the matrix.
H2: Search time and eye movement characteristics depend on the target template format.
H3: Search time and eye movement characteristics depend on impulsivity level of the subjects.

2   Method

Participants. Sixty-three volunteers (with normal or corrected-to-normal vision and good color vision) took part in the experiment; 41 females and 22 males aged 18–48, the mean age being 22 years and 3 months.

Measuring cognitive impulsivity score. Prior to the main experiment, the subjects were to do Matching Familiar Figures Task in order to define their cognitive impulsivity [8]: participants selected the one of six visually presented stimuli, which is identical to an original image (participants completed 20 trials). The task provided a composite Impulsivity score (I-score): response times on this task are inversely related to choice accuracy, providing a measure of cognitive impulsivity in the form of the speed-accuracy trade-off [4]. We classified the subjects into two groups: impulsive and reflective (the Cognitive style factor).

Procedure. During the experiment the participants were seated 0.65 m away from a 19-inch computer screen. The task was to find symbolic images of real-life objects (such as a butterfly, a cactus, a book, etc.) among a variety of other objects.

Stimuli. The icons were provided by Internet icon resources and modified into homogeneous stimulus material (see Fig. 1). A total of 5128 pictures were collected. The stimuli were arranged in a rectangular full screen stimulus matrix 9×9: a 1466×954 pixel rectangle with 300 pixels per square inch resolution. Each matrix contained 1 target item and 80 distractor items. The target stimulus was situated in 8 out of 9 quadrants (all but for the central quadrant). Thirty-two matrices were presented, half black-and-white, the other half colored (the Chromaticity Factor). The icons were framed either by circles or squares. The target template was set either by word, denoting the object, or by an accurate copy of the searched image in grayscale (the Target template factor).

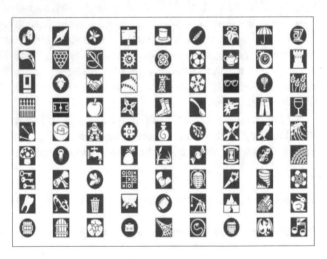

Fig. 1. One of the black and white matrices, presented to the subjects

Recorded data. We recorded the search rate and eye movement data. Eye movements were sampled monocularly at 250 Hz using the SMI iView X RED 4 (FireWire) tracking system with on-line detection of saccades and fixations and a spatial accuracy $< 0.5°$. 32 trials were recorded for each subject, in total 2016 trials. We discarded 33 trials due to a low quality of recorded eye tracking data (leaving 33 trials).

Defining search patterns based on saccade directions and intersaccadic angles. We studied visual search patterns of eye movements in visual search task by analysis of two angles, namely, the angle between a saccadic event and the horizontal direction ("Direction") and the angle between two consecutive saccadic events ("Intersaccadic angle"). We used the data, provided by SMI BeGaze, about the start (Saccade Start Location X, Y) and ending (Saccade End Location X, Y) coordinates of the saccades (saccades being identified by SMI BeGaze by velocity and acceleration thresholds). We used the procedure of saccadic direction angles computation similar to the one described by Amor, Reis, Campos, Herrmann & Andrade [1], where the horizontal angle for the i-th saccade (a_i) is computed as:

$$\arctan(a_i) = \arctan(r_{i,y}/r_{i,x}),$$

where $r_{i,y}$ = |Location End Y - Location Start Y|, and $r_{i,x}$ = |Location End X - Location Start X|.

The obtained angles ($0°–90°$) we attributed to the directions according to the substraction values of y_i and x_i, where y_i = Location End Y - Location Start Y, and x_i = Location End X - Location Start X. We distinguished 8 directions (given in counterclockwise order): Right-upward direction ($0°–45°$), Upward-right direction ($45°–90°$), Upward-left direction ($90°–135°$), Left-upward direction ($135°–180°$), Left-downward direction ($180°–225°$), Downward-left direction ($225°–270°$), Downward-right direction ($270°–315°$), Right-downward direction ($315°–360°$). For example, in case $a_i < 45°$, $y_i > 0$, $x_i > 0$, the direction was Right-upward, in case $a_i > 45°$, $y_i < 0$, $x_i < 0$ the direction was Downward-left. We counted up the percent of saccades in each direction for every trial.

The angle between the i-th saccade and the successive saccade (β_i) was computed as: in case both saccades were directed upwards or both directed downwards

$$\beta_i = 180-(|a_i-a_{i+1}|),$$

in case one saccade was directed upwards and one downwards

$$\beta_i = 180-a_i-a_{i+1},$$

We counted up the percent of saccades in each direction for every trial.

Data processing. The trials were calculated and subjected to factorial ANOVA (two-way ANOVA) using IBM SPSS Statistics 19.

3 Results and Discussion

3.1 The Chromaticity Factor

The assumption that the search effectiveness will differ whether the stimulus matrix is colored or in grayscale did not prove valid. There was little effect of chromaticity on eye movement characteristics, apart from fixation count in the area of target words: on average 3.62 fixations on colored matrices as opposed to 3.33 on grayscale matrices ($F_{(1, 1983)} = 4,127$; $p < 0,05$), whereas dwell time on target words did not change significantly. Therefore, we concluded that the Chromaticity Factor in itself was not important.

This contradicts other data about the chromaticity factor in visual search: for instance, using color maps leads to smaller errors than using gray-scale maps irrespective of the complexity [11].

However, its interaction with the target template format had an effect on the search time. It appears that color might interfere when searching for the target stimulus if it is set in the form of a word and, on the contrary, serve as an aid when the stimulus is presented as an image. It should be noted that even when the stimulus was set as a picture it was in grayscale, so the color did not play a key role in detecting the searched element but rather affected the mechanisms of cognitive processing. We concluded that the color grade of the stimulus provided faster cognitive processing on lower levels of processing rather than on higher levels, when semantic analysis is involved. No significant interaction was found between the two factors in relation to eye movement activity.

3.2 The Target Template Factor

The results indicate that the target template type affects the search rate and the scanning process (Table 1). If the target stimulus is set as a word, it requires a longer search time (as compared to a picture). Moreover, the format of the target template could determine the search strategy, i.e., a specific eye movement pattern. Therefore, the target template format – a high-level factor – regulates eye movement activity. If the stimulus is presented as a word, the search query turns out to be less determined. A subject seemed to use not a mental representation stored in the working memory as a target sample, but rather its semantic field, which could include a number of visual representations. The search in this case requires a deeper and more detailed information processing. It leads to a longer mean fixation duration, shorter-amplitude and slower saccades (see Fig. 2). This eye movement pattern resembles focal information processing. If the stimulus is set as a picture, the search query is highly precise, being a particular mental representation stored in the working memory. The subject's task is to match it with the images he or she comes across. This does not require deep semantic processing and can be accomplished with shorter fixations and faster saccades with longer amplitudes (eye movement pattern, typical for ambient vision).

Table 1. The time of search and eye movement parameters in different formats of the target template – as a word or a picture

	All trials	The target template		F (2; 1983)	Sig
		Word	Picture		
Search time [ms]	12119	13142	11091	12.0	$p < 0.01$
Mean fixation duration [ms]	226.5	230.9	222.1	7.6	$p < 0.01$
Mean saccadic amplitude [°]	3.901	3.736	4.068	6.6	$p < 0.01$
Fixation count on AOI	3.480	3.667	3.287	7.0	$p < 0.01$
Dwell time on AOI [ms]	1745	1821	1666	4.1	$p < 0.05$

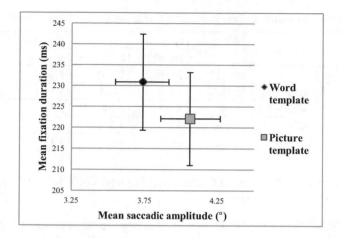

Fig. 2. Mean fixation duration and saccadic amplitude with different templates

The target stimulus was outlined in the matrix as the area of interest (AOI) and eye movement data was registered around that segment. As we can see in Table 1, the mean time spent in the area of interest is longer when the stimulus was set as a word, although the differences are not too pronounced. More apparent differences can be observed when counting the number of fixations in the area of interest, which could be connected with the performed cycles of cognitive processing. This outcome can possibly imply the necessity for the subject to verify the correct stimulus in cases when the format of the search query was incongruent with the element (word vs. picture).

3.3   The Cognitive Style Factor

We divided the subjects into two groups, based on their MFFT score in search time and errors: impulsive and reflective. Impulsive subjects tended to have lower search time, make fewer fixations and have shorter scan path length (the measure was computed in SMI BeGaze program). Reflective subjects tended to have longer search time, consequently, longer scan path length and more fixations. Mean fixation duration was slightly shorter in more impulsive subjects. However, there were no significant distinctions in the mean saccadic amplitude in impulsive and reflective subjects. These results indicate in favor of no interrelation of the Impulsivity score and ambient/focal processing. The most significant distinctions were manifested in AOI dwell time: impulsive subjects tended to spend less time on the area of interest, which indicates a faster decision making process in detection and identification of the target stimulus (Table 2).

The Cognitive style factor proved to have little interrelation with the other factors. However, Scan Path observation led us to the conclusion, that scanning manner differed in subjects with a different Impulsivity score. An additional hypothesis was proposed about the correlation of cognitive style and eye movement patterns in visual search.

Table 2. The time of search and eye movement parameters in impulsive and reflective subjects

	All trials	Cognitive styles		F (2; 1983)	Sig
		Reflective	Impulsive		
Search time [ms]	12119	12758	11417	5.1	p < 0.05
Mean fixation duration [ms]	226.5	229.3	223.4	3.5	p < 0.05
Mean saccadic amplitude [°]	3.901	3.866	3.940	0.3	-
Fixation count on AOI	3.480	3.378	3.582	2.0	-
Dwell time on AOI [ms]	1745	1847	1642	7.1	p < 0.01

3.4 Eye Movement Patterns of Space Scanning and Their Interrelation with Subjects' Cognitive Style

Cluster analysis (Ward's method) was used to categorize the experimental series on the basis of subject's eye movement characteristics, namely, intersaccadic angle and saccadic direction. We defined 8 saccadic directions, 4 upward, 4 downward * 4 left, 4 right, given in counterclockwise order. The percent of the corresponding intersaccadic angles and directions for each trial was counted up. We opted a 3-cluster solution, which is given in details below (see Table 3).

Table 3. Cluster analysis results

	Cluster 1 (370 cases)	Cluster 2 (334 cases)	Cluster 3 (544 cases)
% of the 0°–45° intersaccadic angle	44.39	27.72	24.47
% of the 45°–90° intersaccadic angle	12.58	15.60	24.72
% of the 90°–135° intersaccadic angle	13.37	17.76	22.20
% of the 135°–180° intersaccadic angle	24.80	24.13	22.96
% of the Right-upward direction (0°–45°)	20.08	8.09	13.11
% of the Upward-right direction (45°–90°)	4.29	14.96	9.84
% of the Upward-left direction (90°–135°)	5.11	13.80	11.11
% of the Left-upward direction (135°–180°)	18.06	8.17	15.68
% of the Left-downward direction (180°–225°)	20.56	7.15	15.35
% of the Downward-left direction (225°–270°)	5.83	16.40	9.32
% of the Downward-right direction (270°–315°)	5.27	13.96	9.39

Each cluster was characterized by composite characteristics of "Intersaccadic angle" and "Direction" values:

(1) prevailing horizontally oriented directions (right- and left-upward and -downward), intersaccadic angle values indicating little change of direction (0°–45°), and occasional "backtrack" 135°–180° angles, which can be interpreted as serving "getting back to the start of the next line" purposes
(2) as opposed to Cluster 1, in Cluster 2 vertically oriented directions prevail (upward- and downward-right and -left), as well as intersaccadic angle values, indicating (mostly) little change of direction and occasional "backtrack" 135°–180° angles
(3) a relatively even distribution of saccade directions and intersaccadic angles, with slight prevalence of horizontally oriented saccades.

Therefore, three patterns of eye movements in visual search were identified, denominated as: "Horizontal sequential" (Cluster 1), "Vertical sequential" (Cluster 2) and "Non-sequential" (Cluster 3). The subjects tended to adopt a steady eye-movement pattern (the coefficient of contingency of Subjects * Cluster numbers was 0.61, with $p < 0.05$).

Significant distinctions were also found in the mean saccadic amplitude and Impulsivity score in samples with different patterns (see Table 4).

Table 4. Mean saccadic amplitude and impulsivity score in different patterns

		"Horizontal sequential" pattern (370 cases)	"Vertical sequential" pattern (334 cases)	"Non-sequential" pattern (544 cases)
Mean saccadic amplitude	$F(4; 1248) = 4.7$ ($p < 0.05$)	3.00	3.53	3.40
Impulsivity index (MFFT)	$F(4; 1248) = 4.7$ ($p < 0.05$)	0.79	0.92	1.17

Thus, trials corresponding to "Horizontal sequential" search pattern exhibit shorter saccadic amplitude as compared to trials corresponding to "Vertical sequential" and "Non-sequential" patterns. We assume this difference is due to the similarity of the pattern to eye movements in reading, in which saccades are generally of shorter amplitude, than in scene viewing and visual search. "Horizontal sequential" pattern was also marked by low Impulsivity index, i.e., subjects who were more "reflective" took their time to peruse the matrix in left-right, bottom-top manner; whereas scanpaths of those who were more "impulsive" had no prevailing directions and intersaccadic angles, therefore being described as apparently "chaotic". This difference in eye movement patterns is consistent with the idea of the reflectivity/impulsivity dimension [8], in which the tendency to be reflective or impulsive is described as the relation between the positive value of quick success and the anxiety generated by the possibility of committing an error.

4 Conclusions

Searching for target objects in real or virtual environment is a common task. In order to succeed, one has to choose the right direction and the segment of the environment while disregarding everything that is not related to the target and then confirm the accuracy of the choice. One has to possess a more or less clear representation of the searched object to be able to compare it with the perceived information. Our results showed that the Target template factor has a significant impact on the search time as well as the manner of scanning the environment. It was discovered that a verbally set target stimulus leads to a longer search rate than a picture template. Moreover, it leads to the change in the nature of cognitive processing – longer fixations occur, including slower saccades with shorter amplitudes. This eye movement pattern indicates deeper (semantic) information processing.

It appeared that chromaticity, as a low-level factor, does not play a significant role on its own, although it interacts with other factors. It was illustrated that the color grade helps the subject to detect the target stimulus within a matrix when it is set in the form of a picture and, on the contrary, can hinder the search process when it is presented as a word. If the target stimulus is presented as a picture, the search is conducted through surface-level cognitive processing. At this level, the various physical features of the matrix can play an important role – in particular, the facilitating effects of the color grade become more apparent. If the target stimulus is set as a word, its search and detection requires a deeper semantic analysis. In this case, the role of the physical characteristics is the reverse. What made it easier to detect the target object at the surface-level of processing now has a negative effect.

Significant effects of cognitive style on search process were revealed. Subjects with higher Impulsivity score tended to have faster search rate and exert their effort less, than subjects with low Impulsivity score. They also spent less time in the target AOI and identify their target stimulus faster. The obtained data provide evidence for the possibility of using the combination of intersaccadic angle and dimension measures for identifying visual search patterns. The described patterns, in general, correspond to the patterns obtained in preceding research [1], as well as interrelate with cognitive dimension of impulsivity/reflectivity.

The obtained data can be used in developing search system interfaces with better organization, which will be more interactive and flexible.

Acknowledgement.   This research is supported by the Russian Foundation of Basic Research; Grant № 16-36-00044.

References

1. Amor, T.A., Reis, S.D., Campos, D., Herrmann, H.J., Andrade, J.S.: Persistence in eye movement during visual search. Sci. Rep. **6**(20815), 1–12 (2016)
2. Bawden, D.: Users, user studies and human information behavior. J. Documentation **62**(6), 671–679 (2006)
3. Burmistrov, I., Zlokazova, T., Izmalkova, A., Leonova, A.: Flat design vs traditional design: Comparative experimental study. In: Human-Computer Interaction (INTERACT-2015), pp. 106–114 (2015)
4. Carretero-Dios, H., Macarena, D.S.R., Buela-Casal, G.: Influence of the difficulty of the Matching Familiar Figures Test-20 on the assessment of reflection–impulsivity: an item analysis. Learn. Individ. Differ. **18**(4), 505–508 (2008)
5. Eckstein, M.P.: Visual search: a retrospective. J. Vis. **11**(5), 1–36 (2011)

6. Goonetilleke, R.S., Shih, H.M., On, H.K., Fritsch, J.: Effects of training and representational characteristics in icon design. Int. J. Hum. Comput. Stud. **55**(5), 741–760 (2001)

7. Hout, M.C., Goldinge, S.D.: Target templates: the precision of mental representations affects attentional guidance and decision-making in visual search. Attention Percept. Psychophys. **77**(1), 128–149 (2015)

8. Kagan, J.: Reflection-impulsivity: the generality and dynamics of conceptual tempo. J. Abnorm. Psychol. **71**(1), 17–24 (1966)

9. Kinley, K., Tjondronegoro, D.W., Partridge, H.L., Edwards, S.L.: Human–computer interaction: the impact of users' cognitive styles on query reformulation behavior during web searching. In: Proceedings of Australasian Conference on Computer-Human Interaction (OZCHI 2012), Melbourne, Victoria (2012)

10. Malcolm, G., Henderson, J.M.: The effects of target template specificity on visual search in real-world scenes: evidence from eye movements. J. Vis. **9**(11), 1–13 (2009)

11. Netzel, R., Ohlhausen, B., Kurzhals, K., Woods, R., Burch, M., Weiskopf, D.: User performance and reading strategies for metro maps: an eye tracking study. Spat. Cogn. Comput. **17**(1–2), 39–64 (2017)

12. Pan, B., Hembrooke, H.A., Gay, G.K., Granka, L.A., Feusner, M.K., Newman, J.K.: The determinants of web page viewing behavior: an eye-tracking study. In: Proceedings of the 2004 Symposium on Eye Tracking Research and Applications, pp. 147–154 (2004)

13. Reinecke, K., Yeh, T., Miratrix, L., Mardiko, R., Zhao, Y., Liu, J., Gajos, K.Z.: Predicting users' first impressions of website aesthetics with a quantification of perceived visual complexity and colorfulness. In: Proceedings of the SIGCHI Conference on Human Factors in Computing Systems. ACM, Paris (2013)

14. Velichkovsky, B.M., Joos, M., Helmert, J.R., Pannasch, S.: Two visual systems and their eye movements: evidence from static and dynamic scene perception. In: Proceedings of the XXVII Conference of the Cognitive Science Society, CogSci 2005, pp. 2283–2288 (2005)

15. Wolfe, J.M.: Visual search. In: Pashler, H. (ed.) Attention, pp. 13–73. Psychology Press, Hove (1998)

Using Eye Trackers as Indicators of Diagnostic Markers: Implications from HCI Devices

Thomas D.W. Wilcockson[✉]

Department of Psychology, Lancaster University, Lancaster, UK
t.wilcockson@lancaster.ac.uk

Abstract. Eye tracking allows psychologists to make broad distinctions between groups of participants. Therefore, it may be entirely possible to exploit these distinctions in order to create screening tools for potentially diagnosing various conditions. These diagnostic techniques may have a number of advantages over exiting techniques. However, in order to develop such screening tools it would be beneficial if eye tracking systems were easy to access and use. There are a number of ways to improve accessibility of eye trackers: affordability, transportation, ease of use. This positional paper explores how HCI (Human Computer Interaction) eye trackers can be used for diagnostic purposes of psychological conditions.

Keywords: Eye tracking · Diagnosis · HCI · Cognitive function · Attentional bias

1 Introduction

Eye tracking allows psychologists to make broad distinctions between groups of participants (see Leigh and Kennard 2004). For example, eye movements can be indicative of whether a patient has had a concussion (Samadani et al. 2015a), has schizophrenia (Benson et al. 2012), or Alzheimer's disease (Crawford et al. 2005). Therefore it may be entirely possible to exploit these distinctions in order to create screening tools for potentially diagnosing such conditions. There are a number of neural pathways involved in the generation of eye movements including the cerebrum, brainstem, and cerebellum. Therefore degeneration or damage in any of these areas would affect eye movements in a particular fashion which would be identifiable during eye tracking tasks and may indicate a specific disorder. Thus indicating the potential benefits which eye tracking could offer neurological and cognitive assessment.

At present eye tracking techniques have been used to compliment the assessment of patients (Anderson and MacAskill 2013). Eye movement can be measured in clinical settings without the use of eye tracking equipment. These observations by clinicians may involve simply asking a patient to follow the clinician's finger as it is moved from side to side. Such basic observations are not as accurate as using eye tracking technology but can give an indication of an amplification of a motor impairment of the disorder. Whereas laboratory-measured eye movements (using eye tracking technology) can provide rich and detailed identification of cognitive and/or neurological status and also monitor the progression of a disorder. Therefore there are advantages of using eye

© Springer International Publishing AG 2018
I. Czarnowski et al. (eds.), *Intelligent Decision Technologies 2017,*
Smart Innovation, Systems and Technologies 73, DOI 10.1007/978-3-319-59424-8_29

tracking equipment in the laboratory. This combined with the measurement of saccades enables the measurement of cognition as well as motor impairments associated with certain disorders.

Eye movement research typically aims to measure two types of movements; vestibulo-ocular reflexes (VORs) or saccades. VORs are associated with the stabilising of the eye in relation to motion from the head. The second type of eye-movement is the saccade. This is a shift of the location of the eye in order to focus on areas of interest. It is this second type of eye-movement which is of particular interest due to its close relationship with attention, as attention may be affected by the cognitive impairments associated with, e.g., neurodegenerative disorders. As previously mentioned eye movement deficits can be observed clinically, however, it has generally been found that the recording of eye movements in a laboratory whilst performing an eye-tracking task is much more sensitive.

Traditionally a number of simple eye movement tasks have been used to observe cognitive function within patients. For example during a prosaccade task participants make a saccade (typically from a central location) toward a sudden onset target. Important measurements include the saccade latency, peak velocity, amplitude, and duration. During an anti-saccade task the participants fixate on a motionless target (such as a small dot) and a stimulus is then presented to one side of the target. The patient is asked to make a saccade toward the opposite direction of the stimulus. For example, if a stimulus is presented to the left of the motionless target, the patient should look toward the right. Failure to inhibit a reflexive saccade is considered an error. A reflex paradigm task involves directing the participant to look at a target as soon as it appears on the screen. By manipulating the onset time of the target a measure of inhibition can be obtained. Manipulating the gap can affect activity in the superior colliculus, which is associated with gaze fixation. This task allows the experimenter to change the demands of the task without changing the instructions the participant receives. A memory guided paradigm task requires more cognitive processing than the reflex paradigm. A target briefly appears on the screen, but the participant must suppress the reflexive saccade towards it until a second cue appears. Following the second cue the participant must fixate where the initial target was (which has subsequently vanished), so guided by memory alone.

These tasks have been successfully adopted to measure discrepancies between participant groups. For example Alzheimer's dementia (AD) typically require posthumous diagnosis in order to be confirmed. However, a diagnosis of probable AD is often given to patients with deficits in memory, one other cognitive domain, and everyday functioning. Utilising the antisaccade task with AD has demonstrated key distinctions between AD and non-dementia controls. Patients with AD typically make more uncorrected errors on the task in that they are much more likely to be unable to inhibit their attention toward the salient target (see Crawford et al. 2005). These findings would therefore suggest that the development of AD impairs executive control over eye movements (see Fig. 1).

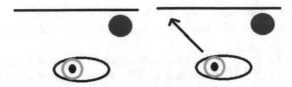

Fig. 1. A diagram of an antisaccade task trial. When a red circle appears on the screen the participant is told to look in the opposite direction of the stimulus.

By using a smooth pursuit task, where participants are told to follow a pattern on a computer screen, Benson et al. (2012) found that 88 schizophrenia cases and 88 controls differed on the task in that the people with schizophrenia were impaired. It was also observed that the eye movement abnormalities appear to be stable traits, independent of current medication or mental states (see Fig. 2).

Fig. 2. The grey line indicates the path which the participant should follow with their eyes. Picture A demonstrates the performance of a control participant. Whereas picture B demonstrates an example of the type of eye movement a Schizophrenic might display.

Samadani et al. (2015b) developed a technique for measuring concussion based upon the clinical examining of a patient who has sustained a head injury. Doctors traditionally would move a finger in front of the patient's eyes to notice any peculiarities in the movement of the eyes. During the eye movement task, participants would watch a video through a moving aperture, which enabled the measurement of ocular motility. It was found that concussion led to a decrease in performance (see Fig. 3).

Fig. 3. A video is displayed on a screen. However, only a portion of the screen can be seen through an aperture.

Another use for eye movement techniques is the diagnosis of substance abuse disorders by measuring attentional biases. Attentional biases are the preferential processing of stimuli within the environment such that, e.g., a heavy alcohol drinker would direct their attention toward an alcohol-related stimulus. Wilcockson and Pothos (2015) observed that, using an eye movement gaze contingent paradigm[1] whereby participants are told to avoid looking at stimuli, once an attentional bias has been developed for a substance, then processing of related stimuli is prioritised but also compulsory to attend to. Substance abuse can also distort perception of an environment so that there is a preoccupation with substance-related stimuli (Wilcockson and Pothos 2016; see Fig. 4). These findings all indicate a robustness of attentional bias: Substance abuse is facilitated by attentional biases which cause a preoccupation with substances, are hard to inhibit, and occur irrespective of context. These attentional biases have also been found to indicate whether substance abuse treatment has been successful or not as they indicate the likelihood of relapse (see Cox et al. 2002). Therefore, the ability to easily measure eye movement toward substance-related stimuli could be useful for predicting treatment success rate.

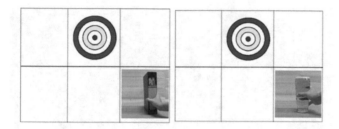

Fig. 4. Example of fixation region and distractor stimuli. The first image is an example of control distractor stimuli depicts a hand reaching for a folder (no grid lines were present in the experiment, they are shown here to represent the 6 sections of the screen which contain the fixation region and distractor stimuli). The second image is an example of a matched distractor stimuli. This example of alcohol distractor stimuli depicts a hand reaching for a pint. The matched control and experimental distractor stimuli are subtracted in order to make a 'difference' score. It is this score that is used to create the measures of attentional biases.

All these findings taken together indicate that the vast array of eye movement techniques can be used to measure distinct clinical characteristics between participant/patient groups.

The importance of using laboratory tasks seems clear. These provide a number of advantages over clinical measure of eye movements. However, it is not also possible to utilise laboratory task on patients. As often patients are seen in clinical settings with relatively little space for equipment, or the time and skills required to set-up eye tracking equipment is lacking. Cost is also a big issue as eye tracking equipment can cost up to $35,000 (compare to $99 for an EyeTribe). Therefore, in order to realise the advantages that laboratory eye movement measures could provide a more practical system would apparently be necessary.

A number of companies produce eye tracking devices for the purposes of human computer interaction (HCI). These eye trackers are generally used as a substitute for

---

[1] A gaze contingency paradigm is a technique which allows a computer screen display to be modified dependent on where the viewer is looking.

using a mouse or for controlling aspects of computer games. These devices are not as powerful as the eye trackers which psychologists would use for data collection. However, some psychological experiments do not require the accuracy and precision provided by expensive eye trackers.

Previous research has indicated that it is indeed possible to use the EyeTribe (Dalmaijer 2014) and Tobii EyeX Controller (Gibaldi et al. 2016) for data collection. These authors considered a number of aspects of the performance of these devices including accuracy, precision, sampling rate, fixation analysis, pupilometry, and saccade metrics. In each respect the HCI eye trackers performed admirably (however, saccade metrics seem to be the weakest aspects of these devices). In each regard the traditional data collection devices out-performed the HCI rivals. However, the HCI tools have the advantage of being cheap, easy to use, easy to transport, and are compact (see Fig. 5).

Fig. 5. A demonstration of the compact nature of the portable eye movement lab system.

Therefore, it has already been demonstrated that this technology would be useful for data collection. The next step is to develop a straightforward and simple way of programming basic tasks which would be useful for the diagnosis of different participant groups in clinical settings. Of importance is developing the appropriate tasks which yield clinical diagnostic markers. But these tasks also need to be simple for clinicians whom are not experts in eye movement data collection and analysis.

The goal of this positional paper was to highlight the utility of HCI eye trackers in diagnosis and demonstrate a software that is very easy to use for this purpose. It was necessary to be able to easily change the stimuli and any demographic questions. The results also need to be instantly available on the screen following the procedure without the need of complicated analyses. This would enable the software and the eye tracker to be used in different situations and address all needs.

2   Method

An EyeTribe device was chosen to develop and explore this viability. The EyeTribe can be used with Windows or Macintosh computers. A Macintosh was used for the programming but a Windows 10 tablet to pilot the tasks. MatLab was used to control the stimulus events. The MatLab task was programmed in such a way that the stimuli could easily be changed by adding pictures to a corresponding desktop folder. This would enable a clinician to easily change the nature of the task e.g. by substituting alcohol-related stimuli for phobia-related stimuli in order to look at a patient with phobias rather than substance abuse. The task also contained demographic questions about the participant. These questions could be easily modified in the MatLab code. The task displayed basic graphical representations of the results on the screen following completion of the task.

Three different eye movement tasks were developed in order to measure different aspects of eye movement task design. The tasks were: a dot probe, an antisaccade, and a gaze contingent task. These tasks were chosen as they broadly represent the different eye movement tasks that have previously been used to demonstrate broad diagnostic distinctions between various patient-types. An example of one of the tasks can be seen in Fig. 6.

Fig. 6. This diagram of an example of a dot probe task. In this example alcohol usage is considered. Participants are first asked a question about their alcohol usage. The participant would respond using either a keyboard or touch screen. The eye movement task would then begin. In our task 10 trials were used. Each trial contained an alcohol and control stimulus. A number of eye movement metrics can be measured. It would be hypothesised that heavy alcohol drinkers would fixate on the alcohol stimulus before the control stimulus. This result can then be shown on the screen following the task. Here a pie chart shows that the participant fixated on the alcohol stimuli more than the control stimuli over the 10 trials.

3   Results

It was found that each of the tasks functioned appropriately with the EyeTribe device. The tasks could easily be modified for different circumstances and the results were clearly available on the screen. No technical ability was needed to set-up the equipment or to perform any analysis.

Specifically, a number of eye movement metrics are measurable with this system. Using the device it is possible to measure coordinates of eye movements on the screen together with timestamps. These can be used to infer a number of variables including fixations, dwell times, and saccades. The software can produce a number of different types of outputs including pie charts, bar graphs, and basic mean comparison statistics. The task could also be modified for different circumstances by changing small parameters related to the MatLab code. For example, changing the pictures in the relevant folders and changing the demographic questions. Therefore computer competency is minimal in order to utilise this system.

4 Conclusions

In conclusion, it would appear that HCI eye trackers could contribute greatly to psychological and medical research. They provide a number of practical and procedural advantages over traditional psychological data collection devices. Of course HCI eye trackers are not able to fully replace the other systems. However, if rich data is not necessarily required then psychologists should consider using HCI devices. Using such systems could potentially enable a clinician to automatically recommend diagnoses for patients with certain conditions.

Following the demonstration that the method can be used simply and efficiently, the next step is to start using this methodology in clinical settings. Such a demonstration would have implications for future diagnosis of patients. Should the methodology be successful in clinical settings then the recommendation would be that clinicians start considering using such devices as part of diagnosing patients.

It should also be acknowledged that web-camera based eye tracking systems may also be a suitable alternative to HCI devices. Such systems may be even cheaper than HCI devices as web-cameras are readily available. Therefore web-cameras have the advantage of collecting data from many devices anywhere in the world with the possibility of downloading clinical tools from app stores. Xu et al. (2015) demonstrated the utility for such devices when fixation measurement is required. However, further development is required before saccades can be measured. This would be integral for some clinical measurements. However, web-cameras have been used in some clinical populations. Chau and Betke (2005) used web-cameras to enable severely paralysed patients to communicate with the computers. Therefore web-camera and HCI device eye tracking systems should both be considered for use by clinicians.

It is a limitation of this paper that there is no verification of our findings. However, the purpose of this paper is proof of concept and is intended to act as a positional paper. Here the theory and a potential methodology are presented. In future research it is hoped that the importance of this method can be demonstrated by conducting diagnostic assessments based on this.

In conclusion, eye movements can provide diagnostic markers for different participant groups. Exploiting these diagnostic markers would enable effective screening tools to be created. A hurdle in the way of deploying such eye movement tasks are the problems inherent with traditional eye tracking devices (e.g. cost, ease of use, time to set up, transportation). However, here it is presented that it is possible to use HCI devices for the measurement of these diagnostic markers. Therefore, it is suggested that with further development clinicians should start to reconsider using eye movement as part of diagnosing in clinical settings.

Acknowledgements. Thank you to Barrie Usherwood for his help in programming the eye movement tasks using the HCI equipment and Colleen McElhatton for her help with an earlier draft.

References

Anderson, T.J., MacAskill, M.R.: Eye movements in patients with neurodegenerative disorders. Nat. Rev. Neurol. **9**, 74–85 (2013)

Benson, P.J., Beedie, S.A., Shephard, E., Giegling, I., Rujescu, D., Clair, D.S.: Simple viewing tests can detect eye movement abnormalities that distinguish schizophrenia cases from controls with exceptional accuracy. Biol. Psychiatry **72**, 716–724 (2012)

Chau, M., Betke, M.: Real Time Eye Tracking and Blink Detection with USB Cameras. Boston University Computer Science Department (2005)

Cox, W.M., Hogan, L.M., Kristian, M.R., Race, J.H.: Alcohol attentional bias as a predictor of alcohol abusers' treatment outcome. Drug Alcohol Depend. **68**, 237–243 (2002)

Crawford, T.J., Higham, S., Renvoize, T., Patel, J., Dale, M., Suriya, A., Tetley, S.: Inhibitory control of saccadic eye movements and cognitive impairment in Alzheimer's disease. Biol. Psychiatry **57**, 1052–1060 (2005)

Dalmaijer, E.: Is the low-cost EyeTribe eye tracker any good for research? PeerJ PrePr **2**, e585v1 (2014)

Gibaldi, A., Vanegas, M., Bex, P.J., Maiello, G.: Evaluation of the Tobii EyeX eye tracking controller and Matlab toolkit for research. Behav. Res. Meth. 1–24 (2016). doi:10.3758/s13428-016-0762-9

Leigh, R.J., Kennard, C.: Using saccades as a research tool in the clinical neurosciences. Brain **127**, 460–477 (2004)

Samadani, U., Farooq, S., Ritlop, R., Warren, F., Reyes, M., Lamm, E., Schneider, J.: Detection of third and sixth cranial nerve palsies with a novel method for eye tracking while watching a short film clip. J. Neurosurg. **122**, 707–720 (2015a)

Samadani, U., Ritlop, R., Reyes, M., Nehrbass, E., Li, M., Lamm, E., Schneider, J., Shimunov, D., Sava, M., Kolecki, R., Burris, P.: Eye tracking detects disconjugate eye movements associated with structural traumatic brain injury and concussion. J. Neurotrauma **32**, 548–556 (2015b)

Wilcockson, T.D.W., Pothos, E.M.: Measuring inhibitory processes for alcohol-related attentional biases: introducing a novel attentional bias measure. Addict. Behav. **44**, 88–93 (2015)

Wilcockson, T.D., Pothos, E.M.: How cognitive biases can distort environmental statistics: introducing the rough estimation task. Behav. Pharmacol. **27**, 165–172 (2016)

Xu, P., Ehinger, K.A., Zhang, Y., Finkelstein, A., Kulkarni, S.R., Xiao, J.: Turkergaze: crowdsourcing saliency with webcam based eye tracking. arXiv:150406755 (2015)

Social Media Analysis and Mining

Social Stream Clustering to Improve Events Extraction

Ferdaous Jenhani$^{(\boxtimes)}$ (ID), Mohamed Salah Gouider,
and Lamjed Ben Said

ISG, SMART Lab, Université de Tunis, Le Bardo, Tunis, Tunisia
jenhaniferdaous@gmail.com,
ms.gouider@yahoo.fr, bensaid_lamjed@yahoo.fr

Abstract. Events extraction from social media data is a tedious task because of their volume, velocity and informality. In a previous work [25], we proposed a successful approach for events extraction from social data. However, messages were processed individually which generates many meaningless events because of missing details scattered within millions of text segments. In addition, many unnecessary texts were analyzed which increased processing time and decreased the performance of the system.

In this paper, we aim to cope with the abovementioned weaknesses and ameliorate the performance of the system. We propose clustering to group semantically-related text segments, filter noise, reduce the volume of data to process and promote only relevant text segments to the information extraction pipeline. We port the clustering algorithm to a stream processing framework namely Storm in order to build a stream clustering solution and scale up to continuously growing volumes of data.

Keywords: Events extraction · Social stream · Clustering · Apache storm · Twitter

1 Introduction

The omnipresent frequent use of social media for business purposes has created a new source of information which analysis has become a real business requirement in order to discover hidden knowledge. Indeed, real time collection and analysis of social data to extract useful information become strongly required to improve decision making in competitive environment. However, social media data are voluminous, dynamic and informal, which hardens their exploitation especially the extraction of the most complete, meaningful and coherent information namely events [17]. Events extraction from streaming data have become the key solution of many problems for a number of streaming applications in which the information should be analyzed as they are generated to make the immediate decision or reactions.

This work is an extension of a previous one [25] for private events extraction from social data. Despite the interesting results we obtained, the overall performance needs to be improved for many reasons; first, because private events are rare compared to public events which hardens the collection of different details scattered within many

I. Czarnowski et al. (eds.), *Intelligent Decision Technologies 2017*,
Smart Innovation, Systems and Technologies 73, DOI 10.1007/978-3-319-59424-8_30

text segments and their individual analysis may lead to incomplete, meaningless and useless information. Second, co-reference resolution requires consecutive text segments which is difficult and may be impossible with dispersed messages generated within different time instants. Third, social media data are short, conversational, informal and contain a lot of noise. We may lose a lot of time and memory by processing text segments that cannot bring any useful information after their long processing. Finally, social text is generated with a very high frequency raising two related challenges velocity and volume. This dataflow exceeds existing processing capacities and may lead to loss of event details necessary to obtain complete and meaningful information which is a major limitation toward obtaining trusted information for better decision making.

In order to overcome these problems, facilitate events extraction from social streaming text and improve its performance, we propose to use clustering as a basic preprocessing work for events extraction. We aim to gather semantically-related texts and remove non relevant ones in order to reduce the volume of data to process and keep only pertinent text segments that could bring coherent events. Interested to private events, we consider semantically-related information the set of messages belonging to the same user and about the same topic. In this paper, we focus on the development of a social stream clustering technique at first which suited better our problem. We use a data stream processing framework called Apache Storm to implement the proposed clustering algorithm and to scale up to larger volumes of data increasingly growing thanks to its parallel and distributed paradigm. Indeed, we combined a simple clustering technique with a big data and streaming technology to obtain an efficient social stream clustering system. Finally, we compare events extraction results before and after clustering to demonstrate how such step is important and we discuss clustering leverage on the performance of events extracted.

This paper is organized as follows; Sect. 2 is a literature review about data stream clustering algorithms and their application for information extraction purposes. In Sect. 3 we present the social stream model. The general clustering framework is explained in Sect. 4. However, the whole clustering process is presented in Sect. 5. In Sect. 6, we explain how we port the proposed stream algorithm to Storm topology and we study its effect on information extraction performance in Sect. 7. Finally, we conclude in Sect. 8.

2   State of the Art

2.1   Stream Clustering Techniques

Clustering data streams is an important task in data stream mining. However, unlike data at rest which had a finite size and could be stored as well as clustered by a batch mode algorithm, data stream clustering imposes several challenges since they can be read only once or small number of times. Many algorithms were proposed in literature where the most important are STREAM [4] which gives an approximation for the k-Median problem to fit the stream model. However, it requires a set of parameters in advance mainly a fixed number of maximum clusters. CluStream [3] uses the two-phase paradigm proposed by Aggrawal et al. [24]. It creates micro-clusters in the online phase by compressing raw data streams, to be used in the offline phase for clustering with K-means. But, it is incapable to handle larger volumes of data since it requires multiple scans of the data. Density-based clustering algorithms are the best suited for data stream clustering since they are one-scan techniques. Moreover, they don't require the number of clusters in advance in addition to their efficiency in handling outliers. Therefore, many adaptations to streaming model were proposed mainly DenStream [21], D-Stream [22] and recently ClusTree [6] which is a parameter-free algorithm since it doesn't require an assumption on the number or the size of clusters. The clustering problem we handle is much simple and doesn't require the use of any of state of the art algorithms. However, we inspire from density-based clustering logic and its conditions to propose a clustering solution.

2.2   Clustering-Based Information Extraction

Information extraction as any discipline was tried to be solve with many techniques within which clustering as an efficient data mining technique. However, no interesting results were realized compared to their use for information retrieval. In [5], contributors used DBSCAN clustering to group similar tweets into homogenous groups. Each group is represented compactly with the most representative words. A subsequent module consists on finding possible relationships between words using association rules. Data handled are data at rest and not streaming. Aggrawal and Subbian [23] used the content, the network structure and the temporal horizon information as clustering features for event detection from social streams. They proposed to create k clusters associated with their k summaries. Major limitations of the proposed approach are the number of clusters K required in advance and the inability to detect outliers. In [7], authors used hierarchical clustering of Twitter stream in order to detect different topics like presidential elections and political events in Syria. They used cosine similarity between keywords and fast cluster library to compute hierarchical clusters. The proposed approach doesn't require a k number of clusters. Nevertheless, it requires the time window parameter in input. In [8], authors used clustering approach for hyperlink recommendation application to group users, tweets and hyperlinks with similar interests. They proposed three-way tensor model called collaborative tensor constructed of three variables namely the users, tweets and hyperlinks then they proceed to cluster output tensors using spectrum clustering. Other approaches used clustering to extract information such as entities and semantic relationships among them [10, 13, 15]. Many contributions are based on linguistic processing to generate triggers and arguments used as clustering features in events extraction tasks [11, 12, 14, 16]. It seems a good solution for well-formed and grammatically correct text. Nevertheless, for unstructured text, it needs many cleaning and pre-processing work. On the other side, few works believe on the importance of clustering as preprocessing work to improve information extraction results. Tanev et al. [9] and Piskorski et al. [20] used news clustering as a preprocessing step for events extraction process. They promote the use of topic-based clustering to the use of lexical lookup to gather relevant data about some subjects in online news. The proposed approach guarantees better precision of information extracted.

3 Social Stream Model

A data stream is a large volume of data arriving continuously and it is either unnecessary or impractical to store them in some form of memory [4]. Some or all of the input data are not available for random access from disk or memory. In [1], a data stream is an ordered collection of data generated continuously by one or more sources that can be read only once. In this work, we are interested to social streams like Twitter and Facebook. A streaming social text is the set of user generated text segments continuously evolving in very high frequency. We borrow the appellation social stream introduced by Aggrawal and Subbian [23] and defined as the continuous and temporal sequence of objects $S_1,...,S_n$ such that each object S_i is the tuple (T_i, A_i, t_i, l_i):

- Contains a text T_i which corresponds to the content published and shared by a participant in the social network mainly a text;
- Authored by a unique author A_i
- Published at a given date time t_i
- Posted from a given location l_i.

4 Social Stream Clustering Approach

Clustering is the problem of grouping a set of data points or to partition them into one or more groups of similar objects based on a distance measure. Clustering data streams is used to overview the data distribution and as a pre-processing task for other algorithms [2]. In this work, we propose to group online streaming social text in order to promote only relevant information to the events extraction pipeline, as well as a reduced volume of data in order to improve the input and eventually the output of events extraction pipeline. In a streaming environment, the clustering problem is much harder because data points are evolving and in continuous change. So, we implement online clustering process in a Strom cluster which is a scalable stream processing framework (Fig. 1):

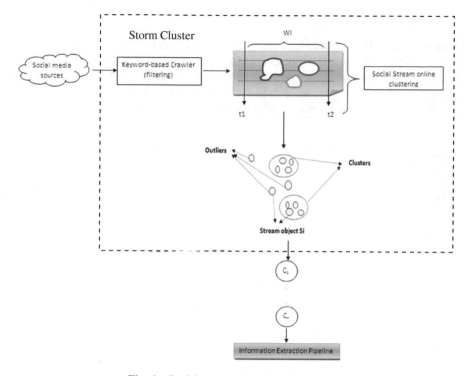

Fig. 1. Social stream clustering framework

5 Online Clustering Process

Many complex and time consuming treatments at the heart of all state of the art clustering algorithms are not required in our problem such as computing the distance measure between objects and the update of cluster centers. In fact, we implement a simple and successful clustering based on the following features:

- The number of clusters is not given in advance,
- Noisy messages are removed in order to save processing time and memory space;
- The size and the number of clusters change since data are continuously evolving;
- The center of clusters is the identifier of the user; so we don't need to update cluster centers each time;
- The distance measure is the exact matching between the user identifier of each message and the center of the cluster.

The clustering process is composed of two steps: cluster detection and outlier detection explained below.

5.1 Cluster Detection

Clustering consists on grouping text segments belonging to the same user and having the same topic. To solve the question of topic, we proposed to use a keyword-based crawler. So, all data collected belong to the same topic. To cluster text segments belonging to the same user, we compare each time the user identifier of the message with the identifier of the cluster using an exact matching. For cluster detection, we assume that:

– C_i: a cluster represents the set of text segments authored by the same user.
– C_{id}: The cluster identifier.
– S_i: a social Stream
– Dist $(C_i, S_j) = \text{Sim} (C_{id}, A_i)$

```
Cluster-Detection Algorithm
Input: Social Stream S
Output: set of clusters
begin
Initialize nbc to 0
Initialize k to 1
Repeat
Receive next stream Sᵢ
Initialize find to false
Repeat
Read next Cluster
compute Dist(Sᵢ, Cₖ)
if (C_id= Aᵢ)
add Sᵢ to Cₖ
find ← true
else
decrement k
until find or k=0
if not find do
create a new cluster
C← Sᵢ
C_id←Aᵢ
Increment nbc
K←nbc
Until end of Stream
End.
```

5.2 Outlier Detection

Each generated cluster corresponds to a potential event. However, not all generated clusters correspond to real and trusted events to be used as basis of future decisions. Therefore, outlier detection is necessary. We consider outliers the clusters containing a number of streaming objects under a given threshold value. The threshold is computed dynamically in function of the number of text segments in each cluster:

```
Outlier-Detection Algorithm
Input: set of clusters SC
Output: final clusters FC
Initialize O to null;
Initialize max, min to 0;
create a unique list of clusters LC from SC
Max= max number of streaming objects in all clusters of LC
Min= min number of streaming objects in all clusters LC
th = (max+min) / nb clusters
For each cluster C_k do
If nb streaming objects <th do
Add C_k to O
Else
Add C_k to FC
End if
End for
```

6 Storm Topology for Clustering Implementation

In order to succeed social stream clustering, reduce its complexity and scale up to larger volumes of data; we propose to use a parallel, distributed and streaming data framework to implement the proposed algorithm. The latest stream processing framework is Storm[1] which is a distributed real-time big data processing system developed to process vast amount of data in a fault-tolerant and horizontal scalable method. It is capable to process thousands messages per second on cluster and guarantees that every message will be processed through the topology at least once. Thus, we propose to divide the whole stream clustering approach into sub-tasks to be distributed among Storm workers as shown in Fig. 2:

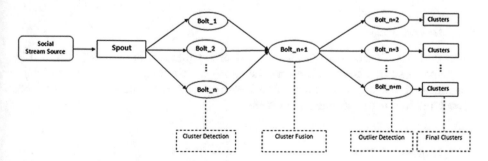

Fig. 2. Storm cluster for social stream clustering

---

[1] http://storm.apache.org/

In fact, a Storm cluster is composed basically of Spouts and Bolts. Spouts are sources of the streaming data and Bolts are tasks distributed among workers.

6.1 Spout

Storm accepts input data from streaming data sources such as Twitter API and performs their real time analysis tuple by tuple. Spout is the first component in the Storm topology charged of collecting streaming data and building social stream objects according to the specification above. At the heart of the spout we implemented a key-word based crawler.

6.2 Bolts

Parallel computing is based on breaking down a big problem into smaller jobs distributed between a given number of workers. In this work, cluster and outlier detection tasks are distributed among bolts. When implementing a Storm cluster on more than one node, we have many bolts so many cluster and outlier detection bolts. A fusion step should be added to the clustering process in order to perform coordination between cluster and outlier detection. This work corresponds to another bolt in Storm topology which algorithm is given below:

```
Cluster-Fusion Algorithm
Input: Initial set of clusters IC
Output: Fused Clusters FC
Read first IC from the initial set of clusters
for each C in IC do
For each IC in the rest of the set
Look for clusters having the same identifiers to C
Fuse C with its similar
Add C to FC
```

6.3 Data Storage

Generated clusters are stored in files for further exploitation. In our case, they will be promoted to an events extraction pipeline as ready data for processing using Natural Language Processing tools. Storm offers many facilities for data storage and even connectors to other platforms and systems.

6.4 Implementation

We created a local Storm cluster composed of cluster detection and outlier detection bolts. We used eclipse java development environment to which we import Storm libraries. The spout implements a keyword based crawler, therefore we imported Twitter4j library as well.

We collected 1000000 tweets about drug abuse and addiction in order to extract drug abuse events. Data collection and clustering details are depicted in Table 1. Collected data belong to three data sets of different sizes collected within different periods. We applied the clustering approach on each one in order to measure the average clustering time. The size of generated clusters corresponds to the number of tweets by cluster:

Table 1. Clustering details of 1000000 tweets

Data set	NB tweets	Extraction period	NB clusters	Size cluster	Clustering time (min)
DS1	504000	30 days	102	388000	15
DS2	300000	23 days	42	126000	5
DS3	196000	12 days	22	65000	3
total	1000000	65 days	166	579000	23

7   Social Stream Clustering Influence on Information Extraction Performance

Our baseline is the system proposed in [25] in which we used Stanford Core NLP [19], for linguistic processing as well as entity extraction, and Open Domain INformer-ODIN [18] to detect possible semantic relationships between entities using linguistic rules. The solution was enhanced with a classification process. In this study, we are limited to the rule-based component to show how clustering ameliorated its performance. We run the events extraction pipeline on the raw set of tweets collected above without clustering and we noted some performance measures. After clustering, we repeated the process on generated clusters and we made the comparison in Table 2.

Table 2. Events extraction performance before and after clustering

	Before clustering	After clustering
Number of tweets	1000000	579000
Information extraction time (minutes)	240	172
Precision	53%	61%
Recall	74%	77%
F-measure	51%	68%

By removing outliers, we remove text segments that could bias the results which decreased the number of false positives and false negatives as well where the improvement of precision, recall and consequently the F-measure. Moreover, the use of clustering as a pre-processing work is an effective way to reduce the volume of data to process thanks to outlier removal which saves memory and processing time by treating only relevant text segments in information extraction phase. Clustering improved also the quality of information extracted by collecting and grouping together semantically-related information and important details that are dispersed at first which resulted in more coherent, meaningful and complete events really useful for further decisions.

8 Conclusion

In this paper, we demonstrate how important and efficient social stream clustering as a preprocessing work to improve events extraction results by grouping semantically-related text segments at first dispersed within social streams. The performance of this approach was strengthened by porting the proposed algorithm to a big data streaming technology namely Storm which succeeded clustering in streaming environment and supported the velocity and volume of data. Nevertheless, events extraction time needs to be improved. Therefore, in future work, we will focus on the use of distributed big data technologies to implement the events extraction pipeline.

References

1. Guha, S., Meyerson, A., Mishra, N., Motwani, R., O'Callaghan, L.: Clustering data streams: theory and practice. IEEE TKDE **15**(3), 515–528 (2003)
2. Gama, J.: Knowledge Discovery from Data Streams. Chapman and Hall Book, Boca Raton (2003)
3. Aggarwal, C., Han, J., Wang, J., Yu, P.S.: A framework for clustering evolving data streams. In: Proceedings of VLDB, pp. 81–92 (2003)
4. Guha, S., Mishra, N., Motwani, R., O'Callaghan, L.: Clustering data streams. In: IEEE Symposium on Foundations of Computer Science, pp. 359–366. IEEE Computer Society (2000)
5. Baralis, E., Cerquitelli, T., Chiusano, S., Grimaudo, L., Xiao, X.: Analysis of Twitter data using a multiple-level clustering strategy. In: Third International Conference on Model and Data Engineering (MEDI 2013), Amantea, Italy, 25–27 September, pp. 13–24 (2013)
6. Kranen, K., Assent, I., Baldauf, C., Seidl, T.: The ClusTree: indexing micro-clusters for anytime stream mining. Knowl. Inf. Syst. **29**, 249–272 (2011). doi:10.1007/s10115-010-0342-8
7. Ifrim, G., Shi, B., Brigadir, I.: Event detection in Twitter using aggressive filtering and hierarchical tweet clustering. In: Second Workshop on Social News on the Web (SNOW), Seoul, Korea. ACM Publisher (2014)
8. Gao, D., Zhang, R., Li, W., Hou, Y.: Twitter hyperlink recommendation with user-tweet-hyperlink three-way clustering. In: CIKM 2012, Maui, HI, USA (2012)
9. Tanev, H., Piskorski, J., Atkinson, M.: Real-time news event extraction for global monitoring systems. In: Joint Research Center of the European Commission, Web and Language Technology Group of IPSC, T.P. 267, Via Fermi 1, 21020 Ispra, VA, Italy (2008)
10. Zhou, D., Chen, L., Yulan, H.: An unsupervised framework of exploring events on Twitter: filtering, extraction and categorization. In: Proceedings of the Twenty-Ninth AAAI Conference on Artificial Intelligence (2015)

11. Georgescu, M., Kanhabua, N., Krause, D., Nejdl, W., Siersdorfer, S.: Extracting event-related information from article updates in Wikipedia. L3S Research Center, Appelstr. 9a, Hannover 30167, Germany (2012)
12. Li, H., Li, X., Ji, H., Marton, Y.: Domain-independent novel event discovery and semi-automatic event annotation (2010)
13. Zhang, Y., Xu, C., Rui, Y., Wang, J., Lu, H.: Semantic event extraction from basketball games using multi-modal analysis (2006)
14. Rusu, D., Hodson, J., Kimball, A.: Unsupervised techniques for extracting and clustering complex events in news. In: Proceedings of the 2nd Workshop on EVENTS: Definition, Detection, Coreference, and Representation, Baltimore, Maryland, USA, 22–27 June, pp. 26–34. Association for Computational Linguistics (2014)
15. Zhang, C., Soderland, S., Weld, D.: Exploiting parallel news streams for unsupervised event extraction (2013)
16. Mehryary, F., Kaewphan, S., Hakala, K., Ginter, F.: Eliminating Incorrect Events from Large-Scale Event Networks by Trigger Word Clustering and Pruning. The University of Turku Graduate School (UTUGS), University of Turku, Finland (2013)
17. Poibeau, T., et al. (eds.): Multi-source, Multilingual Information Extraction and Summarization. Theory and Applications of Natural Language Processing. Springer, Heidelberg (2013). doi:10.1007/978-3-642-28569-1. Chapter 2, J. Piskorski and R. Yangarber
18. Valenzuela-Escarcega, M., Hahn-Powell, G., Hicks, T., Surdeanu, M.: A domain-independent rule-based framework for event extraction. In: Proceedings of the 53rd Annual Meeting of the Association for Computational Linguistics and the 7th International Joint Conference on Natural Language Processing of the Asian Federation of Natural Language Processing: Software Demonstrations (ACL-IJCNLP) (2015)
19. Manning, D., Mihai, C., Bauer, S., Finkel, J., Bethard, J., McClosky, D.: The Stanford CoreNLP Natural Language Processing Toolkit (2014)
20. Piskorski, J., Tanev, H., Atkinson, M., Van der Goot, E.: Cluster-Centric Approach to News Event Extraction. Joint Research Centre of the European Commission Institute for the Protection and Security of the Citizen Via Fermi 2749, 21027 Ispra, Italy (2010)
21. Cao, F., Ester, M., Qian, W., Zhou, A.: Density-based clustering over an evolving data stream over noise, pp. 326–337 (2004)
22. Chen, Y., Tu, L.: Density-based clustering for real-time stream data. In: Proceedings of the 13th ACM SIGKDD International Conference on Knowledge Discovery and Data Mining, KDD 2007, pp. 133–142. ACM Press (2007)
23. Aggrawal, C.C., Subbian, K.: Event Detection in Social Stream. IBM T. J. Watson Research Center, Hawthorne, NY, USA, †Department of Computer Science & Engineering, University of Minnesota, Twin Cities, MN, USA (2011)
24. Aggarwal, C.C., Han, J., Wang, J., Yu, P.S.: A framework for on-demand classification of evolving data streams. IEEE TKDE 18(5), 577–589 (2006)
25. Jenhani, F., Gouider, M.S., Ben Said, L.: A hybrid approach for drug abuse events extraction from Twitter. In: 20th International Conference on Knowledge-Based and Intelligent Information and Engineering Systems (ICKIIES 2016), York, United Kingdom, pp. 1032–1040 (2016)

A New Social Recommender System Based on Link Prediction Across Heterogeneous Networks

Manel Slokom$^{(\boxtimes)}$ and Raouia Ayachi

LARODEC, Institut Supérieur de Gestion Tunis, 2000 Le Bardo, Tunisia
manel.slokom@live.fr, raouia.ayachi@gmail.com

Abstract. Nowadays, the wide use of the Internet around the world allows people to socialize and connect together. This results of the explosion of the Web 2.0 giving rise to a growing demand for *Social Recommendation Systems*. Social recommendation systems are introduced to rescue users from searching and choosing by predicting users' preferences. In this paper, we will focus on recommendation via link prediction across heterogeneous social network. The main objective is to recommend items by predicting the missing or unobserved interactions between actors within a social network while pinpointing different types of objects and links. Probabilistic relational models will be used for prediction of new interactions in a citation network.

Keywords: Social recommendation systems · Link prediction · Heterogeneous network · Probabilistic relational model

1 Introduction

Social networks are dynamic structures evolving over time by addition of new social entities and new social relationships. With the fast growing of the world wide web, online social networks such as Facebook, Twitter, Foursquare have become more and more popular [6]. Thus, web-based social networks are flourishing and several researches have been proposed in order to deal with the *social information overload problem*. An active research field of social network analysis is the *social recommendation systems* [10]. In fact, social recommender systems are designed to solve the information overload in social media and create original opportunities in order to help users better understand what they really want.

A panoply of recommendation methods were proposed to deal with social information overload but the performance was not satisfactory. Therefore, researchers proposed a variety of solutions or techniques in order to improve the recommendation accuracy such as triadic closure [15] and link prediction [7,13], etc. In this paper, we are in particular interested to social recommendation via link prediction since it is widely used in recent years [1,7,10]. In fact, link prediction plays a fundamental role in the analysis of complex networks. It was first introduced by Liben-Nowell et al. [14] and defined as the task of identifying and predicting missing or new interactions within a social network.

© Springer International Publishing AG 2018
I. Czarnowski et al. (eds.), *Intelligent Decision Technologies 2017*,
Smart Innovation, Systems and Technologies 73, DOI 10.1007/978-3-319-59424-8_31

Most of the existing link prediction methods are designed for only simple networks which have *homogeneous* relations [1,3,7]. However, in most of the cases, these approaches only focused on information about the network structure and ignore the important role related to the relational knowledge. To improve the recommendation performance, few recent studies [4,9,11] have introduced the probabilistic relational model (for short PRM) in the recommendation process. However, these methods only consider simple recommender systems that could not be applied for heterogeneous networks.

In this paper, we will develop a new social recommendation approach based on link prediction using PRM. In fact, our approach is based on three steps, namely, *PRM generation* which consists in generating a random PRM, *prediction step* to predict unobserved or missing relations between social entities, and *recommendation step* to generate the most relevant relationships in the network. To illustrate our approach, we provide a case study from the citation domain. The rest of this paper is organized as follows. In Sect. 2, we will briefly present basic concepts related to social recommendation via link prediction and the probabilistic relational model. Section 3 is dedicated to related works. We describe our social recommendation approach in Sect. 4. Finally, we conclude the paper in Sect. 5.

2   Basic Concepts

This section gives a brief overview on both social recommendation via link prediction and probabilistic relational models.

2.1   Social Recommendation via Link Prediction

A Recommender System (RS) is a software application capable of suggesting items to its users based on their preferences [13]. Social Recommender Systems (for short SRSs) are defined as any recommender system that targets social media domains [10]. SRSs play an important role at coping with the social information overload by suggesting the most relevant and attractive content to the user. Many researchers proposed a variety of techniques in order to improve the accuracy of recommendation such as taking explicit and implicit information into account, analyzing the user's social behaviors, link prediction [7,13], social tie recommendation and triadic closure [15]. In fact, link prediction is a fertile research area [1,10] closely related to social recommender systems and has been often offered to enhance it [13,15]. Specifically, given a set of potential links, *link prediction* is a binary classification problem which aims to predict unobserved relationships between actors of a social network [14]. Then, recommendation step consists in suggesting to each actor a list with whom he might create new connections. Most of the recent link prediction methods [1,14] are designed for *homogeneous* networks, where only a single type of link exists in the network. Meanwhile, real world applications are modeled as heterogeneous networks. Thus, link prediction across heterogeneous network assumes that there exist various relationships holding between various social entities, such as friendships, business relationships, and common interest relationships [6,7]. Such network are called *multi-relational* or *heterogeneous social network*.

Example 1. *The DBLP citation is a heterogeneous network containing different interactions between nodes as shown in Fig. 1. Various interactions from different types are present in such network, we cite:* **contain:** *a paper contains different terms/keywords,* **citation:** *a paper is citing/cited by another paper,* **write:** *an author writes a paper,* **present:** *an author is present in ACM conference.*

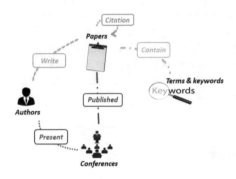

Fig. 1. The DBLP bibliographic network.

Definition 1. *Link prediction problem. Let us consider a heterogeneous network modeled as $G = (V_1 \cup V_2 ... \cup V_M, E_1 \cup E_2 ... \cup E_N)$, $V_u(u \in N)$ and $E_j(j \in M)$ where V_u represents the set of nodes or social entities of the same type u and E_j represents the set of links having same type j. Links in these networks can occur between social entities having the same or different types. The link prediction task is to predict whether there* **exist** *or* **will exist** *a link of type w between a pair of nodes i and j where $(i \in V_x)$ and $(j \in V_y)$. Therefore, the problem is to assign a $score(i, j, w)$ that indicates the existence of a link between social entities i and j where w presents the link type. Formally:*

$$Score(i, j, w) = \begin{cases} 1 & if \ \ w = (i, j) \in E \\ 0 & Otherwise \end{cases} \tag{1}$$

2.2   Probabilistic Relational Model

Bayesian Networks (namely BNs) represent an effective tool for reasoning under uncertainty and representing knowledge [5]. Despite their success, BNs are often inadequate for representing complex domains since they are developed for flat data. To address these limitations, Probabilistic Relational Models (noted PRMs)

[8] have emerged as an extension of BNs. In fact, a *PRM* specifies a unified template for a probability distribution that brings together the strengths of probabilistic graphical models and the relational presentation. The fundamental assymption is that PRM considers not only properties about a particular object but also properties about the related objects and the relationships among them. We will follow the same notation used in [8]. A *relational schema* $\mathcal{R}$ describes a set of classes $\mathcal{X} = \{X_1...X_n\}$. Each class is associated with *a set of descriptive attributes* denoted $\mathcal{A}(X)$ and *a set of reference slots* denoted $X.\rho$ allowing objects to refer to another object. In the context of a relational database, a class refers to a single database table, descriptive attributes refer to the standard attribute of tables and reference slots are equivalent to foreign keys. A *Slot chain* κ is a sequence of reference slots. The *aggregate function* γ is a function which takes a multi-set of values and produces a single value as a summary of the input values. A *relational skeleton* σ_r of a relational schema is a partial specification of an instance of the schema. It specifies the set of objects $\sigma_r(X_i)$ for each class and the relations that hold between the objects.

Definition 2. *Probabilistic relational model.* Π *is defined for each class* $X \in \mathcal{X}$ *and each descriptive attribute* $A \in \mathcal{A}(X)$, *we have:*

- *a set of parents* $Pa(X.A) = \{U_1, ..., U_l\}$, *where each* U_i *has the form* $X.B$ *or* $\gamma(X.\tau.B)$, *where* B *is an attribute on which* A *depends.*
- *a conditional probability distribution (CPD),* $P(X.A|Pa(X.A))$.

Given a relational skeleton σ_r, the PRM Π defines a distribution over the possible worlds through a *Ground Bayesian network* (GBN). It consists of:

1. A qualitative component:
   - A node for every attribute of every object $x \in \sigma_r(X)$, $x.A$
   - Each $x.A$ depends probabilistically on a set of parents $Pa(x.A) = u_1, ..., u_l$ of the form $x.B$ or $x.\kappa.B$, where each u_i is an instance of U_i.
2. A quantitative component relative to the conditional probability distribution (for short CPD) for $x.A$ denoted by $P(X.A|Pa(X.A))$.

Example 2. *Let's consider a relational schema for the citation domain with five classes* $\mathcal{X} = \{Paper, cites, author, writes, co-author\}$ *as described by Fig. 2. The class paper may have descriptive attributes such as topic and words and the class author may have research area and institution as descriptive attributes. The relation cites has descriptive attribute cites. Exists and two reference slots cites. CitingPaper and cites. CitedPaper representing the citation of one paper, the **Cited** paper, by another paper, the **Citing** paper.*

An example of a slot chain would be $writes.author.author^{-1}.paper$ *which could be interpreted as all the papers that can be written by a particular author. Paper.topic* $\rightarrow$ *cites.paper.citingPaper is an example of a probabilistic dependency derived from slot chain of length 1 where paper.topic is a parent of cites.paper.citingPaper as shown in Fig. 2. Also, varying the slot chain length may give rise to other dependencies.*

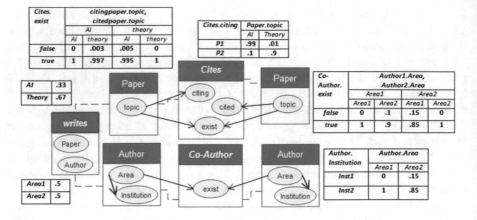

Fig. 2. The PRM structure for the citation domain.

Another important feature of PRMs is their ability to represent not only uncertainty about properties of the attributes but also about *link uncertainty* which is uncertainty over the relational structure of the domain. Typical link uncertainty algorithms are based on reference uncertainty and existence uncertainty in order to model the probability that certain relationships can hold between objects. PRM with **existence uncertainty** *PRM-EU* makes no assumptions about the number of links that exist and is able to model the existence of relations between objects too. Under this extension, the uncertainty of existence of relationships is modeled between objects by introducing a special existence attribute $X.Exists$ whose values are from $V(E) = \{true, false\}$. While, PRM with **reference uncertainty** (PRM-RU) assumes that the objects are prespecified, but relations among them, i.e., reference slots, are subject to random choices [8]. The objective is then to find the probability that a link points to some specific objects.

3   Related Work

Several link prediction based recommendation algorithms have been proposed in literature. In fact, in [7] authors presented a new link prediction and recommendation approach across heterogeneous social networks using a new ranking factor graph model. In addition, in [16] authors proposed a link recommendation method based on both user attributes and graph structure. In the same context, authors in [2] developed a new algorithm based on supervised random walk that combines information from the network structure (node and edge level attributes) in order to predict and recommend links in social networks. In [3] authors proposed a novel link prediction approach based on supervised machine learning in order to improve the academic collaboration recommendation.

Most of existing link prediction and social recommendation approaches give good results in the context of heterogeneous networks while focusing on structural and topological information about the networks or based on probabilistic and sto-chastic models in order to obtain better performance. However, in most of the cases, these approaches only focused on information about the network structure and ignore the important role related to the relational knowledge. Indeed, few recent studies have introduced the probabilistic relational model in the recommendation process. For instance, authors in [4] proposed a novel approach to build a personal-ized PRM-based recommendation model with the help of users' preferences where content-based, collaborative filtering as well as hybrid models can be achieved from the same PRM. In [9], authors described how PRMs can be applied to the task of col-laborative filtering. In [11], authors proposed a unified recommendation framework based on PRMs. While authors in [12] proposed a new recommendation approach that explores the relational nature of the data in hand using relational Bayesian net-works. However, these methods only consider simple RS that could not be applied for heterogeneous networks. To the best of our knowledge, no research studied *link prediction* across heterogeneous networks using PRM in the context of social rec-ommendation.

4 A New Social Recommendation Approach Based on Link Prediction Using PRM

A social recommender system recommends for each user the most relevant and attractive data. This is typically done based on modeling the probabilistic and relational interactions between social entities [7]. To assure this task, we propose a new social recommender system based on link prediction using PRM in order to accommodate users with content that best fit their tastes. The whole process is composed of three steps, as illustrated by the diagram of Fig. 3. The first step concerns the PRM generation. The second step is prediction using existence and reference uncertainty algorithms. The last step is dedicated to recommendation.

Notations

- $\mathcal{I}$ is an instantiation of the database.
- σ_r is the relational skeleton.
- σ_e (resp. σ_e) is the relational entity skeleton (resp. relational object skeleton).
- ψ refers to the partition function.
- S_ρ is a selector attribute within the class $\mathcal{X}$.

4.1 The PRM Generation

The PRM generation is a primordial step in our approach. It consists at first in generating a random PRM, then learning parameters. In fact, the relational schema should be at first generated, then probabilistic dependencies between classes and attributes have to be defined. Our PRM dependency structure depicted by Fig. 2(b) represents two types of links: (i) *Intra-class links* connect-ing attributes of the same class (i.e., Paper.topic → Paper.words) (ii) *Inter-class links* connecting attributes of different classes (i.e., Paper.topic → Cites.Citing).

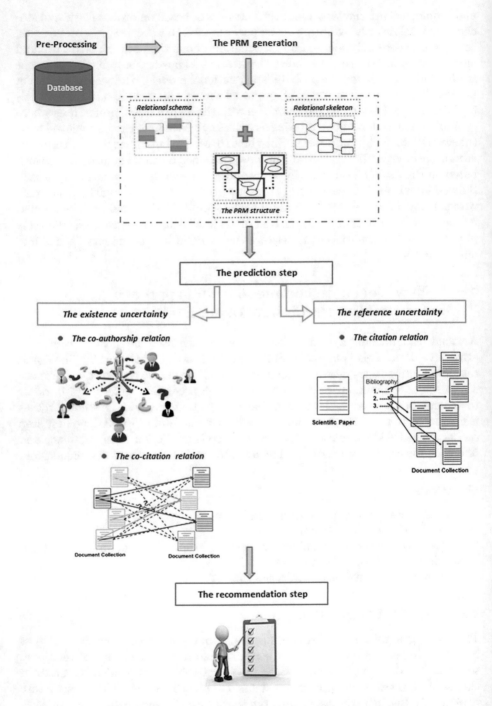

Fig. 3. The proposed social recommendation process

4.2   The Prediction Step

The learned PRM model have certainly missing or unobserved links (relations) between objects, so a prediction step seems to be primary to enrich the model. Our main idea in this step is to basically predict three relations, namely:

- *Co-authorship prediction* (or author-author relation): In a such prediction, we will apply a recent algorithm proposed by Getoor et al. in [8], the so-called *existence uncertainty* proposed to predict whether exist or not a relation between authors. To this end, we introduce a new special binary variable called Exists that tells whether an authorship relation actually exists or not between two authors. We define a new $PRM\,\Pi^*$, in which $Co\text{-}Author.Exists = \{true, false\}$ is a standard descriptive attribute and we specify in the other hand the parents of $Co\text{-}Author.Exists$ to be $Pa^*(Co\text{-}Author.Exists)$ and the associated conditional probability distribution to be $P^*(Co\text{-}Author.Exists|Pa^*(Co\text{-}Author.Exists))$. As a result, a PRM with existence uncertainty for the co-authorship prediction specifies a distribution over a set of instantiations $\mathcal{I}$ given an entity skeleton σ_e as follow:

$$P(\mathcal{I}|\sigma_e,\Pi) = P(\mathcal{I}|\sigma_r\,[\sigma_e]\,,\Pi^*) = \prod_{X\in\mathcal{X}}\prod_{\mathbf{Co\text{-}Author}\in\sigma_r[\sigma_e](X)}\prod_{Exists\in\mathcal{A}(x)}$$

$$P^*(Co\text{-}Author.Exists|Pa^*(Co\text{-}Author.Exists)) \qquad (2)$$

- *Co-citation prediction* (or paper-paper relation): Similarly to the co-authorship prediction, we will apply the existence uncertainty algorithm between two papers in order to predict for each co-authorship relation ($Co\text{-}Author.Exists = True$), the existence of new relation of type *co-citation* between two papers based on the attribute topic. Formally:

$$P(\mathcal{I}|\sigma_e,\Pi) = P(\mathcal{I}|\sigma_r\,[\sigma_e]\,,\Pi^*) = \prod_{X\in\mathcal{X}}\prod_{\mathbf{Cites}\in\sigma_r[\sigma_e](X)}\prod_{Exists\in\mathcal{A}(x)}$$

$$P^*(Cites.Exists|Pa^*(Cites.Exists)) \qquad (3)$$

- *Citation prediction:* is based on the reference uncertainty algorithm while assuming that papers are prespecified, but relations among them, i.e., reference slots, are subject to random choices. In fact, for each *co-citation relation* ($Co\text{-}Citation.Exists = True$) we will predict the topic of papers that can *be cited by* a particular paper, as well as the topic of papers that can *be citing* this paper. To define the semantics of this extension we treat reference slots as random variables:

$$P(\mathcal{I}|\sigma_o,\Pi) = \prod_{x\in\sigma_o(X)}\prod_{A\in\mathcal{A}(X)} P(x.A|Pa(x.A)).\prod_{\rho\in\mathcal{R}(X),Range[\rho]=Y}\frac{P(x.S_\rho = \psi\,[y]\,|Pa(x.S_\rho))}{|\mathcal{I}(Y_{\psi[y]})|}$$

$$(4)$$

Then, once we have papers sorted according to the topic attributes into clusters and we have computed for a target paper the percentage of $P(CitingPaper.Topic)$ and $P(CitedPaper.Topic)$. The next step is to apply existence uncertainty in order to approximately find papers (among the target cluster) that can be citing or cited by the target paper.

4.3   The Recommendation Step

Once unknown or unobserved relationships are predicted based on PRM with existence uncertainty and PRM with reference uncertainty. The recommendation step is then vital in order to classify different relationships into two basic categories, i.e., 'highly recommended' and 'not recommended'. So, we fix a threshold:

- **Co-citation/co-authorship recommendation:** has two percentage values ranging from $\{true, false\}$. We fix a threshold S such that we highly recommend relations having $P(relation = true) \geq S$.
- **Citation recommendation:** in this relation, we have two basic phases. At first, we fix a threshold S' capable of choosing the best citing or cited papers for the target paper based on the partition attribute (= topic). Then, we will recommend only the citation relations where $P(citation = true) \geq S''$.

Illustrative Example

Let us illustrate the different steps of our recommender system by an example. We will consider the random PRM structure depicted in Fig. 2. We consider an instantiation from the DBLP database containing information related to 5 papers and 3 authors. Our goal is to predict new or missing relationships between either two papers or two authors as follows:

1. *Firstly, we predict the existence of relation between two authors a_1 and a_2 based on existence uncertainty. To this end, we will compute $P^*(co - author.Exists = true|a_1.area, a_2.area) = 0.93 \geq S$ and $P^*(co - author.Exists = false|a_1.area, a_2.area) = 0.07 \leq S = 0.8$ meaning we highly recommend a new co-authorship relation between a_1 and a_2.*

2. *Second, in the same manner we compute the existence of relationship of type co-citation between papers related to a_1 and a_2 as follows: $P^*(cites.Exists = true|p_2.topic, p_3.topic) = 0.999 \geq S = 0.8$ and $P^*(cites.Exists = false| p_2.topic, p_3.topic) = 0.001 \leq S = 0.8$. Thus, we highly recommend the existence of co-citation between p_2 and p_3.*

3. *Third, based on both predicted relationships, we will find for a target paper p_5 all citing and cited papers. So, using reference uncertainty, we start by classifying papers into clusters based on the partition function ψ_{topic}. We compute for each citing (respectively cited) paper, the probability of belonging to AI or theory as follow: $P(CitedPaper.topic = theory) = 0.355 \leq S' = 0.6$, $P(CitedPaper.topic = AI) = 0.645 \geq S' = 0.6$, $P(CitingPaper.topic = theory) = 0.38 \leq S' = 0.6$ and $P(CitingPaper.topic = AI) = 0.62 \geq S' = 0.6$. Thus, we can say that 64.5% of papers cited by p_5 are from the theory field ($\in Cluster_{theory}$) and 62% of citing papers $\in Cluster_{AI}$.*

4. *Finally, we suppose a new threshold $S'' = 0.8$ and we compute based on existence uncertainty whether paper p_5 will be citing or not paper p_3, where $p_3 \in Cluster_{AI}$. Therefore, we find that $P(Citation(p_5, p_3).Exists = true) = 0.92 \geq S'' = 0.8$. Consequently, we will highly recommend the appearance of new citation relationship between paper p_5 and p_3.*

5   Conclusion

In this paper, we proposed a new social recommendation method which extends the idea of social recommendation to link prediction in heterogeneous network. We designed the link prediction task using a PRM. Then, we resorted to probabilistic inference to predict the appearance of new relationships among the network. Our approach allows the integration of relational data into the social recommendation process. We took a case a study of our approach in the citation domain. Thus, information related to authors and papers are used to predict new or missing interactions of type: Co-authorship, co-citation and citation.

In our ongoing work, we will focus on the implementation of the social recommendation based PRM framework in order to apply it to real world dataset and validate the efficiency of our theoretical proposal.

References

1. Al Hasan, M., Zaki, M.J.: A survey of link prediction in social networks. In: Aggarwal, C.C. (eds.) Social Network Data Analytics, pp. 243–275. Springer, Heidelberg (2011)
2. Backstrom, L., Leskovec, J.: Supervised random walks: predicting and recommending links in social networks. In: Proceedings of the Fourth ACM International Conference on Web Search and Data Mining, pp. 635–644. ACM February 2011
3. Benchettara, N., Kanawati, R., Rouveirol, C.: A supervised machine learning link prediction approach for academic collaboration recommendation. In: Proceedings of the Fourth ACM Conference on Recommender systems, pp. 253–256 (2010)
4. Chulyadyo, R., Leray, P.: A personalized recommender system from probabilistic relational model and users' preferences. Procedia Comput. Sci. **35**, 1063–1072 (2014)
5. Daly, R., Shen, Q., Aitken, S.: Learning Bayesian networks: approaches and issues. Knowl. Eng. Rev. **26**(02), 99–157 (2011)
6. Davis, D., Lichtenwalter, R., Chawla, N.V.: Multi-relational link prediction in heterogeneous information networks. In: 2011 International Conference on Advances in Social Networks Analysis and Mining, pp. 281–288. IEEE (2011)
7. Dong, Y., Tang, J., Wu, S., Tian, J., Chawla, N.V., Rao, J., Cao, H.: Link prediction and recommendation across heterogeneous social networks. In: 2012 IEEE 12th International Conference on Data Mining, pp. 181–190 (2012)
8. Getoor, L., Friedman, N., Koller, D., Taskar, B.: Learning probabilistic models of link structure. J. Mach. Learn. Res. **3**, 679–707 (2002)
9. Getoor, L., Sahami, M.: Using probabilistic relational models for collaborative filtering. In: Workshop on Web Usage Analysis and User Profiling (1999)
10. Guy, I.: Social recommender systems. In: Ricci, F., Rokach, L., Shapira, B. (eds.) Recommender Systems Handbook, pp. 511–543. Springer, New York (2015)
11. Huang, Z., Zeng, D. D., Chen, H.: A Unified Recommendation Framework Based on Probabilistic Relational Models (2005). SSRN 906513
12. Ishak, M.B., Amor, N.B., Leray, P.: A RBN-based recommender system architecture. In: 5th International Conference on Modeling, Simulation and Applied Optimization (ICMSAO), pp. 1–6. IEEE (2013)
13. Li, Z., Fang, X., Sheng, O.R.L.: A survey of link recommendation for social networks: methods, theoretical foundations, and future research directions. In: Theoretical Foundations, and Future Research Directions (2015)

14. LibenNowell, D., Kleinberg, J.: The link prediction problem for social networks. J. Am. Soc. Inform. Sci. Technol. **58**(7), 1019–1031 (2007)
15. Schall, D.: Social Network-Based Recommender Systems. Springer, Switzerland (2015)
16. Yin, Z., Gupta, M., Weninger, T., Han, J.: Linkrec: a unified framework for link recommendation with user attributes and graph structure. In: Proceedings of the 19th International Conference on World Wide Web, pp. 1211–1212. ACM (2010)

Social Market: Stock Market and Twitter Correlation

Ivo Bernardo, Roberto Henriques[(✉)], and Victor Lobo

NOVA IMS, Universidade Nova de Lisboa, 1070-312 Lisbon, Portugal
ivopbernardo@gmail.com,
{roberto, vlobo}@novaims.unl.pt

Abstract. The text shared on social networks and interactions resulting from all virtual activities have been gaining a great impact on society. In this work, we investigate if Twitter data can be used to predict or describe stock market prices by using sentiment polarity (positive or negative). Using a Bayesian classifier and making two causality models (one with the Stock Market and another with the Twitter sentiment as dependent variable) we could relate the data from Twitter with intra-day and day-to-day stock prices. We reached four significant conclusions. First, the relationship between twitter and the stock market is, in both cases, strongly dependent on the time grouping of the twitter data. Second, using Granger Causality Analysis, we found some companies where we can use tweets to predict the stock price, and others where we can't. Amongst those where we can, there are some where the delay between tweets and changes in price are small (Cisco, American Airlines and Microsoft), and others where those changes take a longer time (LinkedIn). Third, companies with a high number of tweets show a weaker relationship amongst the two variables. Forth, in some cases (British Petroleum), we can predict changes in Twitter sentiment using stock prices.

1 Introduction

The stock market is a core focus of economic and social studies and the capability of developing models that can accurately predict the stock market prices have been thoroughly explored by researchers [1–4]. One of the most recurrent questions that investors and companies make on a daily basis is which data can be used to predict the volatility and the changes of the prices of some specific stock. Financially, stock prices should measure the expected growth and cash-flow that the company will obtain in the future. This expectation is based on some specific set of information from various sources that consist of economic, reputation, financial and social data. According to the Efficient Market Hypothesis (EMH) [1], stock market prices are influenced by news and happenings rather than from past prices, and changes of stock prices can only be predicted with an accuracy of approximately 50%. EMH follows the assumption that news are totally unpredictable and therefore it is not possible to foresee future prices of stock [1]. However, this assumption does not invalidate the fact that stock prices do depend on news, and thus when we have news we can predict stock prices in the near future with good accuracy. Moreover, social media can give an early indication of

© Springer International Publishing AG 2018
I. Czarnowski et al. (eds.), *Intelligent Decision Technologies 2017*,
Smart Innovation, Systems and Technologies 73, DOI 10.1007/978-3-319-59424-8_32

public news facts, further increasing the time over which stock prices can be predicted [5]. Thus, even though stock prices may be difficult to predict in the long run, we may be able to use social networks to predict them in the short run.

With the growth of social networks, the speed of information that travels through the globe has increased exponentially. At the distance of a click you can access the thoughts and opinions of many individuals as soon as they formulate them. Various papers have shown [6] how opinions in social media have influenced things such as company reputation [6, 7], products and brands consumer opinions [8–10], audience sentiment on sports debates [11] or political elections [12]. Social media opinions regarding companies can be categorized in terms of polarity (positive vs. negative) [6] or public mood (calm, alert, etc.) [5]. Duric and Song [13] classify sentiment analysis applications in four categories: (1) those that automatically analyze online customer reviews, (2) those that improve search engines by integrating sentiment analysis, (3) marketing and business intelligence applications and (4) those that perform early detection of negative text and cyber-bullying.

Various techniques for sentiment analysis have been developed [14]. These techniques have different behaviors when applied to different types of texts (tweets, newspaper news, reviews, etc.).

One example of such work was proposed in [15] where a machine learning approach was followed to predict prices by learning from historical price changes.

On Twitter, a social network for micro-blogging [16], users can only post messages with 140 characters. At first glance, this characteristic should imply a simpler sentiment analysis than the one used for longer texts such as news or entire documents. However, tweets' based sentiment analysis poses some challenges [17]. In fact, most tweets are neutral, while using other sources it is usual to have a predominance of negative or positive texts. Also, associated with the size limitation of the text, new and challenging linguistic representations are often used and limited sentiment is usualy provided. As for the relationship between sentiment analysis on news (including tweets) and stock prices, research is still ongoing but some studies found some quantitative correlation between financial news and stock prices [2, 3]. In a paper by Schumaker and Chen [3], stock prices were predicted with 58,2% accuracy using financial news. On another related paper [5], the Dow Jones Industrial Average (DJIA) value was predicted with 86,7% accuracy using tweets processed with a psychology tool called Profile of Mood States. In that paper the general mood state and sentiment was extracted from tweets, and the sentiment alone (positive or negative) was not enough to predict the DJIA. However, the authors only studied the effect of one day sentiments on the next day's prices, in what we call "day to day predictions", and did not try to predict prices during the same day, in what we call "intraday predictions".

In our study, we focused on predicting individual stock prices (as opposed to general averages) using tweets concerning those specific companies as input variables. The key motivation for this work is that Twitter's streaming messages reflect very recent news which are likely to be correlated with stocks in a short time frame. For this, we computed the average sentiment over 1 h periods and 3 min periods. For day to day predictions (which we performed to compare with previous studies) the daily average sentiment was also used. Besides the influence of tweets on stock prices we also tested the inverse relationship, i.e., how tweet sentiments are influenced by stock prices. Sentiment analysis

employs statistical and machine learning algorithms to build classifiers to model sentiment. Some of the most applied classifiers are *Naïve* Bayes [18, 19], Support Vector
Machines [20, 21], Logistic Regression [22] and Random Forest [23]. A more detailed
survey on the use of classifiers for tweet sentiment analysis can be found in [24].

We compared two sentiment analysis techniques: a keyword based classifier; and a
Bayesian classifier. A pre-classified test set of tweets was used to choose the best
sentiment analysis technique. A Granger Causality Test was then applied, as done in
previous studies [5]. Granger Causality [25] is a test that evaluates if we can use the
values of one time series to predict the other with some given lag. It must be noted that
this does not imply actual causality (correlation does not imply causality), but it does
imply that tweets' sentiment of some specific company at moment t can be used to
predict stock prices at moment $t + \delta$.

The rest of the paper is structured as follows: Sect. 2 explains the methodology
used, Sect. 3 presents results, and Sect. 4 discusses the results and draws the main
conclusions.

2 Overview the Methodology

Our aim is to predict individual stock prices using tweets concerning specific companies as input variables. The rising popularity of twitter gives researchers a novel way
of capturing the collective mind up to the last minute. In this paper, we analyse the
positive and negative sentiment over 1 day, 1 h, and 3 min periods.

Our research framework is depicted in Fig. 1.

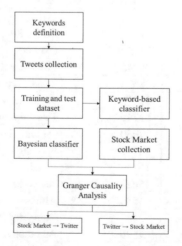

Fig. 1. Starbucks stock price evolution (hourly averages)

First, tweets were extracted using a list of names and related names from 16 main
US companies. In parallel, we obtained data from stock market for the same days
tweets were collected with a price interval of 3 min, gathering 130 stock prices per day

for each company [26]. Next, we built a training and test dataset by manually reading and classifying the tweets as positive and negative. The training dataset is used in the Bayesian classifier, while the test dataset is used in both classifiers to evaluate its accuracy.

Finally, we used Granger Causality Analysis to test the relationship between stock market prices and twitter data sentiment. We tested different lags to relate stock price variations with twitter sentiments using different time periods as well as different time lags. So, as to be clearer, the detailed methodology is explained step by step in the following section where we show the results at each step.

3   Results

3.1   Data Extraction

Twitter data extraction occurred from February 11th to February 28th 2014, between 13:30 GMT and 21:00 GMT, excluding weekends and financial holidays. These times are the opening and closing hours of the stock market in the USA, and thus the times when stock prices can vary, so that we can test the intraday effects of Twitter on the stock market. We only selected the tweets that explicitly mention one of the companies considered in the study, and obtained 1,535,167 tweets. The companies that we considered are presented in Table 1.

Table 1. Companies chosen for analysis, respective industry and tweets gathered

Company	Number of Tweets	Industry
Amazon	455156	E-Retail
Starbucks	311945	Coffee Shop
Nike	273552	Clothing
Microsoft	128045	Computer Soft and Elect.
LinkedIn	108620	Social Network
Sony	99002	Technology
BlackBerry	61246	Telecommunications
BP	37950	Oil and Gas
Barclays	16775	Banking
Sears	16174	Retail
Cisco	15628	Networks
American Airlines	4886	Airline
Logitech	3441	Computer Peripherals
General Motors	1620	Automotive
Marriot	620	Tourism
Quiksilver	507	Clothing

We only gathered Tweets written in English for two reasons:

- Multilingual sentiment analysis research is still in its infancy [27].
- The stock prices used are from the USA, where English is the standard language.

Regarding stock market, we obtained the data from [26] for the days above with a price interval of 3 min for all companies chosen for the study, obtaining a total of 130 stock prices per day for each company. We used this time series in some tests, but we used hourly averages and day averages of these values for others. It must be noted that these prices do not constitute a stationary process, *i.e.*, the average price varies in time, and trends can clearly be seen. As an example, the Starbucks stock price over the period analyzed shows a decreasing trend as can be seen in Fig. 2a.

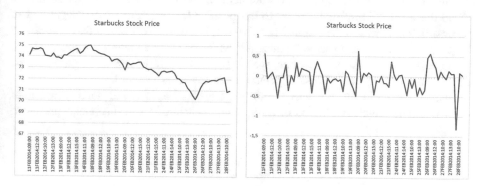

Fig. 2. Starbucks stock price evolution: (a) (hourly averages) and (b) price change in the stock price between hour t and hour t − 1

To overcome this effect, we differentiated this series, obtaining changes in the stock price, as was done by [5]. Using this approach, the stock market series used for Starbucks (using hourly averages) is presented in Fig. 2b.

As done by Miner *et al.* [28] we first pre-processed our tweets removing user names (in twitter, "user mentions" are preceded by @), HTML links, and retweet mentions (RT). To determine sentiment (positive, negative, or neutral) we used two models: a simple keyword-based classifier; and a Bayesian classifier.

3.2   Sentiment Analysis

Keyword-Based Classifier

In line with the work done by [7] a set of "positive words" and "negative words" was used. If more "positive" words appear in the tweet, the tweet is considered positive, if more "negative" worlds appear, it is considered negative, and if no such words appear, or if there is a tie between these two groups, then the tweet is considered "neutral". The list of the 6800 words used in this study (such as "bad" for negative, and "good" for positive) is presented in [7].

Table 2. Proportion of positive/neutral/negative tweets

Sentiment	Proportion
Positive	34%
Neutral	54%
Negative	12%

This method has the advantage of being very simple to apply and of benefiting from short length of Twitter messages. On the other hand, the use of a static word list for classification may decrease the accuracy of the method as time goes by [7]. Internet language has its own inaccuracies and misspells and every year new expressions and sentiment related words appear. To complement the list of words presented in, we added the words "Want" and "Party" to our positive list and "Cut" to our negative list. The word "Free" was in the original positive list, but when we realized that it appeared frequently due to the expression "Amazon Free Super Saver Delivery", we removed it. Using this keyword-based classifier, we obtained the distribution amongst the three classes presented in Table 2. As expected, most tweets are neutral.

Just out of curiosity, the vast majority of tweets has a rather small number of keywords. The actual distribution of positive and negative keyword scores is presented in Fig. 3.

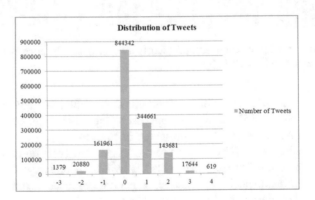

Fig. 3. Positive and negative scores on the training set with the keyword-based classifier

To evaluate the accuracy of this classifier, we used the test set of manually pre-classified tweets mentioned earlier. This test set had 10.000 randomly selected tweets that were personally read and rated by one of the authors of this paper.

The confusion matrix of the keyword-based classifier on the test set is presented in Table 3. The error rate obtained was 31,2%.

Bayesian Classifier

We also applied a supervised learning machine learning method to try and obtain a more accurate classifier. We chose to use a standard Bayesian Classifier, using all the words that appeared in the tweets as inputs. Thus, all tweets are treated as a "bag of

Table 3. Confusion matrix of keyword based classifier (error rate = 31.2%)

	Sample		
	Positive	Negative	Neutral
Positive	1500	171	1632
Negative	42	632	555
Neutral	434	285	4749

words" [29], each one producing a dimensional vector were each component is the number of times a given word appears in that tweet. As expected, most components are zero, and thus the training set is a very sparse matrix.

To obtain a classified training set, we followed the same procedure we used for the test set, *i.e.*, we personally read and rated tweets until we had a set of 100.000 manually classified tweets. To obtain a balanced set, we explicitly chose 50.000 neutral tweets, 25.000 positive, and an equal number of negative tweets.

Instead of using a single Bayesian classifier with three outputs ("positive", "negative" and "neutral") we followed the approach used in [6], and used two binary classifiers. The first separated "neutral" (or what we called "objective") from "positive" and "negative", that we called "subjective". The second classifier was used only on the "subjective" tweets, separating the "positive" from the negative". Again, out of curiosity, the distribution of tweets obtained with the Bayesian classifier on the original set of 1,535,167 tweets is presented in Fig. 4.

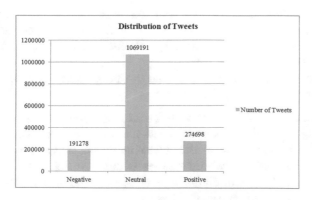

Fig. 4. Distribution of tweets – Bayesian classifier model

Using the same test set that was used for the keyword-based classifier (that was completely independent from the training set), we obtained the confusion matrix presented in Table 4. The error rate is 18.3%, which is marginally better than the error rate of 18.7% obtained by [29].

As would be expected, the supervised learning Bayesian classifier had a better performance than the simple keyword-based classifier. However, it requires a manually

Table 4. Confusion matrix of the Bayesian classifier (error rate = 18.3%)

	Sample		
	Positive	Negative	Neutral
Positive	1333	170	537
Negative	140	769	261
Netural	126	375	6289

produced pre-classified training that requires a lot of work, and it is computationally more demanding than the keyword based method.

3.3   Relating Stock Market and Twitter Data

As mentioned in the introduction, and similarly to what was done in [5] we chose Granger Causality Analysis as the method to estimate the relationship between stock market prices and twitter data. Granger Causality Analysis assumptions and theoretical framework are explained in [25]. Basically, this method assumes that the predicted variable, Y_t is obtained by a difference equation of the previous $Y_{t-\delta}$ values and the values of an independent X_t series, plus a residual random error ε_t and a constant k [2]:

$$Y_t = \sum_{j=1}^{m} \alpha_j Y_{t-j} + \sum_{i=1}^{n} \beta_i X_{t-i} + k + \varepsilon_t \qquad (1)$$

Using as our dependent variable both the stock market and Twitter sentiment (therefore having 2 distinct models where the Y an X series are interchanged) we want to test if our β coefficients are significantly different from 0. Thus, our null hypothesis is $H_0 = \beta_i = 0$ (for $i = 1, 2, \ldots, n$).

If we reject our null hypothesis we can use prior values of X to predict Y thus establishing Granger Causality from X to Y.

Predicting Stock Price Variations with Twitter Sentiments

When in Eq. (1) we use the stock price variation as variable Y, and the twitter sentiment score as variable X, we can test whether these sentiments are good predictors of price. A graphical example of the two time series is given in Fig. 5, where hourly price variations of the stock price of Starbucks is compared with hourly average sentiments estimated with a Bayesian classifier of tweets.

We started by using 3 min averages for both time series, and used a least square regression to adjust Eq. 1 to the different time series for the 16 companies, obtaining the required β values. We repeated this adjustment using increasing values of m and n, i.e., using ever greater length time series. We always considered $m = n$, i.e., used the same time lag for both series. We then computed the p-values of assuming these β were all 0 (i.e. confirming the null hypothesis). The results are presented in Table 5. Very low p-values ($p < 0.05$) are shown in bold, and indicate situations where the null hypothesis is rejected. Thus, bold values show that in some cases twitter sentiment is a good predictor of stock price variations. A quick inspection of these tables show that

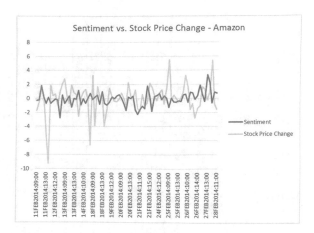

Fig. 5. Amazon vs. stock price change for Amazon

for American Airlines, Cisco, and Microsoft, and providing we consider large time lags (at least 3 to 6 periods of 3 min), this relation holds. As we shall see later, for Cisco there is another effect in place that makes this relation less relevant.

Next we used hourly averages of both time series, and repeated the Eq. (1) adjustment for all 16 companies. The results are presented in Table 5. Here we can see that the previous relations no longer hold (except for Cisco that we will deal with later), but the stock price variations of LinkedIn and Quiksilver can be predicted with some certainty. This means that LinkedIn sentiment helps to predict stock market prices 1 and 2 h later, a pattern that did not appear with the 3 min' time grouping. This may seem do indicate that Twitter and Stock Market relationship is connected to the time grouping of data and that there are stocks prices that adjust quicker than others. (Table 6)

Finally, we tested with daily averages of stock price variations and sentiments. The results are presented in Table 7. As shown in those tables, all p-values are rather large, revealing that the dependency of price variations with twitter sentiment is close to zero. This result is in line with previous work presented in [5]. We may thus conclude that while day-to-day predictions of price variations is not possible, we can for some companies predict intraday price variations by observing twitter sentiments concerning those companies.

Predicting Twitter Sentiments with Stock Price Variations

If we use the stock price variations as the X variable in Eq. (1), and twitter sentiments as variable Y, we can estimate the inverse relation of the previous case. Again, we used 3 min, 1 h, and daily averages for all 16 companies, estimated β coefficients using least square regressions, and computed p-values of assuming these coefficients were 0. The results are presented in Tables 8, 9 and 10, using different length time series.

Regarding 3-minute time series, there is one unexpected result: the sentiment regarding BP can be predicted with some certainty only if we consider a short time series. Besides that, we can only predict sentiment for Cisco. However, as was mentioned before, for this company there is also significant Granger causality in the

Table 5. Statistical significance (*p*-values) for Granger Causality test between stock market and Twitter sentiment (explanatory variable) using an interval of 3 min.

Lag	Amazon	BP	Barclays	American Airlines	BlackBerry	Cisco	General Motors	LinkedIn	Logitech	Marriot	Microsoft	Nike	Quiksilver	Sears	Sony	Starbucks
0	0.489	0.296	0.88	0.117	0.508	0.941	0.278	0.994	0.828	0.966	0.347	0.251	0.411	0.89	0.806	0.185
1	0.739	0.425	0.886	0.14	0.687	0.983	0.428	0.158	0.376	0.574	0.271	0.415	0.786	0.946	0.979	0.127
2	0.814	0.622	0.966	0.101	0.495	0.339	0.124	0.204	0.281	0.71	**0.008****	0.62	0.569	0.915	0.996	0.26
3	0.938	0.531	0.992	0.184	0.545	**0.026****	0.139	0.328	0.296	0.74	**0.000****	0.743	0.574	0.954	0.997	0.345
4	0.348	0.643	0.905	0.299	0.645	**0.028****	0.176	0.396	0.4	0.444	**0.001****	0.702	0.693	0.945	0.99	0.43
5	0.413	0.714	0.945	0.427	0.744	**0.025****	0.222	0.458	0.467	0.43	**0.001****	0.809	0.693	0.903	0.994	0.39
6	0.427	0.664	0.964	**0.020****	0.709	**0.019****	0.303	0.565	0.384	0.526	**0.001****	0.881	0.671	0.897	0.998	0.342
7	0.423	0.713	0.959	**0.039****	0.788	**0.031****	0.38	0.655	0.419	0.62	**0.002****	0.829	0.748	0.872	0.998	0.486
8	0.404	0.801	0.798	**0.063***	0.853	**0.043****	0.463	0.615	0.432	0.516	**0.003****	0.88	0.2	0.965	0.998	0.455
9	0.502	0.852	0.21	**0.039****	0.898	0.066*	0.447	0.722	0.46	0.38	**0.005****	0.903	0.742	0.927	0.998	0.589
10	0.598	0.904	0.231	**0.024****	0.529	0.076*	0.455	0.743	0.25	0.409	**0.006****	0.937	0.839	0.963	0.996	0.561
11	0.663	0.931	0.295	**0.012****	0.571	0.087*	0.544	0.717	0.285	0.556	**0.012****	0.963	0.354	0.892	0.998	0.452
12	0.556	0.928	0.37	**0.026****	0.584	**0.020****	0.59	0.775	0.349	0.582	**0.012****	0.691	0.399	0.925	0.999	0.318
13	0.553	0.956	0.411	**0.041****	0.457	**0.023****	0.554	0.798	0.351	0.515	**0.019****	0.712	0.49	0.878	1	0.127
14	0.592	0.956	0.422	0.059*	0.515	**0.030****	0.608	0.787	0.396	0.576	**0.031****	0.773	0.588	0.898	1	0.196
15	0.481	0.941	0.497	**0.036****	0.567	0.051*	0.233	0.829	0.395	0.639	**0.035****	0.8	0.695	0.925	1	0.191
16	0.538	0.955	0.566	**0.021****	0.554	0.068*	0.064*	0.764	0.439	0.833	**0.039****	0.849	0.654	0.878	0.963	0.216
17	0.53	0.917	0.633	**0.028****	0.613	0.056*	0.073*	0.623	0.382	0.661	**0.035****	0.837	0.756	0.898	0.971	0.191
18	0.405	0.929	0.715	**0.038****	0.637	0.070*	0.057*	0.5	0.438	0.725	**0.046****	0.867	0.571	0.933	0.935	0.23
19	0.46	0.949	0.759	0.051*	0.496	0.066*	0.071*	0.547	0.798	0.764	**0.046****	0.815	0.567	0.96	0.945	0.289
20	0.53	0.96	0.738	**0.014****	0.553	**0.024****	0.075*	0.223	0.524	0.73	0.058*	0.709	0.476	0.962	0.943	0.277

**p-value < 0.05, *p-value < 0.1

Table 6. Statistical significance (*p*-values) for Granger Causality test between stock market and Twitter sentiment (explanatory variable) using an interval of 1 h

Lag	Amazon	BP	Barclays	American Airlines	BlackBerry	Cisco	General Motors	LinkedIn	Logitech	Marriot	Microsoft	Nike	Quiksilver	Sears	Sony	Starbucks
0	0.604	0.483	0.415	0.794	0.308	0.093*	0.919	0.311	0.643	0.371	0.887	0.612	**0.042****	0.789	0.373	0.196
1	0.897	0.656	0.659	0.959	0.288	0.099*	0.894	**0.012****	0.675	0.336	0.853	0.853	0.16	0.958	0.467	0.25
2	0.99	0.689	0.682	0.929	0.452	0.081*	0.944	**0.024*****	0.693	0.25	0.622	0.84	0.205	0.987	0.555	0.319
3	0.999	0.808	0.785	0.266	0.917	**0.018****	0.926	0.059*	0.265	0.248	0.731	0.938	0.282	0.943	0.569	0.25
4	0.999	0.882	0.403	0.417	0.865	**0.033****	0.801	0.13	0.31	0.324	0.822	0.854	0.424	0.975	0.759	0.302

**p-value < 0.05, *p-value < 0.1

Table 7. Statistical significance (*p*-values) for Granger Causality test between stock market and Twitter sentiment (explanatory variable) using an interval of 1 day

Lag	Amazon	BP	Barclays	American Airlines	BlackBerry	Cisco	General Motors	LinkedIn	Logitech	Marriot	Microsoft	Nike	Quiksilver	Sears	Sony	Starbucks
0	0.915	0.571	0.661	0.69	0.99	0.32	0.099*	0.235	0.769	0.847	0.587	0.812	0.566	0.126	0.547	0.605
1	0.997	0.478	0.357	0.971	0.884	0.756	0.193	0.386	0.378	0.987	0.425	0.777	0.097*	0.424	0.718	0.661

**p-value < 0.05, *p-value < 0.1

Table 8. Statistical significance (p-values) for Granger Causality test between Twitter sentiment and stock market (explanatory variable) using an interval of 3 min.

Lag	Amazon	BP	Barclays	American Airlines	BlackBerry	Cisco	General Motors	LinkedIn	Logitech	Marriot	Microsoft	Nike	Quiksilver	Sears	Sony	Starbucks
0	0.929	**0.023****	0.99	0.117	0.508	0.147	0.733	0.874	0.785	0.537	0.195	0.327	0.646	0.771	0.13	0.094*
1	0.972	**0.008****	0.744	0.14	0.687	0.129	0.748	0.595	0.817	0.533	**0.046****	0.392	0.67	**0.049****	0.334	0.162
2	0.993	**0.016****	0.865	0.101	0.495	0.1	0.83	0.717	0.281	0.255	0.068*	0.56	0.162	0.903	0.522	0.184
3	0.899	**0.023****	0.914	0.184	0.545	**0.005****	0.892	0.868	0.779	0.221	0.139	0.769	0.102	0.907	0.709	0.112
4	0.93	**0.044****	0.812	0.299	0.645	**0.014****	0.848	0.699	0.876	0.33	0.090*	0.891	0.141	0.866	0.733	0.123
5	0.961	0.065*	0.868	0.427	0.744	**0.001****	0.976	0.574	0.923	0.422	0.17	0.955	0.059*	0.925	0.716	0.137
6	0.937	0.072*	0.825	**0.020****	0.709	**0.001****	0.989	0.507	0.954	0.508	0.156	0.902	0.121	0.956	0.811	0.189
7	0.943	0.101	0.791	**0.039****	0.788	**0.002****	0.891	0.609	0.912	0.459	0.176	0.95	0.13	0.876	0.876	0.235
8	0.848	0.158	0.839	0.063*	0.853	**0.002****	0.855	0.711	0.943	0.527	0.128	0.914	0.106	0.917	0.903	0.219
9	0.879	0.172	0.841	**0.039****	0.898	**0.006****	0.661	0.786	0.894	0.595	0.063*	0.935	0.11	0.926	0.899	0.265
10	0.917	0.185	0.752	**0.024****	0.529	**0.006****	0.501	0.826	0.928	0.562	0.11	0.873	0.13	0.914	0.93	0.219
11	0.939	0.203	0.793	**0.012****	0.573	**0.008****	0.552	0.801	0.958	0.601	0.135	0.888	0.24	0.947	0.925	0.256
12	0.943	0.268	0.839	**0.026****	0.584	**0.002****	0.371	0.854	0.804	0.625	0.114	0.806	0.291	0.895	0.874	0.317
13	0.969	0.307	0.856	**0.041****	0.457	**0.002****	0.62	0.9	0.834	0.72	0.135	0.874	0.276	0.921	0.861	0.401
14	0.977	**0.036****	0.867	0.059*	0.515	**0.004****	0.693	0.89	0.82	0.786	0.164	0.903	0.262	0.894	0.896	0.415
15	0.983	0.301	0.897	**0.036****	0.567	**0.005****	0.752	0.817	0.791	0.824	0.222	0.938	0.315	0.925	0.844	0.285
16	0.99	0.365	0.934	**0.021****	0.554	**0.008****	0.764	0.743	0.831	0.737	0.23	0.954	0.168	0.949	0.741	0.303
17	0.995	0.39	0.887	**0.028****	0.513	**0.010****	0.857	0.765	0.872	0.806	0.28	0.968	0.168	0.962	0.737	0.488
18	0.997	0.303	0.908	**0.038****	0.637	**0.014****	0.596	0.818	0.915	0.816	0.218	0.983	0.338	0.948	0.797	0.492
19	0.996	0.324	0.924	0.051*	0.496	**0.018****	0.585	0.717	0.955	0.737	0.262	0.966	0.343	0.956	0.699	0.543
20	0.996	0.35	0.933	**0.014****	0.553	**0.024****	0.637	0.769	0.969	0.665	0.301	0.976	0.417	0.959	0.71	0.648

**p-value < 0.05, *p-value < 0.1

Table 9. Statistical significance (*p*-values) for Granger Causality test between Twitter sentiment and stock market (explanatory variable) using an interval of 1 h

Lag	Amazon	BP	Barclays	American Airlines	BlackBerry	Cisco	General Motors	LinkedIn	Logitech	Marriot	Microsoft	Nike	Quiksilver	Sears	Sony	Starbucks
0	0.618	0.288	0.096*	0.838	0.604	0.918	0.427	0.034**	0.328	0.769	0.25	0.62	0.869	0.58	0.475	0.838
1	0.903	0.556	0.12	0.871	0.897	**0.007****	0.364	0.113	0.627	0.543	0.57	0.912	0.969	0.784	0.752	0.871
2	0.893	0.85	0.239	0.462	0.99	**0.010****	0.237	0.095*	0.797	0.695	0.523	0.642	0.336	0.779	0.834	0.462
3	0.95	0.87	0.335	0.442	0.999	**0.019****	0.156	0.169	0.559	0.681	0.549	0.675	0.451	0.893	0.677	0.442
4	0.956	0.903	0.243	0.473	0.999	**0.038****	0.295	0.236	0.589	0.757	0.357	0.564	0.533	0.923	0.193	0.473

**p*-value < 0.05, *p*-value < 0.1

Table 10. Statistical significance (*p*-values) for Granger Causality test between Twitter sentiment and stock market (explanatory variable) using an interval of 1 day

Lag	Amazon	BP	Barclays	American Airlines	BlackBerry	Cisco	General Motors	LinkedIn	Logitech	Marriot	Microsoft	Nike	Quiksilver	Sears	Sony	Starbucks
0	0.434	0.982	0.434	0.267	0.931	0.283	0.199	0.905	0.985	0.366	0.986	0.336	0.568	0.054*	0.397	0.223
1	**0.045****	0.298	0.679	0.561	0.502	0.614	0.65	0.826	0.511	0.508	0.947	0.508	0.889	**0.040****	0.551	0.22

**p*-value < 0.05, *p*-value < 0.1

opposite direction. So, we may ask whether sentiment drives price or price drives sentiment. A closer inspection reveals that it is more consistent to assume the latter. However, since both the price variations and sentiments of Cisco had a very consistent trend over the period in question, it is impossible to determine which one is the driver of the other.

Regarding hourly averages, the same two companies (LinkedIn and Cisco) showed good predictability.

Regarding day-to-day predictions, as expected, no truly significant relationships where identified. However, some relationship was found in the case of Amazon and Sears.

Summarizing, it is rarely possible to predict twitter sentiment with stock price data (people's moods seem to only be sensitive to short time variations in the value of BP).

Twitter negative *vs.* positive daily sentiment of companies does not show any relationship at all with the stock market prices for all companies. This is the same conclusion reached by the original authors and strengthens the idea that inter-day positions are not predictable using Twitter polarity sentiment [5] and the price adjustment regarding sentiment is more notorious during the day.

Curiously, the three companies with more tweets, and that have significantly more the others, do not show any significant relationship amongst variables.

4 Discussion and Conclusions

Regarding sentiment analysis, Bayesian classifier models are more accurate at classifying twitter sentiment than keyword based models. The error rate for this classification (using human classifiers as ground truth) is only 18%, as compared with almost 30% for keyword based classifiers.

Regarding the predictability of stock prices using twitter sentiments, it was shown that some companies show a significant relationship between these variables. This is not contradictory to the results obtained in [5] since in that case stock indexes (that average individual companies) were used. Furthermore, Bollen *et al.* [5] used only day-to-day predictions, whereas we showed in this study that the relationship depends a lot on the time interval used. For some companies the reaction time is very short, and for others it is longer, but the dependency holds only for intraday predictions, and never for day-to-day.

We also showed that the inverse relationship (stock prices influencing twitter sentiment) is rare but does occur for some companies. Finally, companies with more mentions on Twitter tend to show less relationship between the variables in question.

For future work we intend to search for companies with stronger relationships amongst variables, since the predictability depends strongly on the characteristics of individual companies. It would thus be interesting to find which are the characteristics that do induce predictability.

We also feel that some improvement may be done on the sentiment classifiers, namely by using some preprocessing to deal with company names that occur in unrelated tweets (such as "apple" referring to the fruit and not the company) or that do not occur in relevant tweets (such as those that mention iPod but not Apple).

Augmenting our time window for 3 or 4 months and testing shorter time predictⁱ (such as minute-to-minute positions) is another recommendation for future studie This would, however, take considerably more resources.

The fact that companies with many tweets have a weak relationship between tweets and stock prices can have two interpretations: very popular companies have many tweets by people that have little or no influence in the stock prices, or, because of the way we conducted our tests (a relationship was established only when p-values were low thus rejecting the nulls hypothesis) datasets with a very high number of instances allowed for higher p-values even where the null hypothesis was in fact weak.

References

1. Fama, E.F.: Efficient capital markets - a review of theory and empirical work. J. Financ. **25**, 36 (1970)
2. Qian, B., Rasheed, K.: Stock market prediction with multiple classifiers. Appl. Intell. **26**, 25–33 (2007)
3. Schumaker, R.P., Chen, H.: Textual analysis of stock market prediction using breaking financial news: the AZFin text system. ACM Trans. Inf. Syst. (TOIS) **27**, 1–19 (2009)
4. Shiller, R.J.: From efficient markets theory to behavioral finance. J. Econ. Perspect. **17**, 83–104 (2003)
5. Bollen, J., Mao, H., Zeng, X.: Twitter mood predicts the stock market. J. Comput. Sci. **2**, 1–8 (2011)
6. Go, A., Bhayani, R., Huang, L.: Twitter sentiment classification using distant supervision. Processing **150**, 1–6 (2009)
7. Hu, M., Liu, B.: Mining and summarizing customer reviews. In: Proceedings of the 2004 ACM SIGKDD International Conference on Knowledge Discovery and Data Mining, KDD 2004, vol. 4, p. 168 (2004)
8. Jansen, B.J., Zhang, M., Sobel, K., Chowdury, A.: Twitter power: tweets as electronic word of mouth. J. Am. Soc. Inform. Sci. Technol. **60**, 2169 (2009)
9. Bai, X.: Predicting consumer sentiments from online text. Decis. Support Syst. **50**, 732–742 (2011)
10. Salehan, M., Kim, D.J.: Predicting the performance of online consumer reviews: a sentiment mining approach to big data analytics. Decis. Support Syst. **81**, 30–40 (2015)
11. Diakopoulos, N.A, Shamma, D.A: Characterizing debate performance via aggregated twitter sentiment. In: Proceedings of the SIGCHI Conference on Human Factors in Computing Systems, CHI 2010, p. 1195 (2010)
12. Tumasjan, A., Sprenger, T., Sandner, P., Welpe, I.: Predicting elections with Twitter: what 140 characters reveal about political sentiment. In: Proceedings of the Fourth International AAAI Conference on Weblogs and Social Media, pp. 178–185 (2010)
13. Duric, A., Song, F.: Feature selection for sentiment analysis based on content and syntax models. Decis. Support Syst. **53**, 704–711 (2012)
14. Pang, B., Lee, L.: Opinion mining and sentiment analysis. Found. Trends Inf. Retrieval **2**, 91–231 (2008)
15. Schumaker, R.P., Zhang, Y., Huang, C.N., Chen, H.: Evaluating sentiment in financial news articles. Decis. Support Syst. **53**, 458–464 (2012)
16. Kwak, H., Lee, C., Park, H., Moon, S.: What is Twitter, a social network or a news media? In: The International World Wide Web Conference Committee (IW3C2), pp. 1–10 (2010)

. Hassan, A., Abbasi, A., Zeng, D.: Twitter sentiment analysis: a bootstrap ensemble framework. In: Proceedings of the SocialCom/PASSAT/BigData/EconCom/BioMedCom 2013, pp. 357–364 (2013)

18. Asiaee T., A., Tepper, M., Banerjee, A., Sapiro, G.: If you are happy and you know it... tweet. In: Proceedings of the 21st ACM International Conference on Information and Knowledge Management, CIKM 2012, p. 1602 (2012)

19. Spencer, J., Uchyigit, G.: Sentimentor: sentiment analysis of Twitter data. In: Proceedings of the CEUR Workshop, pp. 56–66 (2012)

20. Agarwal, A., Xie, B., Vovsha, I., Rambow, O., Passonneau, R.: Sentiment analysis of Twitter data, pp. 30–38. Association for Computational Linguistics (2011)

21. Cho, S.W., Cha, M.S., Kim, S.Y., Song, J.C., Sohn, K.: Investigating temporal and spatial trends of brand images using Twitter opinion mining. In: 2014 International Conference on Information Science and Applications (ICISA), pp. 1–3 (2014)

22. Stieglitz, S., Dang-Xuan, L.: Political communication and influence through microblogging - an empirical analysis of sentiment in Twitter messages and retweet behavior. In: Proceedings of the Annual Hawaii International Conference on System Sciences, pp. 3500–3509 (2011)

23. Gokulakrishnan, B., Priyanthan, P., Ragavan, T., Prasath, N., Perera, A.: Opinion mining and sentiment analysis on a Twitter data stream. In: International Conference on Advances in ICT for Emerging Regions (ICTer), pp. 182–188. IEEE (2012)

24. da Silva, N.F.F., Hruschka, E.R., Hruschka, E.R.: Tweet sentiment analysis with classifier ensembles. Decis. Support Syst. **66**, 170–179 (2014)

25. Granger, C.W.J.: Investigating causal relations by econometric models and cross-spectral methods. Econometrica **37**, 424–438 (1969)

26. Google Finance: Google Finance: stock market quotes, news, currency conversions & more. https://www.google.co.uk/finance

27. Denecke, K.: Using SentiWordNet for multilingual sentiment analysis. In: Proceedings of the International Conference on Data Engineering, pp. 507–512 (2008)

28. Miner, G., Elder, J., Fast, A., Hill, T., Nisbet, R., Delen, D.: Practical Text Mining and Statistical Analysis for Non-structured Text Data Applications, 1st edn. Elsevier, Amsterdam (2012)

29. Pak, A., Paroubek, P.: Twitter as a corpus for sentiment analysis and opinion mining. In: LREC, pp. 1320–1326 (2010)

Author Index

© Springer International Publishing AG 2018
I. Czarnowski et al. (eds.), *Intelligent Decision Technologies 2017*,
Smart Innovation, Systems and Technologies 73, DOI 10.1007/978-3-319-59424-8